Food Packaging

Food Packaging

EDITED BY

Takashi Kadoya
Kanagawa University
Hiratsuka, Japan

ACADEMIC PRESS, INC.
Harcourt Brace Jovanovich, Publishers
San Diego   New York
Boston   London   Sydney   Tokyo   Toronto

This book is printed on acid-free paper. ∞

Copyright © 1990 by Academic Press, Inc.
All Rights Reserved.
No part of this publication may be reproduced or transmitted in any form or by any means, electronic or mechanical, including photocopy, recording, or any information storage and retrieval system, without permission in writing from the publisher.

Academic Press, Inc.
San Diego, California 92101

United Kingdom Edition published by
Academic Press Limited
24–28 Oval Road, London NW1 7DX

Library of Congress Cataloging-in-Publication Data

Food packaging / [edited by] Takashi Kadoya.
    p. cm.
  ISBN 0-12-393590-3 (alk. paper)
  1. Food--Packaging. I. Kadoya, Takashi, Date.
TP374.F648 1990
664'.119--dc20   89-18381
                                                                            CIP

Printed in the United States of America
90 91 92 93   9 8 7 6 5 4 3 2 1

CONTENTS

Foreword   xi
Masao Fujimaki

Preface   xiii

Part I Fundamentals of Foods

1. Physical Properties and Microbiology of Foods   3
   Toshimasa Yano
   - I. Physical Properties of Foods   3
   - II. Microbiology of Foods   14
     - Nomenclature   21
     - References   22

2. Oxidation of Foods   25
   Setsuro Matsusuhita
   - I. Oxidation of Lipids   25
   - II. Secondary Products   34
   - III. Factors Affecting Lipid Oxidation   35
   - IV. Antioxidant and Singlet Oxygen Quenchers   40
   - V. Reactions between Lipid-Oxidized Products and Other Food Components   41
     - References   42

Part II New Food Packaging Materials

3. New Food Packaging Materials—An Introduction   47
   Akira Kishimoto
   - Text   47
   - References   51

4. Paper and Paperboard Containers   53
   Mitsuhiro Sumimoto
   - I. Characteristics   53
   - II. Materials   54

v

III. Classification of Paper and Paperboard Containers   58
IV. Manufacturing Processes for Cartons   74
    References   83

5. Metal Containers   85
   Hiroshi Matsubayashi

   I. Introduction   85
   II. Features of Metal Containers   85
   III. Metals   86
   IV. Metal Containers and Easy-Open Ends   93
       References   104

6. Glass Containers   105
   Yoshihiro Yamato

   I. Introduction   105
   II. Forming Glass Bottles   105
   III. Properties of Glass Bottles   107
   IV. Trends in Glass Bottles   108
   V. Afterword   115
       References   115

7. Plastic Containers   117
   Koji Kondo   117

   I. Introduction   117
   II. Polystyrene   118
   III. Butadiene–Styrene Copolymer   122
   IV. Acrylonitrile—Styrene Copolymer   124
   V. Acrylonitrile—Butadiene—Sytrene Copolymer (ABS)   125
   VI. Polyvinyl Chloride (PVC)   125
   VII. Polyethylene   129
   VIII. Polypropylene   133
   IX. Nylon   137
   X. Saponified Ethylene—Vinyl Acetate Copolymer (EVOH)   140
   XI. Polycarbonate   141
   XII. Polyethylene Terephthalate   142

Part III Food Packaging and Energy

8. Food Packaging and Energy in Japan—Energy Analysis of Consumer Beverage Containers   149
   Takashi Kadoya

   I. Introduction   149

Contents

      II. Calculation of the Energy of the Transportation for Reference Containers   151
     III. Calculation of Manufacturing Energy of Basic Packaging Materials   152
     IV. Energy to Manufacture the Container   154
      V. Total Energy Consumption   154
     VI. Conclusion   159
         References   161

Part IV Packaging Systems and Technology of Food Materials

9. Recent Development of Packaging Machinery in Japan   165
   Hidekazu Nakai

       I. Computer-Controlled Packaging Machines   165
      II. Computer-Controlled Optimum Accumulating and Packaging System for Multisized Products   169
     III. Heat-Sealing Device Using a Heat Pipe   170
     IV. Industrial Robots and Unmanned Operation   173

Part V New Trends in the Technology of Food Preservation

10. New Trends in the Technology of Food Preservation— An Introduction   179
    Michio Yokoyama

       I. Food Packaging Technology and the Behavior of Microorganisms   179
      II. Sterilization and Control of Microorganisms in Packaged Foods   182

11. Retortable Packaging   185
    Kanemichi Yamaguchi

       I. Introduction   185
      II. History of Retortable Packaging in Japan   186
     III. Types of Retort Foods   186
     IV. Production Systems of Retort Food   188
      V. Microorganism Control in Retort Food   204
     VI. Standards for Packages   208

VII. Shelf Life of Retort Food   209
VIII. Conclusion   210
     References   210

12. Aseptic Packaged Foods   213
Michio Yokoyama

I. Recent Trends in Aseptic Packaged Foods In Overseas Countries and Japan   214
II. Aseptic Food Packaging Systems   217
III. Manufacturing Methods of Aseptic Packaged Foods   221
IV. Future Trends in Aseptic Packaged Food   227
    References

13. Free Oxygen Scavenging Packaging   229
Yoshihiko Harima

I. Introduction   229
II. Summary of Oxygen Absorber   230
III. Types of Oxygen Absorber   235
IV. Advances in Related Technology   244
V. New Knowledge in Microbiology   252

14. Frozen Food and Oven-Proof Trays   253
Kentaro Ono

I. Oven-Proof Tray Development   253
II. Varieties of Dual-Oven-Proof Trays   254
III. Lidding of Trays   264
IV. Usage of Trays   265
    References   267

15. Gas-Exhange Packaging   269
Koji Satomi

I. Gas-Exchange Packaging and Its Aim   269
II. Gases for Usage and Their Properties   270
III. Form and System of Gas-Exhange Packaging   270
IV. Gas-Exchange Packaging and Packaging Materials   272
V. Cases of Gas-Exchange Packaging in Food   272
    References   277

16. Vacuum Packaging   279
Naohiko Yamaguchi

I. Vacuum Packaging Machinery   279
II. Packaging Material   282

Contents

Part VI Packaging Fresh and Processed Foods

17. Fruits   295
    Akiyoshi Yamane

    I. Packaging for Freshness Preservation and Storage   295
    II. Shock-Absorbing Packaging   299
        References   301

18. Vegetables   303
    Masutaro Ohkubo

    I. Old Concepts Requiring Reexamination   304
    II. New Concepts   305
    III. Future Trends   307
        References   307

19. Fresh Meat   309
    Yoshihiko Tomioka

    I. World Meat Production   309
    II. Brief Historical Overview   310
    III. Wholesale Packaging   313
    IV. Retail Packaging   318
        References   321

20. Meat By-Products   323
    Yoshiyuki Tohma

    I. Simple Wrapping   324
    II. Vacuum Packaging   324
    III. Double-Sterilization Packaging   325
    IV. Boil-and-Steam Cooking Packaging   326
    V. Retort Sterilized Packaging   327
    VI. Air-Containing Packaging   328
    VII. Oxygen-Absorbing Agent Packaging   329
    VIII. Aseptic Packaging   329
    IX. Packaging of Typical Product Lines   332
    X. Summary   333

21. Seafood Products   335
    Takao Fujita

    I. Fresh Fish   336
    II. Frozen Fish   337
    III. Dried, Salted, and Other Types of Seafood Products   338
    IV. Canned Fish   339

22. Fish Meat By-Products   341
   Shigeyuki Sasayama
       I. Fish Paste Products (Kamaboko)   341
       II. Fish Ham and Fish Sausage   344
       III. Specially Packaged Kamaboko   348

23. Dairy Products   349
   Kentaro Ono
       I. Yogurt and Fresh Cheese   349
       II. Natural Cheese   351
       III. Processed Cheese   353
       References   355

24. Cakes and Snack Foods   357
   Shinichi Minakuchi and Hachiro Nakamura
       I. Introduction   357
       II. Classification and Required Quality for Packaging of Cakes   358
       III. Quality Deterioration of Cakes   358
       IV. Moisture-Proof Packaging and Gas Barrier Packaging   364
       V. Vacuum Packaging and Gas Substitution Packaging   367
       VI. Packaging with Oxygen Absorbent and Packaging with Gas Substitution Agent   371
       VII. Packaging with Alcohol-Generating Agent   377
       VIII. Future Prospects   378
       References   378

Part VII Physical Distribution and Food Packaging

25. Physical Distribution of Packaged Foods   381
   Yoshio Hasegawa
       I. Packaging for Preservation of Foods in the Distribution Process   381
       II. Distribution of Packaged Foods and Temperature Environment   391
       III. Physical Distribution and Unit Load System   392
       IV. Distribution Package Dimensions by Modular Coordination   395
       V. Examples of Distribution Package Dimensions Selected by Modular Coordination   405

Index   415

FOREWORD

Humans consume other organisms and their constituents as food. All food organisms go through their own life cycles until harvested. The preharvest process comprises these features in chronological order: genesis, growth, maturation, and reproduction. Various environmental factors affect the life process in whole or in part. These include ambient temperature, humidity, light, nutrients, composition of the atmosphere, and so forth. Failure to control the factors will affect the quality of food organisms and sometimes jeopardize their efficient distribution for human consumption. The same may be true for the postharvest process.

Most food organisms after harvest receive artificial treatments for preservation. One main purpose of the treatments is essentially to lengthen the time from harvest to consumption. The longer the time, the farther the food can be distributed from the site of the harvest to the consumer. This relationship is particularly important in the effective feeding of people living in highly sophisticated societies.

Modern science has developed a high level of technology for preserving food organisms postharvest. The technology in the forms of drying, cooking, heating, refrigerating, freezing, pasteurizing, sterilizing, irradiating, and packaging is a major part of the food industry in economically developed areas of the world. If for any reason any one of these treatments fails to function properly, it will affect the quality of food products and their distribution to markets. Packaging is extremely important as a final process of food preservation. It can maximize the time from harvest to consumption and therefore warrant the longest shelf life of any food product. Actually, almost all food items are packaged prior to being sold on the market.

A great many countries have developed new materials and methods for packaging a variety of foods in recent years. Japan may be one of the leading countries in the field of the science and technology of food packaging. With such a background, the editing of this book was carried out by a Japanese team headed by Dr. Takashi Kadoya, Former Professor of the University of Tokyo, who is also the President of the Japan Packaging Research Association.

The planning of the book was first suggested by the late Dr. B. S. Schweigert, former Professor at the University of California, Davis, who acted as the chairman of the Academic Press Editorial Board of the Food Science and Technology Series. The suggestion was then introduced to

Dr. Kadoya by way of Dr. Soichi Arai, Associate Professor at the University of Tokyo. Both of them are gratefully acknowledged.

I am particularly grateful to the Japan Packaging Institute, which has given every possible support during the time from planning to publication, and to Mr. Junichiro Minagawa, President, Academic Press Japan, who has kindly assisted the chief editor Dr. Kaydoya in many respects.

Masao Fujimaki
Professor Emeritus
The University of Tokyo
Ochanomizu University

PREFACE

It is well known that the foods we eat and drink have been changing increasingly to fit a changing life style. In response to new demands, food packaging has become an increasingly important area in modern society. However, few books specifically addressed to this subject have been published anywhere in the world.

The purpose of food packaging is to safely preserve the foods and distribute them, still nice, sweet, tasty, and delicious to the consumer, preventing deterioration under various environmental conditions. In order to achieve this purpose, it is necessary to learn the fundamental properties of food, packaging materials, sanitation control (of food, food packaging materials, and packaging systems), and technology for food materials.

Recently, new packaging materials and their application technologies have been developed for food packaging. Retortable packaging, aseptic packaging, free-oxygen scavenging packaging, and vacuum and gas-exchange packaging are among the new trends in the technology of food preservation. In additional, fresh and processed food packaging technologies have made remarkable progress.

Under these worldwide conditions, the eating habits of Japanese people especially have been dramatically revolutionized, because Japanese people consume foods supplied as both Western style and Japanese, with various origins.

Therefore, an editorial committee on food packaging was made up of many Japanese scientists, representing viewpoints of Japanese food packaging. A close relationship exists between the food industry and the packaging industry to determine packaging materials, technologies, and physical distribution, as well as to meet energy-saving or environmental regulations.

Total consumption of packaging materials in Japan in 1987 was estimated to be about 19,048 thousand tons, at a cost of 40.09 billion U.S. dollars (at ¥130 to the dollar). This amount is second in world consumption to that of the United States. A considerable amount of these packaging materials has been used for food packaging. About 51% of corrugated board is now used for food packaging materials, and considerable amounts of plastic films and containers have been supplied for food packaging. Metal cans also have an important role in food packaging: about 74% of metal cans are destined for food packaging in Japan, and 90% in the United States. Further, glass bottles are widely used for food, with about 77% slated for food packaging. As packaging materials that have played a very important part in the food packaging, several kinds of typical packaging materials are introduced in this book.

Our modern eating habits depend not only on our food materials but also on the packaging technology of food preservation and fresh and processed food packaging. Accordingly, it is necessary to have information on new food preservation technologies. This book also describes representative information on new application technologies. As mentioned earlier, Japanese people consume varied foods from outside Japan as well as their own foodstuffs, so this book also describes kinds of food packaging that may be particular to Japan.

This book is intended as a reference work for researchers and technicians with specialist experience in the field of food packaging, as a textbook for undergraduate university students, and as a reference work for student seminars. Care was taken in selecting contributors from various fields so that each chapter would be compiled from a global viewpoint by mainly Japanese researchers involved in a wide range of food packaging. The following is a short introduction to each section of the book.

Part I takes up fundamental issues in food packaging, such as moisture absorption, dehydration, and thermal qualities of foodstuffs. It also discusses how microorganisms react toward low temperatures, heating, sterilization, microwaves, ultraviolet rays, and other influences. On the subject of food oxidation, it deals with the degeneration of oil, vitamins, browning qualities, and so on.

Part II goes beyond a mere explanation of packaging materials to include comments on the special qualities required in each material when packaging food. It opens with a history of packaging in paper and paperboard, metallic materials, glass, and plastics. Fundamental properties and reactions of each of these packaging materials are described. For paperboard containers, for instance, this section considers the printing suitability for display, and the resistance of containers to crushing and contamination, while in the case of metallic containers it also covers the relationship between the metallic surface and the food.

The total amount of energy consumption needed for the packaging of several beverages, such as milk and juice, is considered in Part III. Energy costs are traced from the cost of the raw materials used in packaging to shipping, manufacturing, and eventual transportation to the supermarket. These energy costs are discussed mainly in terms of Japanese conditions: the amount of energy consumed in packaging is considerably different in each country. In order to calculate the energy consumption of food packaging, milk cartons, glass bottles, metal cans, and retort pouches are selected as reference containers.

Preface

Part IV introduces individual food packaging machines and shows how they combine with other packaging machinery in a total production system, dealing first with computer-controlled packaging machines. Computer-controlled optimum accumulating and packaging systems for multisized products, heat-sealing devices using heat pipes, and industrial robots and unmanned operation for packaging processes are then introduced.

New trends in the technology of food preservation are explored in Part V. Packaging that is retortable or aseptic is discussed. Methods of free oxygen scavenging, gas exchange, and vacuum are applied to packaging. These new trends also include frozen foods on oven-safe trays.

Part VI classifies representative fresh and processed foodstuffs and discusses each category, also taking up a wide range of liquids and other foods. First the packaging of fruits and vegetables is discussed by two authors, and two others describe packaging of fresh meat and meat products. Seafood products and fish, typical marine products, are also introduced here as packaging processes. Fermented milk products and the processed products and yogurt and fresh cheese as examples of packaging of dairy products are considered. As packaging of cakes and snack foods is very important, a chapter in this section deals with them. The final chapter in this section describes the packaging of Japanese foods such as rice, soy sauce, *miso*, and Chinese noodles.

Part VII deals with the physical distribution of packaged foods and the unit load system, commenting on actual examples of transportation and packaging network schemes in Japan, Switzerland, and Sweden.

Takashi Kadoya

PART I

Fundamentals of Foods

CHAPTER 1

Physical Properties and Microbiology of Foods

Toshimasa Yano
Department of Agricultural Chemistry,
The University of Tokyo, Tokyo, Japan

I. PHYSICAL PROPERTIES OF FOODS

A. Moisture Sorption Equilibrium and Water Activity

1. Moisture Sorption Equilibrium

Any food material sorbs, or desorbs, moisture in a package until the moisture sorption equilibrium is reached. The sorption includes both the adsorption of moisture onto the surface of solid materials and the absorption of moisture into the solid or liquid materials. Knowledge of the moisture sorption equilibrium is necessary in order to control the physical, chemical, biochemical, and microbiological reactions that may occur during storage and distribution of packed foods. Since the moisture permeability of packaging material is much less than the rates of moisture sorption and desorption of foods, the atmosphere inside the package may be in the moisture sorption equilibrium, although the equilibrium changes slowly due to moisture permeation through the packaging material.

Figure 1 shows schematically the moisture sorption equilibrium as a function of the relative humidity at a constant temperature. Generally, hysteresis is observed for many foods. Upon desorption from a high water content, the equilibrium water content of foods is higher than that upon sorption from a low water content.

The occurrence of hysteresis has been explained from many points of view. Change in food structure is one of the most likely explanations. The

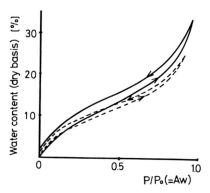

Figure 1. Moisture sorption isotherms of rice flour: ↗↗ sorption, ↙↙ desorption, — 26°C, --- 45°C.

effect of impurities in porous structure is another. However, the most significant and useful explanation may be of an "ink bottle" structure (McBain, 1935). Suppose a water-filled cylindrical space of radius r_2 connected to the atmosphere through a capillary of radius r_1 as in Fig. 2 shows ($r_2 > r_1$). In a capillary, water condensation takes place at a certain vapor pressure lower than the saturation vapor pressure, as the Kelvin equation shows:

$$\ln \frac{P_0}{P'} = \frac{2\sigma V_L \cos \theta}{rRT} \tag{1}$$

where P' is a vapor pressure at which the condensed water fills a capillary of radius r, P_0 the saturation vapor pressure, r the radius of a capillary, R the gas constant, T the absolute temperature, V_L the molar volume of condensed water, σ the surface tension of water, and θ the contact angle. Equation (1) shows that the smaller the capillary, the lower the vapor pressure of condensation.

Upon desorption, the water filling the cylindrical space of radius r_2 cannot be removed until the vapor pressure becomes low enough to desorb the condensed water filling the capillary of radius r_1. Upon sorption, the capillary of radius r_1 is first filled with the condensed water, leaving the cylindrical space of radius r_2 empty until the vapor pressure becomes high enough to fill it with the condensed water. Since radii of capillaries would have a wide distribution in food materials, the hysteresis could be such as Fig. 1 shows.

The hysteresis was found not only for the sorption isotherm as shown in Fig. 1 but also for the sorption isobar obtained by holding the vapor

Chapter 1. Physical Properties and Microbiology of Foods

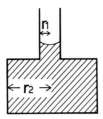

Figure 2. An "ink-bottle" structure.

pressure constant and varying the temperature as Fig. 3 shows (Strasser, 1969).

The hysteresis in the moisture sorption equilibrium is important in microbiological deterioration of packed foods that are exposed to repeated temperature changes. An explanation of the microbiological deterioration is given in Section II,B,4.

2. Water Activity of Foods

The activity of water in foods is much more important than the mere water content of the foods, because the chemical, biochemical, and microbiological reactions occurring in the foods depend on the activity of water rather than the water content.

The activity of water, or simply the water activity, is defined conveniently as

$$a_w = P/P_0 \qquad (2)$$

where a_w is the water activity, P an equilibrium water vapor pressure over a given food, and P_0 the saturation vapor pressure over the pure water at the same temperature and atmospheric pressure. The right-hand side of Eq. (2) is the relative humidity at equilibrium of the ambient atmosphere in Fig. 1. However, Eq. (2) shows that P/P_0 is also the measure of the activity of water contained in the foods. The thermodynamical explanation is as follows.

The thermodynamical definition of the activity is

$$a = f/f° \qquad (2')$$

where a is the activity of a vapor component, f the fugacity of the vapor component at a given condition, and $f°$ the fugacity of the vapor component for a reference state. The fugacity is the thermodynamical pressure corrected for any nonideality of the real vapor component. The reference state could be selected arbitrarily. However, if the reference state is

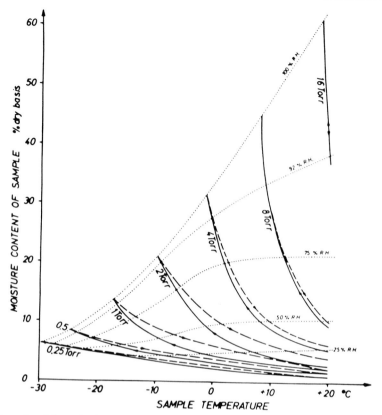

Figure 3. Adsorption isobars (solid lines) and desorption isobars (dashed lines) of cooked, freeze-dried beef. (From Strasser, 1969.)

chosen to be the water vapor pressure over the pure water as mentioned above, the difference between (P/P_0) and $(f/f°)$ is at most about 0.2% (Gal, 1972). In equilibrium at a constant temperature and a constant pressure, the chemical potentials of all components in a closed system must be equal throughout all phases of the system. The chemical potential of the water vapor must be equal to the chemical potentials of the liquid water, the bound water, and all other components in all phases. The chemical potential of the water vapor is related to its activity by

$$\mu_w = \mu_w^\circ + RT \ln (f_w/f_w^\circ)$$
$$= \mu_w^\circ + RT \ln a_w \qquad (3)$$

where μ and $\mu°$ are the chemical potentials for a given state and the reference state, respectively, and the subscript w is for water. Therefore,

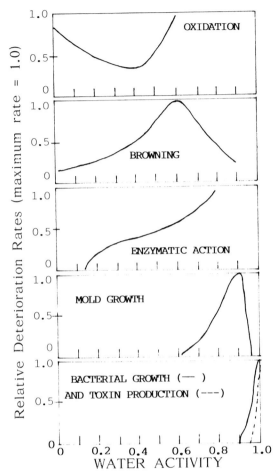

Figure 4. Dependence of relative rates of food deterioration on water activity. Bacterial growth, *S. aureus;* toxin production, *Enterotoxin B;* mold growth, *Xeromyces bisporus;* enzymatic activity, oat lipase acting on monoolein in ground oats; browning, nonenzymatic browning of pork bites; oxidation, lipid oxidation in potato chips. (From Karel, 1975.)

the activity of the water vapor given by Eq. (3) is equal to the activity of liquid or solid water, in equilibrium.

Now Fig. 1, especially the sorption curve, can be read as the relation between the water contents of a given food and the activities of the water. Van den Berg and Bruin (1981) summarized the equations to fit the sorption isotherms.

Figure 4 illustrates some published chemical, biochemical, and microbiological reaction rates as functions of the water activity (Karel, 1975). Note that the lipid oxidation rate has a minimum at a certain a_w value, and a certain microorganism can grow even at a low a_w of 0.65. The effect of temperature on the water activity, assuming ideal solutions, is given as

$$\ln \frac{(a_w)_{T_1}}{(a_w)_{T_2}} = \frac{h_w^E}{R} \left(\frac{1}{T_1} - \frac{1}{T_2} \right) \quad (4)$$

where h_w^E is the excess partial molar enthalpy of water. Equation (4) is a practical form of the Clausius–Clapeyron equation. Although Eq. (4) is applicable for no precise calculation (van den Berg and Bruin, 1981), it suggests the right tendency.

For many food materials the water is sorbed exothermally, that is, h_w^E has a negative value. Therefore the water activity increases with temperature increase. For some cereals the increment in a_w is about 0.0023/°C (van den Berg and Leniger, 1976). For many solutions a_w is not affected significantly by temperature.

For frozen foods P_0 in Eq. (2) is the saturation vapor pressure over supercooled pure water. A relationship between the water activity and subfreezing temperature is shown in Fig. 5 (Fennema, 1978). In the presence of an ice phase, a_w changes about 0.008/°C. For measurements of the moisture adsorption isotherm, refer to Gal (1975) and Spiess and Wolf (1983).

B. Thermal Properties of Foods

For thermal processings of foods, such as sterilization, chilling, and freezing, knowledges on the following thermal properties of foods are essential.

1. Specific Heat

The specific heat capacity, or simply the specific heat, of a mixture of n components is given as follows, if physical or chemical interactions among the components are not significant:

$$C_m = \sum_{i=1}^{n} C_i X_i^w \quad (5)$$

where C_m is the mean specific heat of the mixture, C_i the specific heat of component i, and X_i^w the weight fraction of component i.

Among the components of food materials the specific heat of water is much greater than those of the others, and the specific heats of carbohy-

Chapter 1. Physical Properties and Microbiology of Foods

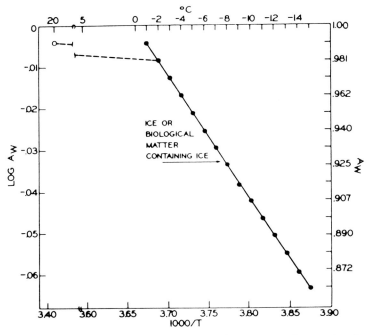

Figure 5. Relationship between water activity and temperature for samples above and below freezing. (From Fennema, 1978.)

drates and proteins are not very different. So many low-fat foods are approximated roughly to be a mixture of two components—a mixture of water and solids—in terms of the specific heat. The classic Siebel's correlation (1892), which is still applicable for many foods at temperatures above 0°C, is one such approximation:

$$C_m = (1.0 X_w^w + 0.2 X_s^w) \times 4187 \quad J/kg \cdot K \tag{6}$$

$$X_w^w + X_s^w = 1.0$$

where the subscripts s and w are for solid and water, respectively, 1.0 and 0.2 are the specific heats of water and solids (kcal/kg · °C), respectively, and 4187 is the conversion factor to SI (Système Internationale) units. Equation (6) may give a little lower value in a low moisture range, as Fig. 6(a) shows. For animal and fish meats Riedel (1957) presented the following correlation:

$$C_m = 0.4 + 0.6 X_w^w \tag{7}$$

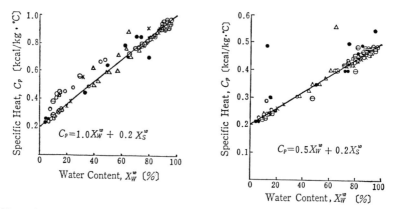

Figure 6. (a) Correlation between specific heat and water content of unfrozen food: ⊖ vegetables, △ meats, ● dairy products, × fruits, ○ miscellaneous. (b) Correlation between specific heat and water content of frozen foods, with same symbols as in (a).

Charm (1963) separated the specific heat of fat and used the following equation in a calculation:

$$C_m = (1.0X_w^w + 0.5X_F^w + 0.33X_s^w) \times 4187 \quad \text{J/kg} \cdot \text{K}$$

$$X_w^w + X_F^w + X_s^w = 1.0 \tag{8}$$

where X_F^w is the weight fraction of fat.

For frozen foods the coefficient of X_w^w in Eq. (6) may be changed to 0.5 as Fig. 6(b) shows, because the specific heat of ice is about 0.5 kcal/kg · °C. However, the ice fraction at a temperature below freezing point differs depending on foods. And the apparent specific heat involves the latent heat of fusion of ice. For the ice fraction the following simplest empirical equation was given by Heiss (1933):

$$X_{ice}^w = X_w^w \left(1 - \frac{T_f}{T}\right) \tag{9}$$

where T_f is the initial freezing point (°C) and T is a given temperature (°C). For more elaborate predictions of ice fractions, refer to Heldman (1974) and Schwartzberg (1976). A thorough compilation of published equations is given by Miles et al. (1983). Figure 7 shows an experimental sample of the apparent specific heats as a function of subfreezing temperature (Sakai and Hosokawa, 1977).

Chapter 1. Physical Properties and Microbiology of Foods

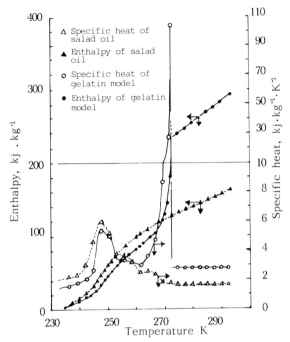

Figure 7. Specific heat and enthalpy (salad oil and gelatin model). (From Sakai and Hosokawa, 1977.)

2. Thermal Conductivity

Thermal conductivity is an important property for prediction and control of the thermal processings of foods. It is defined by Fick's first law for the one-dimensional heat conduction as

$$q = \lambda \frac{dT}{dx} \qquad (10)$$

where q is heat flux (J/m² · s), T is temperature (K), x is the distance in the x direction (m), and λ is the thermal conductivity (J/m · s · K or W/m · K).

The thermal conductivity is a nonadditive property. The thermal conductivity of a composite material is not determined simply from a chemical composition and the physical states of the components, but is affected by a dispersion structure of the composite material. Thus, unlike the specific heat, the mean value of the thermal conductivities of the components does not mean anything for the composite material. The thermal condutivity of such a heterogeneous composite material to be described by Eq. (10) is called the "effective" thermal conductivity.

Although a lot of measurements have been done, the effective thermal conductivity of food materials has not been understood systematically as a function of composition and structure. There have been two types of difficulty that prevented the systematic understanding of the effective thermal conductivity of foods. One of the difficulties was that the thermal conductivities intrinsic to proteins and carbohydrates were unknown, since direct measurements of the intrinsic thermal conductivity are impossible for powdery materials such as proteins and carbohydrates. The intrinsic thermal conductivities of fats and water were measured directly. The other difficulty was that, among more than 10 equations presented so far, the adequate equation to predict the effective thermal conductivity of a given food with a given heterogeneous structure was unknown due to the irregular and complex structures of food materials.

Recently, Yano et al. (1981) and Kong et al. (1982a, 1982b, 1982c) found that for protein gels the following series-model equation was the best among four theoretical equations to explain the experimental data without any irrationality:

$$\lambda_e = \frac{1}{(X_1^v/\lambda_1) + (X_2^v/\lambda_2) + (X_3^v/\lambda_3)}$$
$$X_1^v + X_2^v + X_3^v = 1$$
(11)

where X_i^v is the volume fraction of component i, and the subscripts e, 1, 2, and 3 are for composite material, water, protein, and fat, respectively. The intrinsic thermal conductivity values of proteins to be used in Eq. (11) are shown in Table I together with the values of other food components. The values of potato starch and agar are taken from a review paper (Yano, 1978). The intrinsic thermal conductivity values of the proteins are in the order of their hydrophobicities, which are calculated from their amino acid composition.

For a dispersion of spherical droplets, the following Maxwell-Eucken equation will give the best prediction instead of Eq. (11) in a wide range of volume fraction of the droplets (Lentz, 1961; Yano et al., 1983):

$$\frac{\lambda_e}{\lambda_c} = \frac{\gamma + 2 - 2\phi(1 - \gamma)}{\gamma + 2 + \phi(1 - \gamma)}$$
(12)

where $\gamma = (\lambda_d/\lambda_c)$, ϕ is the volume fraction of the dispersed phase, and the subscripts c and d are for the continuous phase and the dispersed phase, respectively.

For fruit juices and sugar solutions the following empirical equation will be applicable (Riedel, 1949):

Chapter 1. Physical Properties and Microbiology of Foods

$$\lambda = (0.565 + 0.00180t - 0.581 \times 10^{-5}t^2)(1 - 0.54X_s^w) \quad W/m \cdot K \quad (13)$$

for $t = 0-80°C$ where t is temperature (°C).

Other correlations are compiled by Jowitt et al. (1983).

3. Thermal Diffusivity

The thermal diffusivity for a homogeneous material is defined by the Fick's second law as follows. For one-dimensional heat conduction,

$$\frac{\partial T}{\partial t} = \frac{\lambda}{C\rho}\left(\frac{\partial^2 T}{\partial x^2}\right) = \alpha \frac{\partial^2 T}{\partial x^2} \quad (14)$$

where C is the specific heat capacity, t is time, ρ is density, and $\alpha \equiv (\lambda/C\rho)$ is the thermal diffusivity. As Eq. (14) shows, the thermal diffusivity is essential for prediction of temperature distribution and its change.

For a heterogeneous composite material, Eq. (14) is also used macroscopically replacing C, ρ, and λ by C_m, ρ_m, and λ_e, respectively. For C_m refer to Eq. (5) through Eq. (8). For λ_e refer to Eq. (11) through Eq. (13).

For water-rich foods, Riedel (1969) presented the following correlation:

$$\alpha = 0.088 \times 10^{-6} + (\alpha_w - 0.088 \times 10^{-6})X_w^w \quad m^2/s \quad (15)$$

where α_w (m²/s) is the thermal diffusivity of water at the same temperature as that of α. Equation (15) would be applicable for $X_w^w > 0.4$.

Table I Intrinsic Thermal Conductivities of Proteins and Carbohydrates[a]

Component	Unfrozen (1-3°C) (W/m · K)	Frozen (-9 to -11°C) (W/m · K)
Water	0.568	2.30
Gelatin	0.380	0.613
Meat protein	0.342	0.581
Soy protein	0.300	0.488
Egg albumin	0.238	0.403
Wheat gluten	0.219	0.315
Milk casein	0.200	0.273
Potato starch	0.252	0.376
Agar	0.259	0.389
Fats or oils	0.14–0.19	

[a] From Yano (1981).

II. MICROBIOLOGY OF FOODS

A. Microorganisms Associated with Foods

One of the main purposes of food packaging is to isolate the foods from contact with microorganisms that spoil or deteriorate the foods. Groups of microorganisms of importance associated with foods are shown in Tables II and III (Hayes, 1985). In Table II, *Campylobacter,* which causes the

Table II Principal Genera of Bacteria Associated with Foods[a]

Group Description	Genus
Gram-negative	
Spiral or curved rods	*Campylobacter*
Aerobic rods to cocci	*Acetobacter*
	Alcaligenes
	Alteromonas
	Brucella
	Halobacterium
	Halococcus
	Pseudomonas
Facultatively anaerobic rods	*Aeromonas*
	Enterobacter
	Erwinia
	Escherichia
	Flavobacterium
	Proteus
	Salmonella
	Serratia
	Shigella
	Vibrio
	Yersinia
Cocci and coccobacilli	*Acinetobacter*
	Moraxella
Gram-positive	
Cocci	*Leuconostoc*
	Micrococcus
	Pediococcus
	Staphylococcus
	Streptococcus
Endospore-forming rods	*Bacillus*
	Clostridium
	Desulfotomaculum
Asporogenous rods	*Brochothrix* (*Microbacterium*)
	Corynebacterium
	Lactobacillus
	Mycobacterium

[a] From Hayes (1985).

second most common food poisoning enteritis, is a group recognized recently. Among aerobic rods and cocci the most important genus is *Pseudomonas*. Many species of *Pseudomonas* grow at low temperatures, causing food spoilage. Proteinases and lipases are produced by some species of *Pseudomonas* at a low temperature near 0°C even more than at 10°C and above (Aibara, 1980). *Halobacterium* and *Halococcus* grow at a high salt concentration over 12%, producing red or pink patches.

Facultatively anaerobic bacteria grow under either aerobic or anaerobic conditions. Among this group *Salmonella* is the most common causative microorganism of food poisoning. Developments in the detection of *Salmonella* in foods were recently reviewed by Andrews (1985) and others.

Among gram-positive cocci, *Leuconostoc*, *Pediococcus*, and *Streptococcus* produce lactic acid by the fermentation of glucose, causing spoi-

Table III   Principal Genera of Fungi Associated with Foods[a]

	Genus
Moulds	
Class	
Phycomycetes	*Mucor*
(subclass: Zygomycetes)	*Phytophthora*
	Rhizopus
	Thamnidium
Ascomycetes	*Byssochlamys*
	Claviceps
	Neurospora
	Sclerotinia
Fungi imperfecti	*Alternaria*
	Aspergillus
	Botrytis
	Cladosporium
	Fusarium
	Geotrichium
	Penicillium
	Sporendonema
	Sporotrichum
Yeasts	
Class	
Ascomycetes	*Debaryomyces*
	Pichia
	Saccharomyces
Fungi imperfecti	*Candida*
	Rhodotorula
	Torulopsis

[a] From Hayes (1985).

lage of vacuum-packed meats and milk. However, some species of *Streptococcus* are useful as starter organisms for making cheese and yogurt. One species of *Staphylococcus*, *S. aureus*, produces heat-resistant enterotoxins. *Staphylococcus aureus* tolerates low water activity down to 0.86 and grows at fairly high salt concentration.

Endospore-forming rods are very important in food packaging due to the extreme heat resistance of the spores. *Bacillus* species are aerobic or facultatively anaerobic, while *Clostridium* and *Desulfotomaculum* are anaerobic. These organisms cause characteristic spoilage of canned foods. *Bacillus stearothermophilus*, the endospore of which is extremely heat-resistant (some 10 times more resistant than the spores of *Clostridium botulinum*), is responsible for "flat sour" spoilage in low-acid foods—that is, it produces acids from carbohydrates to sour the food but no gas is formed. *Bacillus coagulans* is also responsible for "flat sour" spoilage in acid foods. *Clostridium botulinum*, the causative organism of botulism, produces a toxin that kills an adult at a level of 10^{-8} g. Other *Clostridium* species produce large amounts of carbon dioxide and hydrogen by the fermentation of carbohydrates, causing "swell" spoilage in canned foods. *Desulfotomaculum nigrificans* causes "sulfur stinker" spoilage. It produces hydrogen sulfide, which reacts with the iron of the container to cause blackening of the food. The aroma of the blackened food is very unpleasant. A new type of "flat sour" spoilage of canned coffee was found recently, caused by unidentified organisms similar to *D. nigrificans* (Nakayama and Samo, 1980; Nakayama and Shinya, 1981).

Among asporogenous rods, *Lactobacillus*, like *Streptococcus*, causes spoilage by producing lactic acid, but is used as starter organisms in making cheese and yogurt. *Mycobacterium* comprises *M. tuberculosis*, which at one time caused milk-borne tuberculosis.

Molds shown in Table III are multicellular fungi that form a filamentous branching growth known as a mycelium, while most of the yeasts are unicellular fungi that do not form a mycelium. Most molds grow under aerobic, acidic, and relatively low water activity conditions as compared with bacteria. Molds are more sensitive to heat than the endospore-forming bacteria. However, care should be taken because certain molds produce heat-resistant toxic metabolites. The toxic metabolites include aflatoxins produced by *Aspergillus flavus*, which caused "turkey X disease," ochratoxins by *A. ochraceus*, fusarial toxins by *Fusarium* and possibly *Cladosporium*, and the "yellow rice" toxins of *Penicillium islandicum* and other strains.

Yeasts grow under either aerobic or anaerobic conditions. They are heat-sensitive and readily killed by heat treatment. Osmophilic strains of *Saccharomyces* and *Torulopsis* grow at a high sugar concentration of 65–70%, tolerating low pH conditions less than 4.0.

B. Environmental Conditions and Growth of Microorganisms

1. Temperature

The temperature range in which microorganisms grow is quite different depending on species, even within a genus. Table IV shows a rough classification.

Most of the microorganisms, either bacteria, molds or yeasts, are mesophiles. Psychrophiles include certain species of *Pseudomonas, Vibrio, Fusarium,* and *Cladosporium.*

Bacillus stearothermophilus, the spore of which is extremely heat-resistant, is a thermophile. It grows at 60°C but does not grow below about 35°C. Therefore, if the foods are cooled quickly down to below 35°C after the heat sterilization operation and held at the low temperature, the "flat sour" spoilage caused by *B. stearothermophilus* will be prevented even though spores of the organism may survive. On the other hand, canned hot drinks in vending machines held at about 55°C are media for anaerobic thermophiles to work on. A new type of "flat sour" spoilage caused by anaerobes similar to *Desulfotomaculum nigrificans* is such a case (Nakayama and Samo, 1980; Nakayama and Shinya, 1981).

2. pH

The logarithm of the reciprocal of the hydrogen ion concentration is called pH. The pH value is in the range 0–14, and the pH of pure water is 7. If a pH value is lower than 7, the solution is acidic because the hydrogen ion concentration is higher than that of the neutral pure water. If a pH value is higher than 7, the solution is alkaline. Every microorganism has an optimum pH range for its growth, and in the higher or lower pH range its growth is inhibited.

Most bacteria favor a neutral pH (6.0–7.5). *Clostridium botulinum* cannot grow at a pH below 4.5. That is why the low-acid foods with pH values above 4.5 are sterilized more severely than the high-acid foods with pH

Table IV Relationship between Temperature and Growth of Microorganisms

Group of Microorganisms	Temperature		
	Minimum	Optimum	Maximum
Psychrophiles	−10–5°C	10–20°C	20–40°C
Mesophiles	10–15°C	25–40°C	40–50°C
Thermophiles	35–45°C	50–75°C	60–80°C

below 4.5. However, *C. pasteurianum* and *C. butyricum* grow in a pH range of 4.0–4.5, sometimes causing spoilage with gas production. *Bacillus coagulans* also grows in this pH range and is a causative organism for "flat sour" spoilage (Hayes, 1985).

Yeasts and molds favor an acidic pH (3.5–5.5), with some exceptions.

3. Oxygen

Microorganisms that require free oxygen for their growth are called aerobic organisms, and microorganisms that grow in the absence of free oxygen are called anaerobic organisms. Microorganisms that grow in either the presence or the absence of free oxygen are called facultative anaerobic organisms.

Bacteria involve a few obligate or strict aerobes, many facultative anaerobes, and a few obligate anaerobes. Some species of *Clostridium* are obligate anaerobes. Molds associated with foods are aerobic. Yeasts are facultative anaerobic organisms. Flora of microorganisms will change depending on the oxygen tension inside the packaging.

4. Water Activity

The definition of water activity, a_w, is given in Section I,A,2. It is more useful than the water content of foods since it reflects the physicochemical state of the water involved in the foods.

Most bacteria favor a high water content with a_w above 0.94. However, *Micrococcus* and *Streptococcus* tolerate an a_w of 0.86, and *Halobacterium* and *Halococcus* an a_w of 0.75. Most yeasts grow above an a_w of 0.87, although some osmophilic yeasts grow in high sugar concentrations of jams, syrups, and jellies with an a_w of 0.62. Molds tolerate an a_w of 0.75 or 0.70.

Special attention should be paid to a_w inside the package. The change in a_w may change the microbial flora, permitting growth of unexpected microorganisms. When packaging was not popular, the unpackaged rice cake permitted growth of only harmless molds. Bacteria did not grow on the unpackaged rice cake due to its low a_w. However, after modern packaging became popular, the growth of harmless molds was prevented but it was replaced by the growth of some facultative anaerobic bacteria such as *Staphylococcus* (some species torelate an a_w of 0.86), *Lactobacillus,* and *Bacillus* (Ohta, 1976). This example illustrates that more careful sanitation may be required for a new food technology.

When microbial deterioration is considered in relation to water activity, a great care should be paid to the hysteresis shown in Fig. 1. Suppose that the a_w of a food is adjusted via sorption or desorption of water. Generally speaking, microorganisms can grow and react at lower levels of a_w in the food adjusted via desorption as compared with the food adjusted via sorption (Labuza et al., 1972; Plitman et al., 1973). An explanation of this difference can be given by the ink-bottle structure shown in Fig. 2. Upon desorption, the measured a_w is the a_w of the bottle-neck water, but microorganisms may use the water of higher a_w inside the ink bottle.

For the same reason, exposure of packed foods to repeated temperature change—that is, to repeated sorption and desorption of water in the foods—may accelerate microbial growth and reactions more than in foods stored at a constant temperature, since by the repeated sorption and desorption condensed water may fill part of the ink-bottle structure.

C. Sterilization

The recent requirement for long shelf life for widespread distribution of foods changed the purpose of heat treatment from killing the pathogenic microorganisms involved in foods (so called pasteurization or disinfection) to inducing sterility, that is, permitting no microbial growth at all. The pathogenic organisms are heat-sensitive and killed by mild heat treatments. For sterilization, however, more heat-resistant organism such as endospore-forming bacteria must be killed. The sterilization of foods can be done mostly by one, or combination, of the processes of irradiation by either ultraviolet radiation or γ rays, and heat treatments.

Among ultraviolet rays (UV) with a wavelength of 100–380 nm, the wavelength around 265 nm is most effective for killing microorganisms. It is coincident with the maximum UV absorption spectrum of thymine, one of the four constitutive bases of deoxyribonucleic acid (DNA). With UV absorption thymine forms thymine dimer, which prevents the normal translation of DNA information, causing the death of microorganisms. However, repair of the injured DNA will proceed within the cell to a certain extent.

Gamma rays have a wavelength of 10^{-2}–10^{-8} nm. Due to their high energy level, irradiating γ rays split molecules into ions, including electron, that cause many fatal reactions within the microorganisms. Cobalt-60 γ rays are used for commercial purposes. Three levels of the γ-ray irradiation are recommended: (1) Radappertization: Dose of γ rays 3–5 Mrad. *Clostridium botulinum* will be killed, but unfavorable side reactions may happen. (2) Radicidation: 0.5–0.8 Mrad. Non-spore-forming food-poisoning bacteria such as *Salmonella* will be killed. (3) Radurization: 0.1–0.3 M rad. This only reduces the number of viable microorganisms for longer preservation.

Since the temperature is not raised by irradiation sterilization, the process is called cold sterilization and is used when heat treatments are not applicable.

Heat sterilization is more common than irradiation sterilization. The mechanism of killing microorganisms is different from those above mentioned. Killing by moist heat is caused mainly by the denaturation of proteins, mostly enzymes. The denaturation of enzymes affects the variety of nutrient requirements of heat-treated microorganisms. That is, on more nutritious foods under more favorable pH and temperature, microorganisms are more torelant to heat treatments. Besides the denaturation of proteins, damage is done to membrane functions and ribosomes. These damages are subject to repair to a certain extent. Microorganisms are more tolerant to dry heat than to moist heat. Dry heat is believed to cause oxidation rather than denaturation of proteins, although the effect of the presence of oxygen is not significant (Pheil et al., 1967).

In the heat sterilization of foods, control of side reactions is very important in practice. Many reactions, such as destruction of vitamins, oxidation of lipids, and browning, take place during heat treatments. Some side reactions have to be suppressed and others must be allowed to proceed. A well-known way of controlling side reactions while completing heat sterilization is the high-temperature short-time method (HTST). The HTST principle comes from the difference in the activation energies between the death reaction of microorganisms and the side reactions occuring among low-molecular-weight molecules.

Many related reactions, including the thermal death of microorganisms, are approximately of the first order, and the temperature dependence of the rate constants obeys the Arrhenius equation:

$$\frac{dX}{dt} = \pm kX \qquad (16)$$

$$k = Ae^{-E/RT} \qquad (17)$$

where X is the concentration of a reactant, k the reaction rate constant, t time, A the Arrhenius constant, E the activation energy, R the gas constant, and T the absolute temperature.

The activation energy of thermal death is mostly in the range 270–340 kJ/mol, which is coincident with the activation energy of the denaturation of many proteins. The activation energies of low-molecular-weight reactions such as the destruction of vitamins, the oxidation of lipids, and the browning reactions are in the range 80–110 kJ/mol. As Eqs. (16) and (17) show, the rate of thermal death reaction increases more sharply than that of the low-molecular-weight side reactions, if the sterilization temperature increases. Therefore, the high-temperature short-time heat treatment to complete the sterilization can be with fewer low-molecular-weight side reactions: the higher the temperature, the fewer the low-molecular-weight side reactions. That leads to ultrahigh temperature heat treatment such as 150°C for 0.5 s, which is applicable only to liquid foods. However, the HTST may not be the best way of sterilization if some side reactions must proceed to a certain extent, or some types of protein denaturation have to be depressed.

Superheated steam is used for the sterilization of rather dry powder or particles of foods. Microwaves are also used for sterilization, although the effect is mostly due to temperature rise.

Sterilization of packaging material is discussed in Chapter 3.

NOMENCLATURE

a	Activity
a_w	Water activity
C	Specific heat capacity (J/kg · K)
f	Fugacity (Pa)
$f°$	Fugacity for reference state (Pa)
h^E	Excess partial molar enthalpy (J/mol)
p	Water vapor pressure in equilibrium with food (Pa)
P_0	Saturation vapor pressure of pure water (Pa)
P'	Condensation vapor pressure in a capillary (Pa)
q	Heat flux (J/m² · s)
r	Radius of capillary (m)
R	Gas constant (J/mol · K)
t	Time (s), or temperature (°C)
T	Temperature (K)
T_f	Freezing point (K)
V_L	Molar volume of condensed water (m³/mol)
x	Distance in x direction (m)
X^v	Volume fraction
X^w	Weight fraction
α	Thermal diffusivity (m²/s)
θ	Contact angle (rad)
λ	Thermal conductivity (W/m · K)
μ	Chemical potential (J/mol)
ρ	Density (kg/m³)
σ	Surface tension (N/m)
ϕ	Volume fraction of dispersed phase

Subscripts

c　continuous phase
d　dispersed phase
e　effective value
F　fats
i　ith component
m　mean value
s　solids
w　water

REFERENCES

Aibara, K. (1980). Shokuryo Hozo To Biseibutsu. In "Shokuryo Hozogaku" (M. Fujimaki, ed.), p. 63. Asakura Shoten, Tokyo.

Andrews, W. H. (1985). A review of culture methods and their relation to rapid methods for detection of *Salmonella* in foods. *Food Technol.* **March:**77. Related articles appear also on 83, 90, 95, 103.

Booth, G. H., Cooper, A. W., and Robb, J. A. (1968). Bacterial degradation of plasticized PVC. *J. Appl. Bacteriol.* **31:**305–310.

Charm, S. E. (1963). "The Fundamentals of Food Engineering." Avi, Westport, Conn.

Fennema, O. (1978). In "Dried Biological Systems" (J. H. Crowe and J. S. Clegg, eds.). Cited in Fennema, O. (1981). Water activity at subfreezing temperature. In "Water Activity: Influences on Food Quality" (L. B. Rockland and G. F. Stewart, eds.), pp. 713–732. Academic Press, New York.

Food Engineering Editors (1962). Special report: Plant handbook data. *Food Eng.* **34:**89.

Gal, S. (1972). Über die Ausdrucksweisen der Konzentration des Wasserdampfes bei Wasserdampf-Sorptionsmessungen. *Helv. Chim. Acta* **55:**1752–1757.

Gal, S. (1975). Recent advances in techniques for the determination of sorption isotherms. In "Water Relations of Foods" (R. B. Duckworth ed.), pp. 139–154. Academic Press, London.

Hayes, P. R. (1985). "Food Microbiology and Hygiene." Elsevier Applied Science, London.

Heiss, R. (1933). Z. ges. Kälteind. Bd. 40, S. 97, 122u. 144. Cited by Kato, S. (1971), in "Teion Seibutsugaku Gaisetsu" (T. Nei, ed.), pp. 172. Tokyo Daigaku Shuppankai, Tokyo.

Heldman, D. R. (1974). Predicting the relationship between unfrozen water fraction and temperature during food freezing using freezing point depression. *Trans. ASAE* **17,** 63.

Jowitt, R., Escher, F., Hallström, B. Meffert, H. F. Th., Spiess, W. E. L., and Vos, G. (1983). "Physical Properties of Foods." Applied Science Publishers, London.

Karel, M. (1975). Water activity and food preservation. In "Principles of Food Science, Part II, Physical Principles of Food Preservation" (M. Karel, O. R. Fennema and D. B. Lund, ed.), pp. 237–263. Marcel Dekker, New York.

Kong, J. Y., Miyawaki, O., Nakamura, K., and Yano, T. (1982a). The "intrinsic" thermal conductivity of some wet proteins in relation to their hydrophobicity: Analysis on gelatin gel. *Agric. Biol. Chem.* **46,** 783.

Kong, J. Y., Miyawaki, O., Nakamura, K., and Yano, T. (1982b). The "intrinsic" thermal conductivity of some wet proteins in relation to their hydrophobicity: Analysis on gels of egg-albumin, wheat gluten and milk casein. *Agric. Biol. Chem.* **46,** 789.

Kong, J. Y., Iibuchi, S., Miyawaki, O., and Yano, T. (1982c). Analysis and prediction of the effective thermal conductivities of meats. *Agric. Biol. Chem.* **46,** 1235.

Labuza, T. P., Cassil, S., and Sinskey, A. J. (1972). Stability of intermediate moisture foods. 2. Microbiology. *J. Food Sci.* **37,** 160.

Lentz, C. P. (1961). Thermal conductivity of meats, fats, gelatin gels, and ice. *Food Technol.* **May,** 243.

McBain, J. W. (1935). An explanation of hysteresis in the hydration and dehydration of gels. *J. Am. Chem. Soc.* **57,** 699.

Miles, C. A., van Beek, G., and Veerkamp, C. H. (1983). Calculation of thermophysical properties of foods. In "Physical Properties of Foods" (R. Jowitt et al., eds.), pp. 269–312. Applied Science Publishers, London.

Nakayama, A., and Samo, S. (1980). Evidence of "flat sour spoilage by obligate anaerobes in marketed canned drinks. *Bull. Jpn. Soc. Sci. Fisheries* **46,** 1117.

Nakayama, A., and Shinya, R. (1981). A new type of flat sour spoilage of commercial canned coffee. *J. Food Hyg. Soc. Jpn.* **22,** 30.

Ohta, Y. (1976). Bacteria found in plastic-packed rice cake (in Japanese). *Kagaku To Seibutsu* **14,** 157.

Pheil, C. G., Pflug, I. J., Nicholas, R. C., and Augustin, J. A. L. (1967). Effect of various gas atmospheres on destruction of microorganisms in dry heat. *Appl. Microbiol.* **15,** 120.

Plitman, M., Park, Y., Gomez, R., and Sinskey, A. J. (1973). Viability of *Staphylococcus aureus* in intermediate moisture meats. *J. Food Sci.* **38,** 1004.

Riedel, L. (1949). Wärmeleitfähigkeitsmessungen an Zuckerlösungen, Fruchtsäften und Milch. *Chem.-Ing.-Tech.* **21,** 340.

Riedel, L. (1957). Kalorimetrische Untersuchungen über das Gefrieren von Fleisch. *Kältetechnik* **9,** 38.

Riedel, L. (1969). Temperaturleitfähigkeitsmessungen an Wasserreichen Lebensmitteln. *Kältetechnik* **21,** 315.

Sakai, N. and Hosokawa, A. (1977). Enthalpy changes of foods in freezing and thawing processes (II). *Nogyo Kikai Gakkaishi* **39,** 321 (in Japanese).

Schwartzberg, H. G. (1976). Effective heat capacities for the freezing and thawing of food. *J. Food Sci.* **41,** 152.

Siebel, E. (1892). Specific heats of various products. *Ice Refrig.* **2,** 256.

Spiess, W. E. L., and Wolf, W. R. (1983). The results of the COST 90 project on water activity. In "Physical Properties of Foods" (R. Jowitt et al., eds.), pp. 65–91. Applied Science Publishers, London.

Strasser, J. (1969). Detection of quality changes in freeze-dried beef by measurement of the sorption isobar hysteresis. *J. Food Sci.* **34,** 18.

van den Berg, C., and Bruin, S. (1981). Water activity and its estimation in food systems: Theoretical aspects. In "Water Activity: Influences on Food Quality" (L. B. Rockland and G. F. Stewart, eds.), pp. 1–61. Academic Press, New York.

van den Berg, C., and Leniger, H. A. (1976). Proc. Int. Congr. Engng. Food, Boston, August 9–13. Cited in van den Berg and Bruin (1981).

Yano, T. (1978). Shokuhin no dennetsu bussei. *New Food Industry,* **20,** 55 (in Japanese).

Yano, T. (1981). Effective thermal conductivities of food gels. *J. Jpn. Soc. Food Eng.* **1,** 26–37 (in Japanese).

Yano, T., Kong, J. Y., Miyawaki, O., and Nakamura, K. (1981). The "intrinsic" thermal conductivity of wet soy protein and its use in predicting the effective thermal conductivity of soybean curd. *J. Food Sci.* **46,** 1357.

Yano, T., Matsushita, Y., Shiinoki, Y., and Sakiyama, T. (1983). Yugen Yosoho o mochiita niso-bunsankei no Yuko-netsudendodo kaiseki. Abstract of Kagaku Kogaku Kyokai 17th Shuki Taikai, p. 315.

CHAPTER 2

Oxidation of Food

Setsuro Matsushita[1]
Research Institute for Food Science,
Kyoto University, Kyoto, Japan

When foods have contact with air, food components are oxidized directly or indirectly. The most frequently occurring reaction is lipid oxidation. This results in reduction or destruction of essential fatty acids, formation of off-flavors, formation of brown pigments, and alteration of pigments and flavors. Therefore, control of lipid oxidation reaction is necessary for the preservation of foods. Keeping foods away from oxygen is an ideal procedure. For that purpose, foods are often wrapped with a layer that is impervious to oxygen, and nitrogen gas or a free-oxygen absorber may be enclosed in the package. However, it is difficult to apply such management to all foods, and therefore use of antioxidants cannot be avoided. There are many factors that accelerate lipid oxidation, and a full understanding of those factors may help to find a way to protect food from lipid oxidation.

I. OXIDATION OF LIPIDS

There are two pathways of nonenzymatic lipid oxidation: autoxidation and photosensitized oxidation. These are different in mechanism and products.

A. Autoxidation

Oxidation occurs in the unsaturated fatty acids contained in lipids. Autoxidation is a reaction that occurs slowly at room temperature. When unsaturated fatty acids come into contact with oxygen, oxidation proceeds

---

[1] Present address: Toua University, Shimonoseki, Japan.

slowly at a uniform rate in the first stage. After the oxidation has proceeded to a certain point, the reaction enters the second step, which has an accelerating rate of oxidation. The initial step is called an induction period (Fig. 1). After the induction period, the products work as the catalyst on the reaction, and therefore this reaction is called autocatalytic autoxidation. The reaction is a chain reaction with radicals.

Based on the radical theory, the reaction is explained as divided into three stages: initiation, propagation, and termination (Lundberg, 1962):

Initiation $\quad RH + O_2 \xrightarrow{catalyst} R^\cdot + {}^\cdot OOH$

$\quad\quad\quad\quad\quad\; RH \xrightarrow{catalyst} R^\cdot + {}^\cdot H$

Propagation $\; R^\cdot + O_2 \longrightarrow ROO^\cdot$

$\quad\quad\quad\quad\quad ROO^\cdot + RH \longrightarrow ROOH + R^\cdot$

Termination $\left. \begin{array}{l} ROO^\cdot + ROO^\cdot \\ ROO^\cdot + R^\cdot \\ R^\cdot + \text{Another radical} \end{array} \right\} \longrightarrow$ Nonradical products

A lipid molecule, RH, forms a free radical, $R^\cdot$, in the presence of oxygen and a catalyst; that is, the initiation starts by the action of an external energy source such as heat, light, or high-energy radiation, or by chemical initiating involving metal ions or heme proteins. The mechanisms of the initiation is still not fully understood. In propagation, the free radical, $R^\cdot$, combined with an oxygen molecule, forms a lipid peroxy radical, $ROO^\cdot$. This radical takes a hydrogen atom from another unsaturated fatty acid

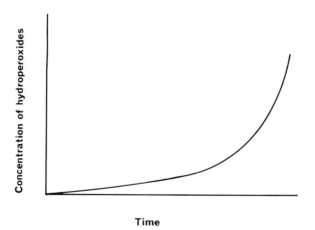

Figure 1. Hydroperoxide development in lipid autoxidation.

molecule and forms the hydroperoxide, ROOH. A new free radical is left on the unsaturated fatty acid molecule from which a hydrogen atom was removed. These reactions are repeated and hydroperoxide molecules are accumulated. The self-propagating chain reaction can be stopped by the termination reaction, where two radicals combine to give nonradical products.

The formation mechanism of hydroperoxides is shown more concretely. Elimination of a hydrogen atom occurs from the α-carbon to the double bond of allyl group, and a free radical is formed. In the case of linoleate, the hydrogen atom at the methylene group between two double bonds is easily eliminated (Fig. 2). The radical moves between two double bonds (pentadiene radical), and two positional isomers are formed after binding with oxygen molecules and taking hydrogen atoms from other linoleate molecules. At the same time, the double bond moves to form conjugated double bonds, followed by formation of hydroperoxides as described above. However, some peroxy radicals cyclize in linolenate and arachidonate, and the molecule is subjected to further oxidation. These are shown later as secondary products. At hydroperoxide formation, cis–trans rearangement occurs at the same time, though most of the double bonds of natural lipids are cis form. The mechanism of formation of monohydroperoxides is schematically shown in Figs. 3 and 4.

Table I shows the hydroperoxide positional isomers formed by the oxidation of methyl linoleate, methyl linolenate, and methyl arachidonate. In the cases of linolenate (Frankel, 1980) and arachidonate (Yamagata et al., 1983), the amounts of the inner isomers (12- and 13-isomers in the case of linolenate and 8-, 9-, 11-, and 12-isomers in the case of arachidonate) are less than those of the outer isomers (9- and 16-isomers in the case of linolenate and 5- and 15-isomers in the case of arachidonate). This ratio is considered to depend on the cyclization of the inner peroxy radicals (Neff et al., 1981). When antioxidants coexist, hydrogen atoms are given to

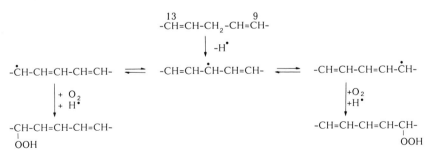

Figure 2. Formation of hydroperoxides in autoxidation of linoleate.

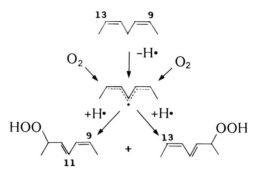

Figure 3. Mechanism of linoleate autoxidation. (From Frankel et al., 1977b.)

peroxy radicals from antioxidants before cyclization. Therefore, the ratio of hydroperoxy isomers formed is almost equal for each isomer (Peers et al., 1981; Yamagata et al., 1983). One example, arachidonate, is shown in Fig. 5.

Positional isomerization in which the oxygen atoms of the hydroperoxy groups exchange with atmospheric oxygen occurs in methyl linoleate monohydroperoxides (Chan, 1978). The reaction occurs at over 20°C and in nonpolar solvents. This isomerization occurs frequently in the case of linoleate hydroperoxides but occurs less often in the case of linolenate due to competing with the reaction of cyclization (Terao et al., 1984). Stereo-

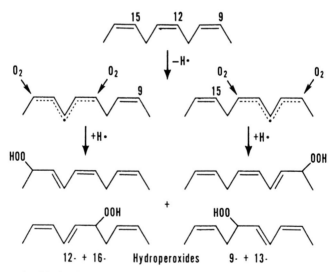

Figure 4. Mechanism of linolenate autoxidation. (From Frankel et al., 1977c.)

Table I GC-MS Analysis of Monohydroperoxide Isomers Formed by Autoxidation of Unsaturated Fatty Acids[a]

Methyl Ester	PV	Isomeric Hydroperoxides											
		5-	6-	7-	8-	9-	10-	11-	12-	13-	14-	15-	16-
													%
Oleate	461				27	23	23	27					
Linoleate	1249					50				50			
Linolenate	495					31		11	11	12			
Arachidonate		26			9	12		13	8			32	46

[a] From Frankel et al. (1977a, 1977b, 1977c), Frankel (1980), and Yamagata et al. (1983).

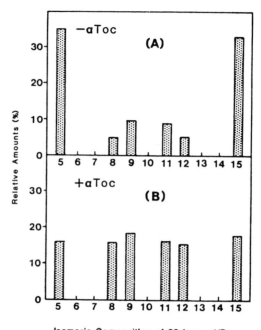

Figure 5. Isomeric compositions of monohydroperoxides on autoxidation of methyl arachidonate with or without α-tocopherol: (a) without α tocopherol, after 12-h incubation; (B) with α-tocopherol, after 120 h incubation. (From Yamagata et al., 1983.)

isomerization, the rearrangement of a cis–trans form to a trans–trans form, also happens at the same time as positional isomerization (Fig. 6) (Chan et al., 1979; Porter et al., 1980; Porter and Wujek, 1984). The structures of linoleate monohydroperoxide isomers are shown in Fig. 7.

B. Photosensitized Oxidation

When a mixture of unsaturated fatty acids and a photosensitizer is irradiated, unconjugated hydroperoxides are formed in addition to the same conjugated hydroperoxides that are produced by autoxidation (Rawls and Van Santen, 1970). This reaction depends on singlet oxygen, $^1O_2$ (type I). Singlet oxygen reacts directly with double bonds electrophilically by addition at either end of the double bond (Foote et al., 1965). That is, the number of hydroperoxide isomers formed is twice the number of the double bonds (Fig. 8 and Table II) (Terao and Matsushita, 1977; Frankel et al., 1979; Porter et al., 1979; Chan and Levette, 1979; Thomas and Pryor, 1980). Singlet oxygen reacts 1500 times faster with methyl linoleate than

Chapter 2. Oxidation of Food

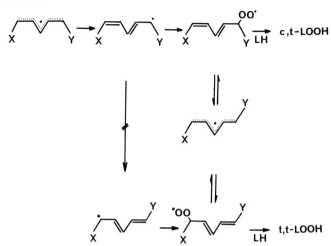

Figure 6. Formation of *cis,trans*- and *trans,trans*-hydroperoxides in the oxidation of methyl linoleate. (From Yamamoto et al., 1982.)

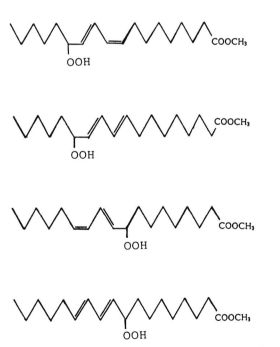

Figure 7. Structures of monohydroperoxides of methyl linoleate.

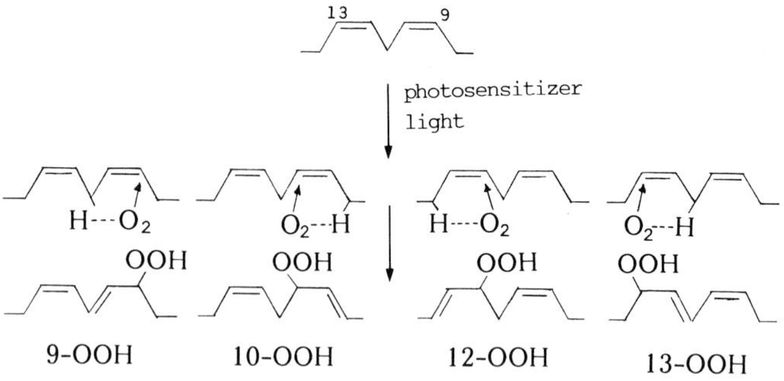

Figure 8. Mechanism of linoleate photosensitized oxidation.

does triplet oxygen (Rawls and van Santen, 1970). Using S to depict the sensitizer and an asterisk for the excited state, we have:

$$^1S + h\nu \longrightarrow {}^1S^* \rightsquigarrow {}^3S^*$$
$$^3S^* + {}^3O_2 \longrightarrow {}^1O_2^* + {}^1S$$
$$^1O_2^* + RH \longrightarrow ROOH$$

Chlorophyll, heme compounds, and their degradation products that are contained in foods act as the photosensitizers (Cobern et al., 1966). Food additives, erythrosine, phloxine, and rose bengale, are also photosensitizers (Umehara et al., 1979). Riboflavin is also considered to be as a photosensitizer, but it accelerates lipid oxidation by forming a radical (type II) (Chan, 1977).

C. Lipoxygenase

Lipoxygenases are distributed widely in the plant kingdom. Beans and wheat are especially rich in the enzyme activity. The reaction mechanism is a radical reaction similar to autoxidation (Fig. 9) (de Groot et al., 1975). Therefore, the products are conjugated hydroperoxides, but they are optically active against the products of autoxidation. Lipoxygenase prefers free fatty acids as substrate, though some can react with triglycerides. Soybean lipoxygenase contains three isozymes: one works at pH 9 and two work at pH 7 (Galliard and Chan, 1980; Axelrod et al., 1981). The one that works at pH 9 forms 13-L-LOOH for linoleic acid, and those that work at pH 7 form 9-D- and 13-L-hydroperoxides in the same ratio. That of corn germ works at pH 7 and forms 9-D-LOOH.

Table II GC-MS Analysis of Monohydroperoxide Isomers Formed by Photosensitized Oxidation of Unsaturated Fatty Acids[a]

Methyl Ester	PV	\multicolumn{12}{c}{Isomeric Hydroperoxides}
		5-	6-	7-	8-	9-	10-	11-	12-	13-	14-	15-	16-
													%
Oleate	1727					48	52						
Linoleate	1124					32	17		17	34		13	
Linolenate	1566					23	13		12	14		19	25
Arachidonate		10	5		15	13		14	15		9		

[a] From Frankel (1980), and Terao and Matsushita (1980, 1981).

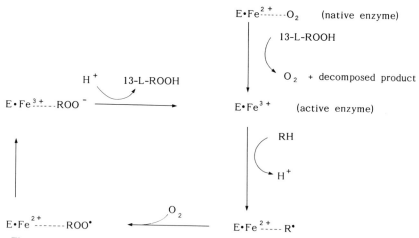

Figure 9. Reaction mechanism of soybean lipoxygenase. (From deGroot et al., 1975.)

D. Monohydroperoxides of Triglycerides and Phospholipids

In the case of both triglycerides and phospholipids, monohydroperoxide isomers formed are the same as described above for unsaturated fatty acids (Park et al., 1981; Terao et al., 1981). However, these lipids consist of several kinds of unsaturated fatty acids, and the fatty acids consist of hydroperoxides differing in degree of oxidation. With the increase of PV, monohydroperoxides from oleic acid increase and those of linolenic acid decrease. That is, this phenomenon shows that the rate of decomposition differs among hydroperoxides formed from different fatty acids.

II. SECONDARY PRODUCTS

Hydroperoxides are considered to be the primary stable products. But they are decomposed with accumulation and form various kinds of lower-molecular-weight compounds. These contain many kinds of compounds having odor, and these are called secondary products in a narrow sense. However, monohydroperoxides are changed in several ways, such as further oxidation, polymerization, and decomposition. These products are all included in the secondary products. Cyclization of peroxy radicals occurs as an alternative pathway prior to becoming hydroperoxides. These compounds are also dealt with in this section.

A. Further Oxygenated Products

Di-OH, tri-OH, or epoxy-hydroxy compounds are found in the chemically reduced mixture of autoxidized products formed from linoleate (Neff et al., 1978; Terao et al., 1975; Pryor et al., 1976) or linolenate (Neff et al., 1981; Toyoda et al., 1982). The unreduced forms of them are deduced from gas chromatography–mass spectroscopy (GC-MS) data, and some of their structures are determined after fractionation by HPLC (Neff et al., 1981; Coxon et al., 1981). More complicated mixtures were obtained by photosensitized oxidation of linoleate and linolenate (Neff et al., 1983) and also arachidonate autoxidation (Yamagata et al., 1984). Those structures are dihydroperoxides, hydroperoxy-epidioxides, hydroperoxy-bis-epidioxides, epoxy-hydroxy compounds, etc. These compounds must also be formed by the oxidation of triglycerides and phospholipids (Sessa et al., 1977; Terao et al., 1981).

Hydroperoxyepidioxide

B. Volatile Compounds

When hydroperoxides and further oxidized products are decomposed, various complex compounds are formed and rancid flavor appears. Diacetyl and 2,3-pentadione show butter flavor, 2,4-pentadienal has potato flavor, *cis*-4-heptenal and 2,4,7-decatrienal have fish flavor, 2-pentyl furan, 2-(1-pentenyl) furan, and 2-(2-pentenyl) furan show rancid flavor, etc. The formation mechanisms of the volatile compounds formed by autoxidation are considered to be fundamentally the same as that of heat decomposition (Terao and Matsushita, 1975; Frankel et al., 1981, 1983). Some examples of compounds produced by heat decomposition are shown in Tables III and IV. Hydrocarbons are also produced.

C. Polymerized Products

Upon lipid oxidation, formation of odor is generally noticed, but the amount of polymers formed is greater than the amounts of volatile compounds, as shown in Fig. 10.

III. FACTORS AFFECTING LIPID OXIDATION

The stability of fats to oxidation is a matter of primary concern in processed foods. Deterioration of fats depends mainly on autoxidation and

Table III GC-MS Analysis of Volatiles from Thermally Decomposed Methyl Linoleate Hydroperoxides[a]

Compound	Autooxidation (rel. %)	Photosensitized Oxidation (rel. %)	Origin
Acetaldehyde	0.3	0.4	?
Pentane	9.9	4.3	13-OOH
Pentanal	0.8	0.3	13-OOH
1-Pentanol	1.3	0.3	13-OOH
Hexanal	15	17	12-/13-OOH
2-Heptenal	Tr	9.9	12-OOH
1-Octen-3-ol	Tr	1.9	10-OOH
2-Pentylfuran	2.4	0.6	?
Me heptanoate	1.0	0.3	?
2-Octenal	2.7	1.5	?
Me octanoate	15	7.6	9-OOH
2-Nonenal	1.4	1.6	9-/10-OOH
2,4-Nonadienal	0.3	0.3	?
2,4-Decadienal	14	4.3	9-OOH
Me 8-Oxooctanoate	1.3	0.9	?
Me 9-Oxononanoate	19	22	9-/10-OOH
Me 10-Oxodecanoate	0.7	0.7	?
Me 10-Oxo-8-decenoate	4.9	14	10-OOH
Unidentified peaks	9.9	12	

[a] From Frankel et al. (1981).

is affected by several factors. It is necessary to know those factors and keep away from them.

A. Composition of Fatty Acids

Highly unsaturated fatty acids are more easily oxidized than lower unsaturated fatty acids. Therefore, the stability of fats can be determined from the iodine value.

B. Temperature

Generally, the higher the temperature, the faster the chemical reaction proceeds. Therefore, fats should be kept in a cold room. But it must be noted that autoxidation proceeds even at low temperature and that the decomposition of hydroperoxides is slow. Thus a rather high concentration of hydroperoxides is accumulated at low temperature.

Table IV GC-MS Analysis of Volatiles from Thermally Decomposed Methyl Linolenate Hydroperoxides[a]

Compound	Autooxidation (rel. %)	Photosensitized Oxidation (rel. %)	Origin
Ethane/ethene	10	3.2	16-OOH
Acetaldehyde	0.8	0.6	?
Propanal/acrolein	7.7	9.0	15-/16-OOH
Butanal	0.1	0.8	?
2-Butenal	0.5	11	15-OOH
2-Pentenal	1.6	1.2	13-OOH
2-/3-Hexenal	1.4	3.4	12-/13-OOH
2-Butylfuran	0.5	0.3	?
Me heptanoate	1.8	1.0	?
2,4-Heptadienal	9.3	8.8	12-OOH
Me octanoate	22	15	9-OOH
4,5-Epoxyhepta-2-enal	0.2	0.2	?
3,6-Nonadienal	0.5	1.1	9-/10-OOH
Me Nonanoate	0.7	0.3	?
Decatrienal	14	4.8	9-OOH
Me 8-Oxooctanoate	0.6	0.4	?
Me 9-Oxononanoate	13	12	9-/10-OOH
Me 10-Oxodecanoate	1.0	1.5	?
Me 10-Oxo-8-decenoate	4.2	13	10-OOH
Unidentified peaks	11	12	

[a] From Frankel et al. (1981).

C. Light

Light is the most effective accelerator for fat oxidation. Ultraviolet light, which has higher energy, affects this more than visible light. However, visible light also accelerates the reaction vigorously when a photosensitizer coexists. Not only sunlight but also light from a window or fluorescent light from the ceiling can accelerate the oxidation. Therefore, fat should be stored in a dark place or kept in a brown glass jar or a tin can.

D. Metals

Transition metals accelerate the oxidation of fats. However, it is not clear that they affect the formation of free radicals in the stage of initiation. But it is obvious that they accelerate the decomposition of hydroperoxides. Ferrous ion is more effective than ferric ion:

$$Fe^{2+} + ROOH \longrightarrow Fe^{3+} + RO^{\cdot} + OH^{-}$$
$$Fe^{3+} + ROOH \longrightarrow Fe^{2+} + ROO^{\cdot} + H^{+}$$

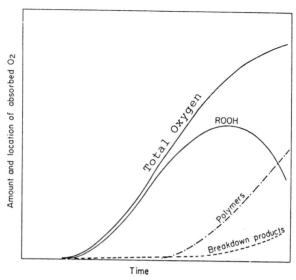

Figure 10. Distribution of oxygen absorbed during lipid oxidation. (From Karel, 1975.)

E. Hematin Compounds

The peroxidation of fats by hematin compounds is a basic deteriorative reaction of meat materials. Hematin compounds coordinate to hydroperoxides and lead hydroperoxides to decomposition (Fig. 11). In this case no valence change of iron ions is observed. When hematin compounds bind to NO or CO and become the low-spin condition, high acceleration is not observed (Taradgis, 1968).

F. Enzymes

The highly active enzyme that catalyzes lipid oxidation, lipoxygenase, is contained in beans and cereals. Therefore, it is necessary to heat such materials as soon as possible in the processing or cooking of such materials.

G. Air Interface Area

Fats are easily oxidized with an increase in the area in contact with air. Oils in plastic bottles are not stable for long periods because oxygen can enter through plastics. Some foods are sealed *in vacuo* or with nitrogen gas. The purpose is to protect oxidation of fats in foods by eliminating the contact of oil with air. Free-oxygen absorber is now often used in packages. When

Chapter 2. Oxidation of Food

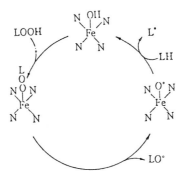

Figure 11. Oxidation of unsaturated fatty acid with hematin catalyst. (From Tappel, 1962.)

fish are frozen, ice crystals grow in the tissue, and air easily penetrates inside. Glaze, an ice cover on the fish surface, is used to prevent oxidation reaction. It must be noted that the level of dissolved oxygen in fats is seven to eight times that in water.

H. Water Activity

In studies with intermediate-moisture food (Fig. 12), water at higher A_w seems to inhibit the contact of air with fats. At an intermediate moisture region, $A_w = 0.6$–0.85, the oxidation rate increases as A_w increases. This

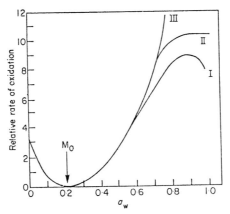

Figure 12. Overall effect of water activity on rates of lipid oxidation: Case I—dilution exceeds viscosity effect; case II—dilution balances viscosity effect; case III—viscosity effect predominates. (From Labza, 1975.)

is deduced to depend on the ease of moving of metal catalysts. In the monolayer region, with A_w around 0.4, fats are most stable toward oxidation. In the drier region, fats come into contact with oxygen directly and are easily oxidized.

IV. ANTIOXIDANT AND SINGLET OXYGEN QUENCHER

A. Antioxidants

Antioxidants are often used in processed foods. Antioxidants are free radical scavengers and cut the chain oxidation reaction. There are many kinds of antioxidants, and their reaction mechanisms are not uniform. However, the most popular antioxidants react as follows:

$$\begin{aligned} ROO^{\cdot} + AH &\longrightarrow ROOH + A^{\cdot} \\ \left. \begin{array}{c} ROO^{\cdot} + A^{\cdot} \\ A^{\cdot} + A^{\cdot} \end{array} \right\} &\longrightarrow \text{Inactive products} \end{aligned}$$

In the preceding section, it was noted that the ratio of monohydroperoxide isomers formed is uniform when antioxidant is present. The change of cis–trans form to trans–trans form of monohydroperoxides (in the case of linoleate) and the formation of cyclic peroxides are also inhibited by antioxidants.

Antioxidants used for foods as the food additives are all phenolic compounds. The synthetic antioxidants BHT (butylated hydroxytoluene) and BHA (butylhydroxyanisole) are very effective, but tocopherols are often used now because they are natural substances.

When antioxidants are used, the mixing of two or more antioxidants has a much better effect than either antioxidant alone. It is also more effective to use antioxidants with citric acid or ascorbic acid. These effects are called *synergism*. Ascorbic acid itself often shows pro-oxidative effects, especially in the presence of water, but it seems to become a hydrogen donor to the antioxidant radicals that have formed by donating hydrogen atoms to fatty acid radicals.

Tocopherols are often used for the stabilization of animal fats against oxidation. They occur widely in most vegetable oils and fats. Those used for food processing are generally mixtures of α-, β-, γ-, and δ-tocopherols, which are derived from molecular distillation of vegetable oils.

B. Singlet Oxygen Quenchers

There are some synthetic singlet oxygen quenchers, but β-carotene and tocopherols are found as natural substances. When a substrate (A) and tocopherol (T) exist, the following reaction occurs (Foote and Denny, 1971; Matsushita and Terao, 1980):

$$A + {}^1O_2 \longrightarrow AO_2$$
$$T + {}^1O_2 \longrightarrow T + {}^3O_2 \quad \text{(physical quenching)}$$
$$T + {}^1O_2 \longrightarrow TO_2 \quad \text{(chemical quenching)}$$
$$^1O_2 \longrightarrow {}^3O_2$$

There are two ways to quench singlet oxygen. One is to liberate energy physically without change of chemical structure of the quencher. The other is to diminish singlet oxygen by combining directly with singlet oxygen.

C. Free-Oxygen Absorber

Ideally, the package should be oxygen-free to prevent the oxidation of packed foods. For that purpose, a free-oxygen absorber is used. The oxygen-removing mechanism of the absorber, specially treated iron powder, is as follows.

$$Fe + 2H_2O \longrightarrow Fe(OH)_2 + H_2$$
$$3Fe + 4H_2O \longrightarrow Fe_3O_4 + 4H_2$$
$$2Fe(OH)_2 + O_2 + H_2O \longrightarrow 2Fe(OH)_3$$
$$Fe_2O_3 \cdot 3H_2O$$

One gram of iron requires 300 ml in volume or 0.43 g in weight of oxygen to be completely converted into ferric hydroxide [$Fe(OH)_3$] under normal conditions.

V. REACTIONS BETWEEN LIPID-OXIDIZED PRODUCTS AND OTHER FOOD COMPONENTS

When protein and lipid hydroperoxide come into contact with hydrophobic interaction and hydrogen bonding, a lipid–protein complex is formed (Kanner and Karel, 1976). Then protein radicals are formed as follows:

$$PH + LOOH \longrightarrow (PH \cdots LOOH) \longrightarrow P^\cdot + LO^\cdot + H_2O$$
$$\longrightarrow PH + LO^\cdot + {}^\cdot OH$$
$$PH + LO^\cdot \longrightarrow P^\cdot + LOH$$
$$PH + LOO^\cdot \longrightarrow P^\cdot + LOOH$$

where PH is protein and LOOH is lipid hydroperoxide.

Proteins in water solution or under high A_w conditions form cross-linking in the presence of oxidized lipids. The cross-linking formed by free radicals is considered to be the termination reaction (Karel et al., 1975). Polymerization of proteins occurs also in freeze-dried foods:

$$P^{\cdot} + P^{\cdot} \longrightarrow P\text{-}P$$
$$POO^{\cdot} + P^{\cdot} \longrightarrow POOP$$

These polymerizations result in (1) enzyme inactivation, (2) reduction of solubility, and (3) formation of coloring matrials. The reactivity of proteins to lipid hydroperoxides differs among protein species (Matsushita, 1975).

Under dried conditions, splitting of peptide bonds occurs on proteins, especially gelatin (Zirlin and Karel, 1969; Matoba et al., 1982).

The secondary products, various aldehydes, react with amino groups, and browning of color occurs (amino-carbonyl reaction).

Not only proteins but also other food components are oxidized with hydroperoxides. The compounds that have conjugated double bonds, such as vitamin A or β-carotene, are easily decomposed and decolorized.

REFERENCES

Axelrod, B., Cheesbrough, T. M., and Laakso, S. (1981). *Methods Enymol.* **71,** 441.
Chan, H. W.-S. (1977). *J. Am. Oil. Chem. Soc.* **54,** 100.
Chan, H. W.-S. (1978). *J. Chem. Soc. Chem. Commun.* 756.
Chan, H. W.-S., and Levett, G. (1977). *Chem. Ind.* **20,** 692.
Chan, H. W.-S., Levett, G., and Matthew, J. A. (1979). *Chem. Phys. Lipids* **24,** 245.
Cobern, D., Hobbs, J. S., Lucas, R. A., and Mackenzie, D. J. (1966). *J. Chem. Soc.* 1897.
Coxon, D. T., Keith, R. P., and Chan, H. W.-S. (1981). *Chem. Phys. Lipids* **28,** 365.
deGroot, J. J. M. C., Veldink, G. A., Vlegenthart, J. F. G., Boldingh, J., Wever, R., and Gelder, B. F. (1975). *Biochim. Biophys. Acta* **377,** 71.
Foote, C. S., and Denny, R. W. (1971). *J. Am. Chem. Soc.* **93,** 5168.
Foote, C. S., Welxler, S., and Ando, W. (1965). *Tetrahedron Lett.* **46,** 4111.
Frankel, E. N. (1980). In "Autoxidation in Food and Biological Systems" (M. G. Simic and M. Karel, eds.), p. 141. Plenum, New York.
Frankel, E. N., Neff, W. E., Rohwedder, W. K., Kahmbay, B. P. S., Garwood, R. F., and Weedon, B. C. L. (1977a). *Lipids* **12,** 901.
Frankel, E. N., Neff, W. E., Rohwedder, W. K., Khambay, B. P. S., Garwood, R. F., and Weedon, B. C. L. (1977b). *Lipids* **12,** 908.
Frankel, E. N., Neff, W. E., Rohwedder, W. K., Khamby, B. P. B., Garwood, R. F., and Weedon, B. C. L. (1977c). *Lipids* **12,** 1055.
Frankel, E. N., Neff, W. E., and Bessler, T. R. (1979). *Lipids* **14,** 961.
Frankel, E. N., Neff, W. E., and Selke, E. (1981). *Lipids* **16,** 279.

Frankel, E. N., Neff, W. E., and Selke, E. (1983). *Lipids* **18**, 353.
Galliard, T., and Chan, H. W.-S. (1980). In "The Chemistry of Plants," Vol. 4 (P. K. Stumpf, ed.), p. 132. Academic Press, New York.
Karel, M. (1975). In "Water Relations of Foods" (R. B. Duckworth, ed.), p. 435. Academic Press, New York.
Karel, M., Schaich, K., and Roy, R. B. (1975). *J. Agric. Food Chem.* **23**, 159.
Labza, T. P. (1975). In "Water Relations of Foods" (R. B. Duckworth, ed.), p. 455. Academic Press, New York.
Lundberg, W. O. (1962). In "Lipids and Their Oxidation" (H. W. Schultz, E. A. Day, and R. O. Sinnhuber, eds.), p. 31. AVI, Westport, Conn.
Matoba, T., Yoshida, H., and Yonezawa, D. (1982). *Agric. Biol. Chem.* **46**, 979.
Matsushita, S. (1975). *J. Agric. Food Chem.* **23**, 150.
Matsushita, S., and Terao, J. (1980). In "Autoxidation in Food and Biological Systems" M. G. Simic and M. Karel, eds.), p. 27. Plenum, New York.
Neff, W. E., Frankel, E. N., Scholfield, C. R., and Weisleder, D. (1978). *Lipids* **13**, 415.
Neff, W. E., Frankel, E. N., and Weisleder, D. (1981). *Lipids* **16**, 439.
Neff, W. E., Frankel, E. N., Selke, E., and Weisleder, D. (1983). *Lipids* **18**, 868.
Park, D. K., Terao, J., and Matsushita, S. (1981). *Agric. Biol. Chem.* **45**, 2071.
Peers, K., and Coxon, D. T. (1983). *Chem. Phys. Lipids* **32**, 49.
Peers, K. E., Coxon, D. T., and Chan, H. W.-S. (1981). *J. Sci. Food Agric.* **32**, 898.
Porter, N. A., and Wujek, D. G. (1984). *J. Am. Chem. Soc.* **106**, 2626.
Porter, N. A., Logan, J., and Kontoyiannidou, V. (1979). *J. Org. Chem.* **18**, 3177.
Porter, N. A., Weber, B. A., Weenth, H., and Khan, J. A. (1980). *J. Am. Chem. Soc.* **102**, 5597.
Pryor, W. A., Stanley, J. P., and Blair, E. (1976). *Lipids*, **11**, 370.
Rawls, H. R., and Van Santen, P. A. (1970). *J. Am. Oil. Chem. Soc.* **47**, 121.
Sessa, D. J., Gardner, H. W., Kleiman, R., and Weisleder, D. (1977). *Lipids* **12**, 613.
Tappel, A. L. (1962). In "Lipids and Their Oxidation" (H. W. Schultz, E. A. Day, and R. O. Sinnhuber, eds.), p. 122. AVI, Westport, Conn.
Taradgis, B. G. (1968). In "Metal Catalyzed Lipid Oxidation" (R. Marcuse, ed.). SKI Rapport, Gotteborg.
Terao, J., and Matsushita, S. (1975). *Agric. Biol. Chem.* **39**, 2027.
Terao, J., and Matsushita, S. (1977). *J. Am. Oil Chem. Soc.* **54**, 234.
Terao, J., and Matsushita, S. (1980). *J. Food Process. Preserv.* **3**, 329.
Terao, J., and Matsushita, S. (1981). *Agric. Biol. Chem.* **45**, 587.
Terao, J., Hirota, Y., Kawakatsu, M., and Matsushita, S. (1981). *Lipids* **16**, 427.
Terao, J., Ogawa, T., and Matsushita, S. (1975). *Agric. Biol. Chem.* **39**, 397.
Terao, J., Inoue, T., Yamagata, S., Murakami, H., and Matsushita, S. (1984). *Agric. Biol. Chem.* **48**, 1735.
Thomas, M. J., and Pryor, W. A. (1980). *Lipids* **15**, 544.
Toyoda, I., Terao, J., and Matsushita, S. (1982). *Lipids* **17**, 84.
Umehara, R., Terao, J., and Matsushita, S. (1979). *Nihonnougeikagaku Zasshi* **53**, 51.

Vliegenthart, J. F. G., and Veldink, G. A. (1980). In "Autoxidation in Food and Biological Systems" (M. G. Simic and M. Karel, eds.), p. 529. Plenum, New York.
Yamagata, S., Murakami, H., Terao, J., and Matsushita, S. (1983). *Agric. Biol. Chem.* **47,** 2791.
Yamagata, S., Murakami, H., Terao, J., and Matsushita, S. (1984). *Agric. Biol. Chem.* **48,** 101.
Yamamoto, Y., Niki, E., and Kamia, Y. (1982). *Lipids* **17,** 870.
Zirlin, G. A., and Karel, M. (1969). *J. Food Sci.* **34,** 160.

PART II

New Food Packaging Materials

CHAPTER 3

New Food Packaging Materials: An Introduction

Akira Kishimoto
Corporate Research & Development,
Toyo Seikan Group,
Yokohama, Japan

A. History of Packaging

The history of the food package goes back to the dawn of human civilization. In ancient times humans may have used perhaps a cup from a large leaf or vegetable shell as a simple container. Egyptians made mummies and preserved their contents in original form in the hope of revival some day. As time passed, packages and containers became tools for subdivision and transportation of goods, as well as elements of ritual, and became rooted deeply in the human life of respective ages.

1. Paper and Paperboard

Papyrus was used in ancient Egypt as a paper material, but was too valuable to be used for packaging. A paper industry was established by the English in the seventeenth century. In 1871 improved corrugated fiberboards were produced and applied as containers for a wide range of products, replacing wooden containers. The manufacture of corrugated board has not basically changed since 1890, but effort has concentrated on how to make a light, moisture-proof, and rigid container. The joint use of shrink plastic film with the corrugated fiber box illustrates such a container.

The paper carton now widely used in liquid packaging is of comparatively recent origin and began to be used for liquid contents in 1935 in Japan. This was not effective without waterproof treatment such as wax coating. With the coming of the plastic age, polyethylene was converted into paper coating and permitted much stronger heat seals and moisture protection. In 1960, Tetra Pak in Sweden developed a paper form, fill, and seal system that was accepted for diary products and fruit juice products successfully. By 1976, the market share of milk paper cartons exceeded 50% in Japan.

2. Metals

The history of canned foods is one of food preservation and also of tinplate manufacturing. All of the world's metal-canned foods were first enclosed in tin-coated steel in the form of tin cans. During World War II tin became unavailable. In 1961 Japanese steel mills developed tin-free steel (TFS). Can makers attempted to make TFS cans by using welding and side-seam cementing techniques. These may be cheaper and may eliminate the soldered side seam. Now the side-seam-cemented TFS can in Japan has the largest production volume in the beverage and food can market.

In 1955 Kaiser Aluminum developed two-piece aluminum drawn-and-wall-ironing can technology, which then spread into the beer and beverage can market. In 1973 the tinplate drawn-and-wall-ironing can was marketed for carbonated beverages. Since the petroleum crisis, developmental efforts have focused on saving material, and the weight of cans has been reduced further and further.

On the other hand, side-seam-cemented TFS cans were modified to withstand hot fill and retort temperatures by Toyo Seikan. Now they are used for hot-fill juice and retorted coffee drinks and are continuing to enlarge a market for low-acid food such as meat and fish.

3. Glass

Glass bottles are known to have been used in Egypt about 1500 B.C., but until recent times glass containers were a work of art. Glass-blowing technique appeared in 300 B.C., which was a great invention for glass packaging and was transmitted to Rome and Europe from Egypt. In Japan, Francisco Xavier brought glasswares in 1573; before then, glass goods were not familiar to Japanese people. The first bottle-making machines were invented in the United States in 1882. Automatic bottle-making machines were introduced into Japan in 1913 and built the foundation of the modern glass packaging industry. Technological efforts have been devoted to getting lighter weight and to protecting from breakage.

4. Plastics

Plastics are modern materials and now have a greater part in the packaging industry. The growth of the petrochemical industry has been so rapid that plastics are used for many fields of packaging. The first polyethylene extrusion blow-molded bottle in Japan appeared on the market in 1955, followed by polyvinyl chloride bottles as edible oil containers. In 1972, high-barrier coextrusion multilayer bottles were developed by Toyo Seikan for mayonnaise and ketchup, using EVAL® (saponified ethylene-vinyl acetate copolymer) as an oxygen barrier material. This was a memorable event in packaging. A recent innovation is the success of the biaxially oriented polyethylene terephthalate (PET) bottle, which is so clear and tough that it became marketed as the larger size container for beer and carbonated beverages instead of glass bottles.

Plastics packages are processed by many methods, such as extrusion blow molding, injection molding, laminating, coating, and thermoforming. Foil, paper, and various plastic films are laminated to make composite materials as flexible packaging. Special combinations of PET/foil/polyolefin film were developed for retort pouches in 1968 in Japan, and could be sterilized under pressurized retort and applied to package low-acid food as a flexible can. The production volume of retort-pouched food in Japan exceeded 600 million units in 1984.

B. Definitions of Packaging and Its Functions

The difficulty in defining packaging is a result of a wide range of functions in materials and technology. According to Japanese Industrial Standards JIS Z 0101, packaging is defined as techniques and states for applying suitable materials or containers to goods in order to protect and maintain product quality during shipping and storage. There are three kinds of packaging. First is primary or item packaging, which is defined as packaging of individual goods and states for applying suitable materials or containers in order to attract the consumer or to protect product quality. Second is inner packaging, which is defined as inner packaging of packaged freight to protect goods or item packaging. Third is outer packaging, defined as outer packaging of packaged freight for the purpose of shipping.

Since packaging includes all operations necessary before the product reaches the consumer, it should be planned to meet distribution processes and the external environment. In the United States, the Packaging Institute, U.S.A. (Beckman et al., 1979), says packaging is the enclosure of products, items, or packages in a wrap, pouch, bag, box, cup, tray, can, tube, bottle, or other container form to perform one or more of the following major functions:

1. Containment for handling, transportation, and use.
2. Preservation and protection of the contents for required shelf life and sometimes protection of the external environment from any hazards of contact with the contents.
3. Identification of contents, quantity, quality, and manufacturer.
4. Facilitation of dispensing and use, including ease of opening, reclosure, portioning, application, unit of use, multipacks, second use or re-use, and working features such as are found in the cook-in pouch.

Packages must be designed to meet internal or external difficulties that may be encountered at respective stages of distribution, to protect goods safely from such difficulties, and to send goods to the consumers satisfactorily. The packaging engineer and researcher must know that the food packer's main aim is not to sell a package but its contents.

C. Classification of Packaging

Packaging is referred to as consumer type, industrial type, shipping type, etc., but this terminology is used loosely. In this chapter, packaging is classified by the materials used. These are paper and paperboard, cellophane, plastics, metals, glass, wood, textiles, and others. Paper and paperboard includes paper bags, laminated paper, paper cartons, corrugated fiberboard, and so on. Cellophane is not a new material, but has served as a base material for laminating with other plastics. Recently cellophane has been replaced by other synthetic films, and the demand is decreasing from year to year. Plastics include films, sheet, blow-molded bottles, injection-molded containers, extrusion lamination resin, stretched flat yarn, and foamed products. Metals are used in two-piece and three-piece steel and aluminum cans, metal tubes, aluminum foil, pails, and drums. Woods include folding wood cartons, wooden boxes, barrels, and wooden frames. Textiles may be larger cloth bags and sacks woven from jute or hemp or cotton. Other materials include composite materials and materials that are difficult to classify in the categories already described.

Using data from the Japan Packaging Institute (1987), consumption of packaging materials in Japan is shown in Table I. Paper and paperboard and metals packaging have stable market share among the various packaging materials. Plastics have gradually increased over time, while glass increases for a while and then decreases. This is due to the boom of light-weight, nonreturnable Plasti-Shield bottles. As of 1986, paper and

Table I Japanese Market for Packaging Materials (Thousands of Tons)

Packaging Materials	1983	1984	1985	1986
Paper and paperboard	8866.1(52.7)[a]	9399.5(52.9)	9641.7(53.7)	9960.2(54.5)
Cellophane	59.4 (0.4)	56.4 (0.3)	50.2 (0.3)	43.2 (0.2)
Plastics	1924.8(11.4)	2184.7(12.3)	2322.3(12.9)	2374.7(13.0)
Metals	1663.5 (9.9)	1763.3 (9.9)	1798.9(10.0)	1802.6 (9.9)
Glass	2301.0(13.7)	2394.6(13.5)	2222.7(12.4)	2159.8(11.8)
Wood	1668.5(10.1)	1660.0 (9.4)	1629.5 (9.1)	1646.0 (9.0)
Textiles	223.9 (1.3)	219.5 (1.2)	211.6 (1.1)	204.8 (1.1)
Others	86.0 (0.5)	83.5 (0.5)	82.0 (0.5)	96 (0.5)
Total	16,793.2	17,761.5	17,958.9	18,271.3

[a] Percentage value given in parentheses.

paperboard, metals, and plastics made up most of the shipping volume for the preceding decade.

REFERENCES

Beckman, H. M., Simms, W. C., and Corvington, J. (1979). "Glossary of Packaging Terms," 5th ed. The Packaging Institute, U.S.A., New York.

Japan Packaging Institute (1987). 1986 statistics of shipping value and mass of packaging materials, containers and machineries in Japan. *JPI. J.* **25,** 485–515.

CHAPTER 4

Paper and Paperboard Containers

Mitsuhiro Sumimoto
Dai Nippon Printing Co., Ltd.,
Tokyo, Japan

I. CHARACTERISTICS

Cartons are generally superior to other containers in such packaging functions as printed explanations, protection of contents, easy using, easy handling, etc. They have been extensively used for many years for the circulation of consumer goods. And they are readily adaptable to modern needs in merchandise circulation that are affected by a changing social environment. Merchandise has taken many forms, the packaging of goods in shops has largely given way to prepackaging, sales methods have changed to large-scale retailing, and the merchandise itself must perform sales functions. The aspects of cartons that offer the largest degree of adaptability are listed below.:

1. Printability is superior. Offset printing, gravure printing, flexographic printing and typographic printing can be used according to the usage, purpose, and quantity of the packages. As the printing is done on paper in general, a wide variety of designs can be effectively printed. This is a great advantage in increasing the merchandise effects.
2. Processing is easy. Carton processes are simple, such as a various coatings, lamination, die-cutting, folding, and carton making. These supplement the water resistance and moisture-proofing, give barrier qualities, increase functions, provide variety, and enable mass production.
3. Proper stiffness. Paper and paperboard containers have suitable stiffness, which is necessary to keep the packaging containers intact and to

Food Packaging Copyright © 1990 by Academic Press, Inc. All rights of reproduction in any form reserved.

protect the contents such as fragile cookies and crackers. This stiffness also makes it easy to transport or pile up the containers, and increases merchandise effects.
4. Automatic packaging machines. Various kinds of automatic packaging machines have been developed to increase productivity, according to the various shapes of the cartons. It is also possible to systematize the packaging process from production to receiving and delivery.
5. Simple waste disposal. The disposal of paper and paperboard containers wastes is simple. As for waste pollution, the burning or burying of these containers is easy, and systems for collection and recovery of paper have been established in some countries.
6. Light and cheap. The materials of paper and paperboard containers are light and collapsible. Their transport is easy because of compactness. They are cheaper than the other materials. These become important factors in view of increasing demands from the market.
7. Laminate materials. These are often made of layers of paper, plastic, and aluminum foil to add to container properties such as waterproofing, moisture-proofing, and grease-proofing.

II. Materials

A. Paperboard

Paperboards are the primary materials for production of containers such as folding cartons and paper cups, and coated boards are used in the largest quantity. The thickness of paperboards is 0.2–1.0 mm, and the weight per surface area of raw paper varies from 120 to 700 g/m^2. Although low-weight paperboard is formed in one layer, the thicker paperboard is made by combination of multiple raw paper sheets.

For most paperboard, the variation of blending raw materials for each layer gives advantages. Therefore, a multilayer structure may be adopted even for relatively low-weight boards. Table I shows the kinds and structures of paperboards used for cartons.

B. Laminate Materials

1. Plastic

Plastics are often laminated to paperboards to add some properties such as heat sealability and waterproofing to the containers. This contributes to appearance and function as follows.

Table I Kinds and Structure of Paper Boards[a]

Kinds of paperboards			Structure[b]	Usage
General	Subtype	Special Type		
Manilaboard	Coated A Manila board (No coat A manila board)		Coating layer SP or bleached KP GP	Item packaging of relatively small goods, such as cakes, cigarettes, etc.
	Coated B Manila board (No coat B Manila board)		Coating layer SP or bleached KP Waste paper GP	
Whiteboard	Gray back	Coated board (no coat board)	Coating layer SP or bleached KP Waste paper	Most general packaging use
	White back	Coated board (no coat board)	Coating layer SP or bleached KP Waste paper SP or bleached KP GP	Package of relatively larger goods, comparing with manilaboard
	Kraft back	Coated board (no coat board)	Coating layer SP or bleached KP Waste paper KP	Packaging of heavy goods, requiring relative strength

(*continued*)

Table I (*Continued*)

General	Kinds of paperboards		Structure[b]	Usage
	Subtype	Special Type		
High-quality paperboard	Ivory board	Both sides (one side) coated ivory (no coat ivory)	Coating layer / SP or bleached KP / Coating layer	Item packaging of high class cakes, food, cosmetics, etc.
	Both sides (one side) coated card (Card)		Coating layer / SP or bleached KP / GP / SP or bleached KP / Coating layer	
Colorboard	One side kraft		SP or bleached KP (colored) / Waste paper	Mostly used for the secondary packaging, such as dozen cases
	Both sides kraft		SP or bleached KP (colored) / Waste paper / SP or bleached KP (colored) / Waste paper	
	Brown board, gray board			
Yellow paperboard			Straw pulp	As the base paper of pasted board carton

[a] Special paperboards (for lower cost, materials of one-side coated ivory board or one-side coated card are partially changed) have been recently much used for the cartons.

[b] Abbreviations: SP, sulfite pulp: bleached KP, bleached kraft pulp (sulfate pulp); GP, groundwood pulp.

a. Gloss Various films such as polyvinyl chloride, polypropylene, and polyester are laminated to paperboards, depending on the packaging design and functions. Sometimes vacuum-metallized films also are laminated.

b. Heat Sealing Polyethylene and polypropylene are extrusion coated to the paperboards to heat seal when the containers are formed, to make, for example, milk cartons or frozen food cartons.

c. Liquid Containers Paper cups that are wax-coated have been used for beverage containers. Recently polyethylene-coated paperboards have been used to make the liquid containers for milk and hot drinks.

In addition to these containers, bag-in-carton and bag-in-box are widely used for liquid containers in the food packaging industry. Bag-in-box and bag-in-carton have an inner bag made of different combinations of paper, coatings, plastic, and aluminum foil to achieve the barrier properties required by each product. These are now being used for wine, ground coffee, and cakes, among others.

d. Spouts Especially for liquid containers, injection-molded plastic spouts are used in the food packaging industry.

e. Combination with Plastic Molding Printed blanks are inserted into the injection mold, and then plastic such as polyethylene and polypropylene is injected into the mold in the melt state to combine the printed blanks and plastic.

2. Aluminum Foil

Aluminum foil is composited with paper and paperboards to achieve barrier properties and surface gloss. The thickness of aluminum foil is generally 7–30 μm. Sometimes thick aluminum foil is used for the lid of a container. Such lids are formed in a special shape and heat sealed at the top of the container. Examples include Japanese sake cartons, juice cartons, and composite cans.

C. Selection of Paper and Paperboard Containers

The following points should be considered in designing and manufacturing the containers.

1. Appropriateness for transportation or display.
2. Suitability for printing, conversion, and filling processes.

3. Reasonable cost.
4. Commodity value and eye appeal.
5. Properties applicable to forming into structures: weight per area (g/m^2), thickness (mm), tensile strength (kg/cm), folding strength (g), elongation (%), and resultant internal tearing strength, dart impact, and suface strength, stiffness.
6. Properties of finished products: size, water absorbency, oil absorbency, water repellency, water vapor permeability ($g/m^2 \cdot 24$ h), odor, brightness, smoothness, and wet strength.

D. Grain of Paper

The fibers in paper materials have a tendency to run along the manufacturing direction of the paper-making machinery. When paper is cut so that its long length lies along the machine direction it is called "paper of machine direction," and the paper cut in within long length across the machine direction is called "paper of cross direction." When the paperboard is torn along the long side or short side of the board, the tearing direction naturally proceeds along the grain of paper (Fig. 1). Another way to distinguish the grain of paper is that the grain runs in the direction in which curls are easily made when the paperboard is slightly curled with the fingers (Fig. 2).

When the cartons are made, if the creasing line is parallel to the paper grain, the creasing will not be made sharply and the paperboard will tend to crack. Therefore, it is important to arrange the grain of paper at a right angle to the major creasing line.

III. CLASSIFICATION OF PAPER AND PAPERBOARD CONTAINERS

Although the term cartons may refer to packaging containers chiefly made of paperboards, the term is sometimes considered to include not only paper cartons but also paper tubes, paper cups, paper holders, or even display cards that are not for packaging but are chiefly made of paperboards.

The manufacturing process for cartons is basically printing, die-cutting,

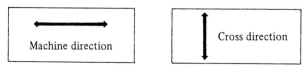

Figure 1. Machine and cross directions.

and carton making. Various coatings on paperboards are chosen according to the usage and functions of the cartons. Cartons are generally superior to other containers in such packaging functions as explanation, protection of contents, easy use, easy handling, etc., and they have been used for the longest time, in the greatest quantity and popularity, for the circulation of consumer goods.

A. Classification

A classification of containers may be outlined as follows:

I 1. Folding carton
   a. Straight carton: tuck-end carton, sleeve, seal-end carton, lock-bottom carton.
   b. Tray carton: collapsible carton, formed carton.
  2. Rigid box.
  3. Special cartons: blister pack, skin pack, strip pack, tray pack, multi-pack, composite can, oven-proof carton or tray, reclosable carton.
  4. Paper cups.
  5. Display card: hanger display, counter display, stand display.
  6. Composite containers: milk carton, sake carton, juice carton, bag-in-box.

B. Folding Carton

The basic shapes of cartons are generally divided into two kinds, based on the completed shapes and the manufacturing processes: the tubelike straight cartons called sack cartons, and the tray cartons, which are the major paper containers at the present.

The folding carton is the most popular paperboard container. The pre-printed and die-cut blanks are folded and glued by the folding machine. These folding cartons are shipped collapsed to the food companies. They can be operated smoothly with an automated filling and packaging machine. They are used widely in the food industry.

Paper of cross direction   Paper of machine direction

Figure 2. Paper of cross and machine directions.

1. Straight Carton

This is the most popular carton style, generally called the sack carton. The straight cartons are widely used for the packaging of mass production merchandise, because of good productivity and the ease of the assembly and packing automation.

a. Tuck-End Carton This is divided into the straight tuck and the reverse tuck, based on the position of the tuck. Figure 3 shows the straight carton that is most widely used for item packaging of soap, for example.

b. Sleeve Although this is the most simple shape of the straight cartons, it can be varied by creasing or cutting method to develop the carton with midboard or the carrier. By increasing the parts, it also can be used for paper containers for cigarettes, caramels, etc. (see Fig. 4).

c. Seal-End Carton Without the tuck part of the tuck-end carton, this is made by gluing the upper cover of the main flap and its bottom part, and is used for the paper container for medicines. For powder containers, a structure to prevent the powder leakage has also been developed (see Fig. 5).

d. Lock-Bottom Carton This has some durability against weight, because the bottom of the flap is made into the lock structure, and it is widely used for packaging bottles of cosmetics, medicines, wines, etc. The form in which the bottom lock is manually assembled is called a semiautomatic lock-bottom carton. Another form, with a lock structure automatically made at high speed, gluing together the glue flap and the bottom flap (after the attachments for gluing are prepared on the lock face), is called an automatic lock-bottom carton (see Fig. 6).

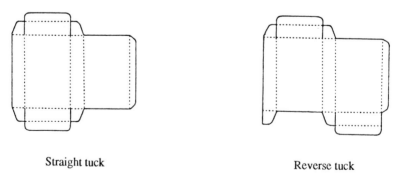

Straight tuck   Reverse tuck

Figure 3. Tuck-end carton.

2. Tray Carton

This is the name for the tray-shaped carton, which includes most cartons other than the straight cartons. There are two kinds of structure: the most simple collapsible structure (paper is die-cut, glued or stitched, then folded) and the assembly type (die-cut parts are assembled at the time of use). They are used in prepacking of products in stores, especially for cakes, textile products, and miscellaneous goods, in various shapes depending on purpose.

Figure 4. Sleeve.

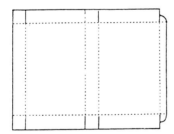

Figure 5. Seal-end carton.

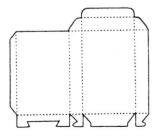 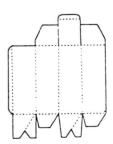

Semi-automatic lock bottom carton (D.O. rock)   Automatic lock bottom carton (Auto-bottom)

Figure 6. Lock bottom carton.

a. Collapsible Carton Because of its production by gluing while holding, the area of this carton becomes small. Therefore, the transportation of cartons and the handling of stock are easy and economical.

Depending on the folding structure of the corners, these cartons are called one-piece, two-piece, with hinge cover, display carton, etc. (see Fig. 7).

b. Assembly-Type Carton This is a carton structure such that the paper parts (die-cut and creased) are manually locked and assembled when it is used. There are two basic structure types: double-wall style and locking style. For some purposes, picture-frame style and double side-wall style (improved work efficiency by partial gluing) have also been used (see Fig. 8).

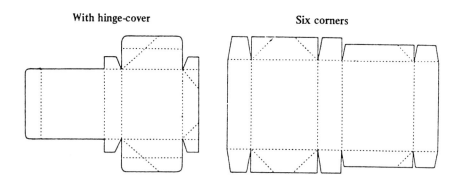

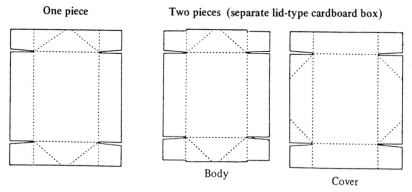

Figure 7. Collapsible carton.

C. Rigid Box

This carton consists of a box body and box cover, made of paperboard on which the cover paper is pasted. Its characteristics are:

A high-quality appearance can be produced by proper details.
It has strength.
Because it is not collapsible, a great deal of space is required for transport and storage.
Because the productivity is lower, the cost becomes higher.

The rigid box must be adapted for use with an understanding of these qualities (see Fig. 9).

They are used in high-quality packages or gift packages of cosmetic, dolls, and cakes.

D. Special Cartons

1. Flip-Top Style

This is the reclosable carton that has been recently developed. It has a structure that allows mass production by machine, and has a shape that lends itself to high-quality appearance (Fig. 10).

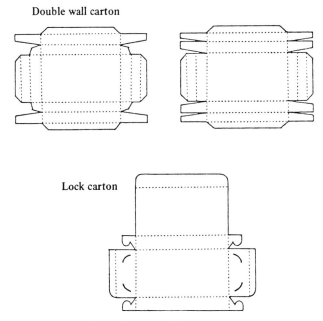

Figure 8. Assembly type carton.

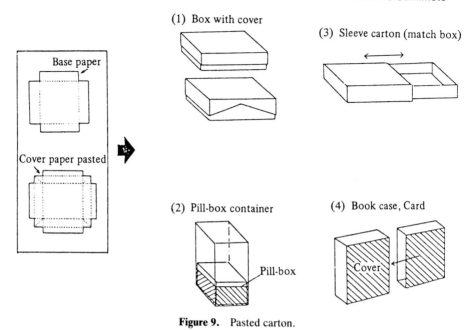

Figure 9. Pasted carton.

2. Cartons with Spouts

Cartons used for packaging of mixes, powdered milk, and cleanser require a spout for convenience. The spout often has been attached at the side of the carton. There are many types of spouts, including an inner carton box for the spout, a die cut at the side of carton, and the metal or plastic spout.

3. Oven-Proof Trays or Cartons

These packages were developed to meet the demand for economical, dual-oven food containers. They are used in many food industries to satisfy the needs of both microwave and conventional oven users. They are made of paperboard that is extrusion coated with polyethylene terephthalate (PET) and polymethylpentene (TPX). They can resist a temperature as high as 200°C. There are two styles, pressure-formed or pressed tray and preformed carton or tray.

4. Multipack and Carrier Cartons

For the easy transporting of bottles, cans, etc., various kinds of carrier cartons have been developed and used.

The carrier is wrapped around the product, and the locks are threaded and positively punched into position on the bottom of the carrier. No glue

Chapter 4. Paper and Paperboard Containers

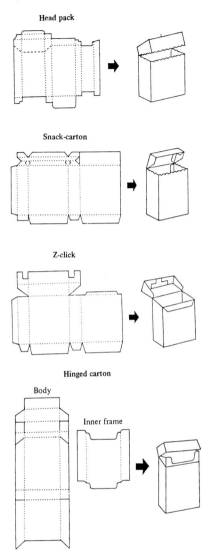

Figure 10. Flip-top style.

is required in closing this carton, as the small stable locks are designed to hold the carrier snug and square.

Full display panels at top and sides provide extensive area for sales messages. The carrier carton also gives full protection to the bottles and cans, and protects against dust.

In selection and design, package strength is preferred. The strong paper of kraftboard is used. Water resistance may be given by polyethylene resin coating (see Fig. 11).

Carrier cartons sometimes compete with multipack shrink films.

5. Composite Cans

These are lightweight packages with strong graphic potential for a wide array of moisture- and oxygen-sensitive products.

They can replace the can in some parts of the food industry. Composite structures are necessary to achieve multifunctional properties economically. Laminate materials are used in the body of the containers. There are two methods of making the body: spiral winding and flat winding.

They are used in the packaging of cookies, green tea, liquid juice, etc.

E. Paper Cups

Paper cups are made of paperboard or polyethylene-coated paperboard and are used for drink cups, ice cream cups, and general food cups. They can be formed by cup-forming machines. They often compete with plastic cups made by thermoforming and injection molding.

Recently aluminum foil and plastic film laminate materials have been used in paper cups, adding barrier properties and heat insulation for hot service.

F. Composite Containers

Composite containers have been made of laminate materials such as aluminum foil, plastic film, and paperboard. The strength and barrier properties of the containers make it possible to packaging many products found at present in glass, plastic, and even metal containers. Various

Example of carrier carton (bottle carrier)

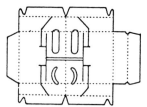

Figure 11. Carrier carton.

Chapter 4. Paper and Paperboard Containers

kinds of composite containers have been used in the food industry (Fig. 12).

1. Milk Cartons

Glass bottles have been used for milk containers. However, due to changing distribution systems, the shortage of manpower for transporting and recovering the bottles, and the increase of labor, cost, etc., conversion to nonreturnable containers, chiefly paper cartons, has been rapid.

Although there is a difference in thickness according to the different size of the container, the water-resistant kraft paper coated on both sides with polyethylene extrusion has been chiefly used. The paperboards are often made of soft wood pulp.

Preprinted, die-cut, and coated paperboard blanks are heat sealed at the milk filling and packaging machine. There are also machines of roll-fed type. For the milk carton, there are three shapes: gable-top style, flat-top style, and tetrahedron style. They are all automatically filled and sealed. Milk cartons can protect milk from losing vitamins and flavor due to exposure to light, especially fluorescent light in store dairy cases. Aseptic packaging can be applied to the milk carton by using hydrogen peroxide as a sterilizer.

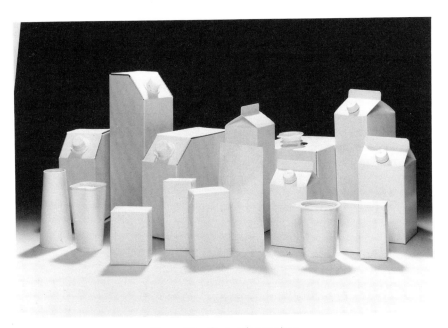

Figure 12. Composit container.

2. Liquid Carton with Spout

This is a gable-topped paper carton that contains liquids. Aluminum lamination of the inner surface improves gas barrier properties and ensures longer shelf life.

The carton features a unique, easy-to-pour spout, which is attached in-line during the carton molding process.

Standard capacities are 500, 900, and 1800 ml, for each of which different filling and sealing machines have been developed.

Filling speed is 500–4000 cartons per hour.

The compactness of the cartons (one-fifteenth the weight and half the bulk of glass bottles) greatly reduces shipping costs.

Handsome designs produced by offset or photogravure techniques enhance the effectiveness of merchandise display.

These cartons are widely used in Japan to contain sake and mineral water (Figs. 13–15).

3. Small Brick-Shaped Paper Containers

A filling and sealing system has been developed for paper containers hot filled with fruit and other beverages (Figs. 16 and 17). The compactness and economy of the filling machine have been appraised and have drawn attention. The system forms small brick-shaped paper containers from printed carton roll paper, and can fill the contents and even seal on-line.

Figure 13. Liquid carton with spout.

Chapter 4. Paper and Paperboard Containers

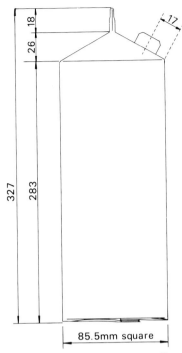

Figure 14. Dimensions of formed carton. Dimensions will vary slightly according to filling machine adjustment.

The application of straws or pull-tabs to the paper container can also be done on-line. Hot filling up to 93°C is possible. Juice and other acid-containing beverages can also be distributed without refrigeration.

Juices, soups, and other products containing pulp or semisolids may be filled and sealed as well. In general, dimensions of the container can be specified when ordering the machine to accommodate any one volume

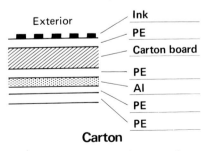

Figure 15. Materials and construction.

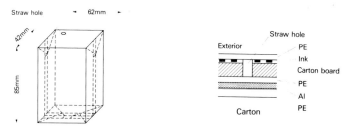

Figure 16. Small brick-shaped paper container. Carton sizes (*left*); materials and construction (*right*).

from 90 to 300 cm³. The speed for filling and sealing for the 200 cm³ size is 2200/h. Each straw is wrapped in polyethylene film supplied in roll form, separated by a cutter, and attached by the straw applicator to the portion, to which hot melt has been previously applied. When a pull-tab is applied instead of a straw, each pull-tab, supplied in roll form, is separated by a cutter and heat sealed at the designated spot.

As multicolor gravure printing is used for the paper containers, beautiful artwork is possible. As the paper container is composed of polyethylene, aluminum foil, and paper, it has high gas barrier properties, which protect the quality of the contents. Since secure fin seals are used, there is no need to worry about the contents staining or leaking out of the paper section. The need for taping at the edges of the paper is also eliminated. This is

Figure 17. Small brick-shaped paper containers.

designed to cut costs for producing the container and making the device more compact.

This system is currently widely used for juices such as orange and apple, sake, and lactic acid beverages that are sold in stores and automatic vending machines.

4. Hot-Filling Cup

In light of the perpetual problems surrounding the disposal of metal cans, the trend toward paper beverage containers is steadily increasing. A hot-filling paper cup has been developed. It ensures content preservation and permits distribution at ambient temperatures.

This uniquely shaped container is already being used on the market for juices. The merits of the hot-filling cup are as follows:

a. It can withstand temperatures up to 93°C, and resists shape distortion despite the pressure drop during cooling. A unique shape sets it apart from other products of its type.
b. The cup is constructed of paper, aluminum foil, and polyethylene for excellent hermetic sealing, allowing the option of long-term distribution of acidic drinks at ambient temperatures.
c. The tape-style pull-tab is easy to open and is not dangerous like the pull-tabs of metal cans.
d. The tapered shape allows for stacking and efficient distribution.
e. The exterior paper surface permits a variety of designs and printing effects.
f. The cups can be easily disposed of by incineration.

Moreover, two types of filling machines have been developed to accommodate variations in processing (Fig. 18). The semiautomatic filler produces 1500 cups/h and the fully automatic filler produces 6000 cups/h.

5. Cylindrical Container

The cylindrical container is easily disposable. It is made of laminated paper, aluminum foil, and plastic. The bottom of the container is constructed to absorb distortions during the pressure change following hot filling. The top, made of molded aluminum foil, features a convenient tape pull-tab, which is much safer than metal can pull-tabs.

These containers can withstand temperatures as high as 95°C. The virtual absence of air within the filled container makes it extremely well suited for long-term preservation.

They are widely used for juices.

Figure 18. Hot filling cup.

6. Bag-in-Box

Bag-in-box or bag-in-carton is a carton in which a flexible bag is attached to the inside (Fig. 19).

When an inner bag is put on the inside of the carton, it is called a lined carton. The inner bag is composed of minimum amounts of paper, aluminum foil, and various plastics to give the exact protection and shelf life each product requires.

These are widely used for snack foods, coffee, sugar, skimmed milk, and liquids.

In the past, sake was exclusively distributed in 1.8-l glass bottles. However, by introducing the more compact, lighter-weight bag-in-box system, shelf life has been extended, distribution cost has been reduced, and carry-home convenience has been enhanced.

Another merit of the bag-in-box system is that it greatly improves the display effect. This system, also applied to the distribution of vinegar, has proved to be highly useful in physical distribution systems.

These containers also are used in aseptic packaging systems. Aseptic packaging systems are already widely used and are considered optional for products such as cream, fruit juice, and seasoned foods.

Chapter 4. Paper and Paperboard Containers

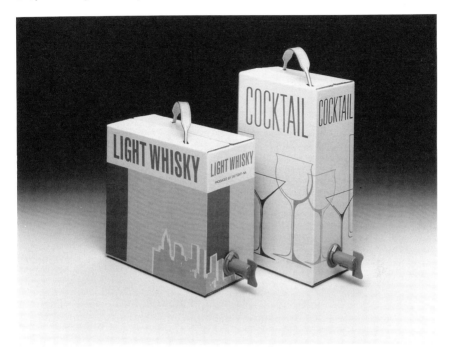

Figure 19. Bag-in-box.

In general, a sterilized inner bag with a spout is fitted to a holder in the filler; the cap is removed, the presterilized contents are injected, and the container is recapped, all under aseptic conditions.

The inner bag is of superior construction, including a leak-resistant four-sides seal with an easily sterilized spout.

The spout's cap and fitted grommet are designed according to very precise specifications.

7. Container with Ultra-Lock

Among recently developed packaging, a new container with a tamper-resistant, easy-to-open, and reclosable top, Ultra-Lock, is receiving attention. The yogurt container as application has this top combined with a paper-based container (Fig. 20).

A polyethylene formed top is ultrasonically sealed to the main body. The V-shaped notch cut in the flange of the container in advance is the special feature that gives both proof of tampering and easy opening. This new package is designed so that a small packet of sugar can be inserted in the polystyrene formed top and then covered with a paper top.

Figure 20. Container with "ULTRA-LOCK."

The body is composed of paper, aluminum foil, and plastic. The container is formed as one by inserting the multi-ply blank in a metal mold and injecting plastic. The container is gravure printed on a paper surface, enabling beautiful, colorful artwork and extremely high visibility in store displays. Moreover, uniqueness can be emphasized with the wide variety in shapes that can be designed. With a tapered container, stacking for transportation is also possible. The container can be used for margarine, jams, and other dessert products in addition to yogurt.

IV. MANUFACTURING PROCESSES FOR CARTONS

The manufacturing processes for cartons are basically printing, die-cutting, and carton making.

Various surface coatings are applied according to the uses of the cartons.

A. Printing

Many systems such as the letterpress, offset press, gravure press, flexographic press, and silk screen printing are used. However, the major ones are the offset press and the gravure press.

In the offset printing system, due to the use of dampening water, the

Chapter 4. Paper and Paperboard Containers

printing is unstable, the tone tends to vary, and drying of ink is delayed. The introduction of UV-hardened ink has made momentary drying available, which results in an advance in productivity. The improvement of quality by excluding the troubles of the offset and insufficient drying and such superior qualities as odorlessness and heat resistance have been recognized by the users. Offset printing will be further used as a powerful method in application and development of the cartons, whose shapes are multiplying.

In gravure printing, many kinds of ink are used for printing various materials. Recently, large-scale, high-speed, web-fed rotary gravure printing machines have been used, and the halftone gravure processing technique has been advanced. Therefore, gravure printing has become the major method of printing cartons in quantity.

In letterpress printing, the ink adhered quantity is better than in offset printing, and there is less instability from dampness. However, the cost of printing plates is higher, and the printing ability is inferior. Therefore, letterpass printing is limited in use.

As for the flexographic press, the printing cost is low, but the delicate printing is difficult. Therefore, it is used chiefly for milk carton printing.

1. Printing Machines: General Considerations for Cartons

Because paperboards are printed, the following points are considered for planning the sheet-fed press.

a. Automatic lift quantity can be increased by raising by making the height of the feeder and the delivery device.
b. In order that the paper sheets flow evenly, the arrangement of the printing drums must be checked.
c. To suit the thickness of the paperboards, up and down adjustment of the gripper pad is possible.
d. A nonstop feeder and a nonstop delivery device are necessary.
e. Cylinder cut-down is different, because cylinder arrangement is different.
f. A device for preventing the rubbed fouls, a sprayer, and a board pick-up device are necessary.
g. In the case of the rotary press, every part must be strong.

2. Letterpress Printing Machines

Letterpress printing machines are classified by type:

a. Platen press: manual printer, treadle printer, roll-fed platen press.

b. Cylinder press: stopped cylinder type (high-bed type, flat-bed type), rotary cylinder type (two-times rotary type, one-time rotary type).
c. Rotary press: sheet-fed type, web-fed type.

For the printing paper cartons, the sheet-fed typographic rotary press is used. In this press system, the paper sheets are passed and printed through the plate drum, on which the wraparound plate with thickness of about 0.8 mm is fixed, and the impression drum. The paper feeding, printing, inking, and delivery structures are its major parts (Fig. 21).

3. Offset Printing Press

On the offset printing press the printing plate is first dampened, then inked. The paper feeding, pressing, and delivery structures are similar to letterpress printing. Although the offset printing press is divided into the flat-bed type and the rotary type, the flat type is not efficient and is used for proof printing. The sheet-fed offset rotary press is used for printing paper cartons.

The printing part structure of this machine is complex, because of the dampening device. The picture on the plate is transferred to the rubber plate, and then the paper is printed. The major parts of the machine are the paper feeding, ink feeding (dampening water feeding and inking), printing, and delivery devices (Fig. 22).

4. Gravure Printing Press

In gravure printing, after the printing plate is inked, excess ink is excluded with a thin steel blade called a doctor, and then the ink left in the concave

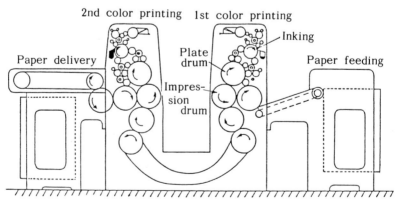

Figure 21. Sheet-fed typographic rotary press.

parts on the plate is removed to the paper. Other features are most similar to letterpress printing and lithography printing. The gravure printing press is divided into the sheet-fed type and the web-fed type. In the sheet-fed system, paper sheets are printed; most of these are one-color machines, but multicolor machines are also used. The sheet-fed type is suitable for the printing of relatively small quantities.

The web-fed type is generally called a gravure rotary press. It prints roll paper and is suitable to the printing a large quantity.

On the sheet-fed gravure press, the paper feeding, paper delivery, and paper removing structures are similar to the offset press. However, the ink feeding and printing structures are unique and are described below.

a. Printing Structure Most press structures are rotary types. As the printing plates, flat plates or cylindrical plates are used. The flat plate is made of copper, and is wrapped around the drum. The cylindrical plate is made of an iron drum on which copper is plated. The printing plate is directly made on the surface of the copper. The impression drum replaces both the rubber drum and the impression drum of the offset press. Therefore the structure is equipped with grippers for wrapping the rubber blanket and for paper feeding.

b. Ink Feeding Structure As with gravure printing, a readily drying ink with high fluidity is used. The ink is stored in an ink pan. There are two methods: surface of the printing plate is directly dipped into the pan, or an ink roller is provided between the ink pan and the plate cylinder (Fig. 23).

c. Web-Fed Gravure Rotary Press Although the paper feeding structure of the web-fed type is simple in structure, high efficiency is required,

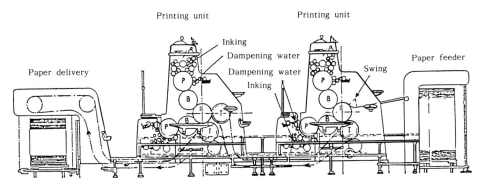

Figure 22. Offset printing press.

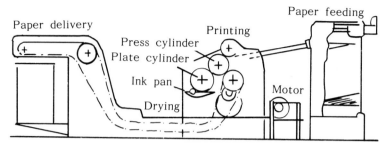

Figure 23. Ink-feeding structure.

and an automatic tension device and automatic paper splicer are provided.

Because of the speedy printing, the required ink quantity becomes large. The ink-feeding system shown in Fig. 24 has been used. The ink pumped up from the ink tank is sprayed onto the plate cylinder, and excess ink is returned to the ink tank through the ink pan.

As the printing plate, the web-fed rotary press uses a cylinder plate. An iron drum with its surface coated with rubber is used for the impression drum.

Because the printing is speedy, a drying device with high efficiency is adopted. The ink for the gravure printing, it is flammable due to its low

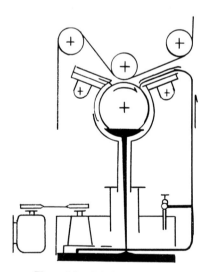

Figure 24. Ink-feeding system.

flash point. Therefore, one set of drum type dryers (using steam heat) is provided for each printing unit.

B. Surface Coating

Various surface coatings are applied to the paperboards used for the cartons. For many uses, fabrication with only printed paper boards would be deficient in the gloss, wearing resistance, water resistance, content resistance, etc.

1. Varnish Over-Printing

As for color printing, varnish is printed onto the required parts, by using plates. The cost is relatively low.

2. Vinyl Coating

The resin solution is coated with a roll coater. Compared to varnish over-printing, it is possible to increase the coating quantity.

Press Coating

Paperboards are first printed with a roll coater. Then the surfaces are coated with solutions of acrylic resin, nitrocellulose resin, vinyl chloride/ vinyl acetate, or others. The surfaces are then heated and pressed with the metallic roller or flat board, which has been glossy mirror coated with chromium, of with the endless steel belt; thus the resin membranes are fused, and the smoothness and the gloss are given. This method is used for high-quality cartons of foods and cartons of medicines and cosmetics.

4. PVDC Coating

A dispersed solution of PVDC resins is applied with an air knife coater or a roll coater. For complete coat formation, high-temperature curing is necessary. This coating is superior in the oil resistance, moistureproofing, gas-barrier function, and heat sealing. It is used for inner coating of cartons in which oily foods such as chocolate are directly packaged.

5. Wax Coating

The wax coating of cartons is generally divided into the paraffin wax coating and the gloss wax (blended petroleum resins etc.) coating. In the latter case, by blending, superior gloss, water resistance, oil resistance, and heat-sealing qualities can be obtained. In all these coating processes, the wax is heated and fused, it is coated on the board with a rubber roller or a gravure roller, it is suddenly cooled, and water is removed by nipping with a rubber roller. Wax coating is used on cartons for frozen foods, butter, and margarine.

6. Film Lamination

On the surfaces of printed paper boards, films of PVC, PP or acetate, etc. are laminated. Film lamination gives smoothness, gloss, and friction resistance.

7. Extrusion Coating

With the extruder, the thermoplastic resins such as polyethylene, polypropylene, polyester, etc. are heated and fused, then extruded, pressed, and coated as a film on the paperboard surface. This coating has been rapidly adopted due to good productivity and low cost. Paperboards that have both sides are coated with polyethylene by this method are used for milk cartons and frozen food cartons.

C. Die-Cutting and Creasing

Die-cutting is cutting the shape of the spread carton from the printed paper board. Creasing is making necessary lines so that the paperboard can be folded when the carton is assembled. Generally, die-cutting and creasing are performed simultaneously with a die-cutter. As to the structure of the cutting machine, the punch is made by combining the cutting edge (thin steel blade) shaped as desired with the creasing edge (thin steel plate without blade) fixed slightly lower than the blade (they are placed in a chase). The female die is made by pasting the solid thick paperboard on the flat steel plate. The printed paper is fed between the punch and the die; then the paper is simultaneously cut and creased by strongly pressing.

Die cutting machines come in two types: (1) platen press type, which includes the sheet-fed die-cutter (which can be either a vertical-type Victoria die-cutter or a platen die-cutter), and the web-fed die-cutter; and (2) rotary type with rotary die-cutter.

1. Platen Die-Cutter

The platen die-cutter consists of the paper-feeding, die-cutting, stripping, and paper delivery parts. The paper sheets piled on the table of the feeding part are picked up by the sacker, and are fed one by one. After the side and front edges are registered, the front of the paper sheet is taken by the

Chapter 4. Paper and Paperboard Containers

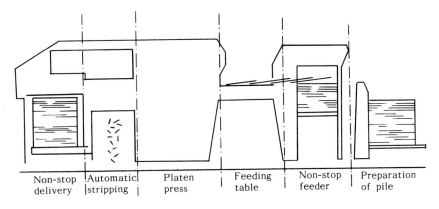

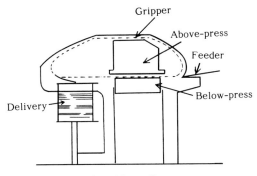

Figure 25. Platen die-cutter.

gripper, and the sheet is sent to the die-cutting part by a chain's intermittent motion. The die-cutting part may have a toggle or crank structure. The sheet is die-cut between the plate beds or chases, which are moving up and down. In the stripping part, the gripper margins are removed by the pins, and the sheets are sent to the delivery part, where the sheets are separated from the gripper bar and stacked (Fig. 25).

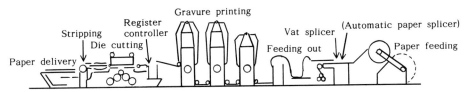

Figure 26. Web-fed die-cutter.

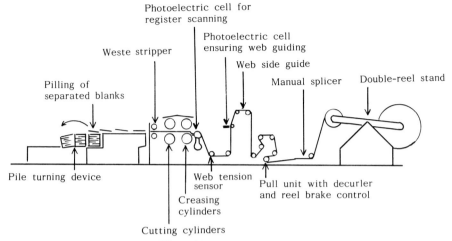

Figure 27. Rotary die-cutter.

2. Web-Fed Die-Cutter

The web-fed die-cutter is the machine to die-cut the paperboard fed from the web. It is generally divided into two kinds, off-line type and in-line type. In the off-type system, the web-fed paperboard is printed by the gravure press, then rewound. This printed and rewound roll is placed on the paper feeding part (unwinder) of the die-cutting machine; then it is die-cut. This system has the merit that high-speed printing is possible, because the printer is not restricted by the speed of the die-cutting. In the in-line system, the die-cutting machine is connected to the printing machine. Manpower can be reduced, because the process from the printing of white paper fed from web to the die-cutting can be performed in one process (Fig. 26).

3. Rotary Die-Cutter

The rotary die-cutter is similar to the web-fed die cutter in every part's function. However, to increase the speed, the die-cutting part is made to a cylinder type. This cylinder consists of the above part and the below part, of which the cutting edge, creasing edge, and female die are made of steel. There are the one-cylinder type and the two-cylinder type (cylinders for the cutting edge and creasing edge). Because the cost of the cylinder is high, this machine is used for mass production of same-dimension products such as milk cartons (Fig. 27).

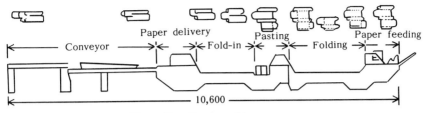

Figure 28. Carton-making process.

D. Carton Making

Carton making is the process in which the raw material that has been finished by printing and die-cutting is folded and pasted to produce cartons. Efficient special machines are used for this, because mass production is impossible manually.

The machine that produces folding cartons by pasting the body of the "blank" (die-cut paperboard) is called a folder-gluer. Carton-making machines are classified as (1) sack gluer, (2) special gluer, or (3) flame sealer.

1. Sack gluer: This machine produces the straight sacks, reverse sacks, and seal cartons by directly pasting their bodies' paste margins. The process speed is fastest, because the timing or folding at a right angle to the flow is not required.
2. Special gluer: This machine pastes the cartons of "autobottom type," and has a structure that pastes the sacks and the bottoms at the same time. For this machine also, the timing is not taken. For the bottom pasting, after the carton has been folded, the paste is directly applied, by utilizing the difference of paper thickness.
3. Flame sealer: This machine seals the blanks, both sides of which are PE coated for frozen food cartons or milk cartons. It is called a flame sealer because the PE is generally fused by a gas burner (Fig. 28).

REFERENCES

Hayashibara, I., and Ohkata, S. (1980). "Paper Containers and Printing." Japan Packaging Institute, Tokyo.

CHAPTER 5

Metal Containers

Hiroshi Matsubayashi
Corporate Research & Development,
Toyo Seikan Group,
Yokohama, Japan

I. INTRODUCTION

In the field of metal containers, various kinds of new metal containers have been developed since the mid 1960s, and the traditional soldered side-seam can, used for about 170 years, has been gradually replaced by them. The new containers are the cemented side-seam can, welded can, DWI (draw and iron) can, and DRD (draw and redraw) can. They have been developed to pursue high quality, high productivity, and cost reduction, and with these developments metal containers have been defending their market position from paper, plastic and glass containers.

II. FEATURES OF METAL CONTAINERS

Metal containers have the following advantages.

1. Metal containers are applicable to almost all foods, because of their rigidity and high thermal resistance. They can withstand outside or inside pressure and high temperature. Various filling and sterilization process can be used. Metal containers are vacuum sealed and heat sterilized under low oxygen pressure. Oxygen in containers accelerates the deterioration of nutritional components of food such as vitamins, fats, and proteins. In metal containers the decomposition of nutrients is very low. Metals perfectly exclude oxygen, moisture, and light, so that correctly processed foods keep for long periods.

2. Double seaming of metal containers is very reliable, and moreover the integrity can be checked by acoustic or optical methods after seaming.
3. High production, high filling, and high sealing speeds can be obtained with the rigid metal containers, and their excellent durability makes long transportation possible.
4. Various kinds of easy-open ends have been developed.
5. Metals' lacquerability and printability are excellent, so that superior decoration can attract consumers' attention.

Though metal containers have many strong points as mentioned above, they are inferior to paper or plastic containers in material cost, container weight, and ease of crush. These disadvantages are gradually becoming less important with the development of low-cost materials and the use of thinner metals.

III. METALS

A. Tinplate

Tinplate is the oldest material, and is still the most important material, for metal containers. Tinplate is low-carbon steel coated with tin by hot dipping or electrodeposition. Now almost all tinplate is made by electrodeposition. The tin coating weight is between 1.12 g/m^2 and 15 g/m^2 on each side of the plate, and differentially coated tinplates are also available.

The use of tin coating generally protects the base steel from rust and corrosion and produces a bright appearance, suitable for printing or coating with organic lacquer and inks.

In the production of tinplate, electroplating is followed by melting, quenching, chemical treatment, and applying a lubricating oil. The most commonly used chemical treatment is a cathodic electrochemical process using sodium dichromate solution (CDC): it provides a surface made very resistant to the formation of tin oxides and has the following attributes (American Iron and Steel Institute, 1979): (1) minimal tin oxide growth after prolonged storage; (2) minimal discoloration during subsequent baking required for organic coatings and lithography; (3) minimal discoloration due to soldering operation; and (4) provision of some resistance to tin sulfide staining of certain sulfur-bearing food products.

Sodium dichromate dip (SDCD) treatment and cathodic sodium carbonate (CSC) are also available for special use. The lubricants mostly used are acetyl tributyl citrate (ATBC) and dioctyl sebacate (DOS). They improve the slip characteristics of plate and reduce surface scratching. Single-reduced tinplate is produced in thicknesses between 0.16 and

0.38 mm, and double-reduced tinplate is between 0.13 and 0.29 mm. A cross section of tinplate is shown diagramatically in Fig. 1.

Containers with lower tin coating levels have recently been developed for cost reduction. The minimal tin coating weight depends on the can manufacturing operation, except for plain cans. That needed for soldering is about 2.8 g/m^2 and for welding is about 0.8 g/m^2. In the case of the tinplate DWI can, tin works as a lubricant during ironing. The minimal tin coating weight is about 0.5 g/m^2.

Corrosion resistance is one of the very important properties of tinplate. It depends on the internal properties of the base steel—kinds and amount of chemical components, nonmetallic inclusions—and on surface properties such as tin coating weight, uniformity, tin crystal size, and passive film. The chemical component of base steel has a very significant effect on the subsequent corrosion resistance and mechanical properties of tinplate. Chemical requirements for tin mill products are shown in Table I (ASTM A623M-79).

Type L steel, low in metalloids and residual elements, has improved corrosion resistance and is used for acidic-aggressive foods. Type MR steel is similar in metalloid content to type L but less restrictive in residual elements; it is commonly used for most tin mill products. Type D steel, aluminum-killed steel, is sometimes required to minimize severe fluting and stretcher strain hazards or for severe drawing applications. Nonmetallic inclusions increase the corrosion rate of steel, and they also cause the surface defects during plastic deformation. The cupping operation is common to both DWI and DRD can manufacture, and nonmetallic inclusions cause microfractures in the cup walls. Such fractures may develop during

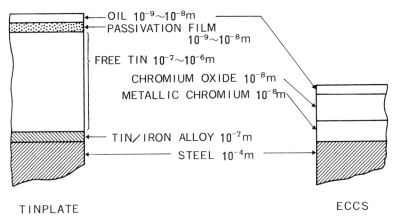

Figure 1. Diagrammatic sections of tinplate and ECCS.

Table I Chemical Requirements for Tin Mill Products

Element	Cast composition (max. %)		
	Type D	Type L	Type MR
Carbon	0.12	0.13	0.13
Manganese	0.60	0.60^b	0.60^b
Phosphorus	0.020	0.015	0.020
Sulfur	0.05	0.05	0.05
Silicon[a]	0.020	0.020	0.020
Copper	0.20	0.06	0.20
Nickel	—	0.04	—
Chromium	—	0.06	—
Molybdenum	—	0.05	—
Other residual elements; each	—	0.02^b	—

[a] Where strand cast steel is furnished, the silicon maximum may be increased to 0.080 unless expressly prohibited by purchaser.

[b] Unless otherwise agreed upon between the manufacturer and the purchaser.

cupping or any subsequent redraw or wall ironing. Surface fractures represent a hazard of pack integrity and corrosion performance. Continuous casting aluminum-killed steel is the most suitable for DWI and DRD manufacture.

Temper is applied to blackplate (base material for tinplate and ECCS): it is the summation of integrated mechanical properties. The combination of properties required for specific applications cannot be measured by any single mechanical test because no one test adequately measures the many characteristics such as drawability, elasticity, stiffness, springiness, and fluting tendency. Ranges of superficial Rockwell hardness are known as temper ranges. Table II shows the temper designations and examples of their usages (ASTM A623M, 1981).

B. Electrolytic Chromium-Coated Steel (ECCS)

Tinplate is traditionally one of the principal materials used for packaging. But rising tin prices and availability of materials such as aluminum and plastics encouraged can manufacturers and steelmakers to develop lower-cost steel base container materials, such as tin-free steel. Against this background, competition for tin-free steel development was waged throughout the world. After studying the feasibility of tin-free steels, such as chemically chromium passivated steel, thin chromium-plated steel, thin nickel-plated steel, aluminum-plated steel, and phosphate–chromium

Table II Temper Designations and Examples of Their Usage

Temper designation	Targeted Rockwell hardness range HR30T	Characteristic	Examples of usage
T-1	46–52	Soft for drawing	Drawn requirements, nozzles, spouts, closures
T-2	50–56	Moderate drawing	Rings and plugs, pie pans, closures, shallow-drawn and specialized can parts
T-3	54–60	Shallow drawing, general purpose, with fair degree of stiffness to minimize flutings	Can ends and bodies, large-diameter closures, crown caps
T-4	58–64	General purpose where increased stiffness is desired	Can ends and bodies, crown caps
T-4-CA	58–64	Moderate forming, fair degree of stiffness	Closures, can ends, and bodies
T-5-CA	62–68	Increased stiffness to resist buckling	Can ends and bodies

Designation	Targeted tensile yield strength (Mpa), longitudinal, at 0.2% offset	Targeted Rockwell hardness HR30-T	Examples of usage
DR-8	550	73	Round can bodies and ends
DR-9	620	76	Round can bodies and ends
DR-9M	660	77	Beer and carbonated beverage ends

film-coated steel, it was concluded that the electrolytically chromium-coated steel (ECCS), developed in Japan in the 1960s, could meet the quality and cost requirements for metal containers.

ECCS is blackplate that is processed by the electrolytic deposition of metallic chromium and chromium oxides and coated with a lubricating oil. The thickness of surface film of ECCS is about 0.04 μm, almost one-twentieth that of no. 50 tinplate. A cross section of ECCS is shown in Fig. 1.

Recently in Japan, ECCS was improved in terms of lacquer adhesion in order to apply the cemented side-seam can to heat-sterilized food. The chemical composition of chromium oxide, uniformity, continuity, and thickness of surface films have very significant effects on the rate of adhesion loss between primer and ECCS during heat sterilization (Matsubayashi, 1982). Chromium oxide contains Cr^{3+} and OH^- as main components. It also contains sulfate anions and fluoride anions from the additives of the plating bath, and Cl^- and Ca^{2+} from the impurities of water used in bath. It is clear that sulfate anion is the most harmful one, and ECCS containing low levels of sulfate anion was developed and commercialized in Japan. The thickness of chromium oxide is preferably between 10 and 30 mg/m^2 from the point of view of lacquer adhesion.

ECCS has the following advantages:

1. Lacquer adhesion on ECCS is superior to tinplate, and resistances to underfilm corrosion and filiform corrosion are excellent.
2. Resistance to sulfide stain is superior to tinplate.
3. Workability of enameled ECCS is superior to tinplate, and lacquer adhesion strength remains even after deep drawing.
4. ECCS does not contain a low-melting-point tin layer (232°C), so that higher lacquer baking temperatures and therefore shorter baking time can be used for the lacquering of ECCS.
5. The cost of ECCS is the lowest in can materials.

On the other hand, disadvantages of ECCS are that surface films of ECCS do not work as a sacrificial layer, unlacquered ECCS is less resistant to corrosion than tinplate, and it is necessary to use a lacquer system for protecting ECCS. ECCS containers cannot be made by traditional soldering, but this disadvantage has been overcome by the developments of the cemented side-seam can, welded can, and DRD can.

C. Aluminum

Aluminum was first used for metal containers in Scandinavia and Switzerland for the packaging of fish, milk, meat, and vegetables. Its widespread adoption in packaging came in the mid 1940s when techniques for rolling and decorating thin aluminum foils were perfected. Now aluminum is widely used for DWI cans for beer and carbonated beverages and for DRD cans.

Pure or commercially pure aluminum types (1000 alloy series) and corrosion-resistant aluminum alloys (3000 alloy series and 5000 alloy series) are used depending on the container design and fabrication method being used. The 1000 series alloys have a minimum content of 99.0% aluminum and are used for the manufacture of extruded containers and foil. The 3000 series alloys contain between 0.3 and 1.5% manganese and are used for manufacture of DRD cans and DWI cans. The 5000 series alloys contain between 0.5 and 5% magnesium and are used for manufacture of DRD cans, can ends, and ring pull tabs.

Magnesium increases the strength of material, but it also reduces its resistance to corrosion by alkalis and acids. Table III shows the mechanical properties and typical uses of some of the most commonly used aluminum alloys (Tsurumaru, 1982).

Aluminum is produced in "O" and "H" tempers. The O temper is produced by subjecting the material to full annealing or complete recrystallization. Aluminum with O temper is the softest form available and has the lowest physical strength. Temper is produced by work hardening the metal as in rolling operations. The H19 is fully strain-hardened by rolling to give maximum physical properties.

Aluminum surfaces are chemically treated to improve the lacquer adhesion and corrosion resistance of aluminum alloys. The chromate–phosphate type of surface treatment was dominant, but recently chromium-free zirconium type or titanium type treatments have gradually increased in popularity.

The inherent advantages of aluminum as a food packaging material are as follows:

1. No taste and odor: Being completely tasteless and odorless, aluminum can be used for the most sensitive products such as beer.
2. No sulfide staining: Aluminum does not produce sulfide staining with sulfur-bearing food products.
3. Workability: Aluminum has excellent workability and requires less strength for draw or draw and ironing operations.
4. Light weight: The specific gravity of aluminum is 2.7, about one-third that of steel. The use of light-weight aluminum containers reduces shipping cost.

On the other hand, disadvantages of aluminum are as follows:

1. Poor resistance to low acid or chloride-containing products: In general, aluminum displays excellent resistance to a wide range of corrosive

Table III Mechanical Properties of Aluminum Alloys

Alloy type	Thickness (mm)	Temper range	Mechanical Properties			
			Tensile yield strength at 0.2% offset (kg/mm^2)	Tensile strength (kg/mm^2)	Elongation (%)	Temper
1080	0.007–0.015	O.H	2.4	7.7–10.4	15–30	O
1100	5–20	O	2.4	7.7–10.4	15–30	O
3003	0.2–0.4	H14–18	11.9	13.2–16.8	1–4	H16
3003	0.2–0.4	H14–18	17.6	18.2	5	H16
3004	to 0.4	H19	27–30	28–31	1–2	H19
		H14–18	23.2	26.7	5	H16
5052	to 0.35	H16–18	24.6	28.1	8	H16
5082	to 0.35	H18–19	38.0	40.1	4	H19
5182	to 0.35	H19	39.2	42.7	4	H19

Chapter 5. Metal Containers

agents. But when the pH of the foods is low, they react with aluminum, and the aluminum surface must be coated to prevent contact. Chloride possesses a strong corrosive action leading to pitting; in case of foods with both low acid and high chloride contents, the highest integrity of protective coating is necessary.
2. The cost of aluminum is the highest in can materials.
3. Mechanical properties of aluminum are weaker than steel.

IV. METAL CONTAINERS AND EASY-OPEN ENDS

A. Three-Piece Can

1. Soldered Side-Seam Can

The soldered can has been used for the packaging of thermally processed foods like meat and vegetables since the late nineteenth century.

The soldered can is made by forming a cylindrical body from a sheet of tinplate and joining the edges of the cylinder with a soldering operation. A cross section of soldered can is shown in Fig. 2 (Toyo Kohan, 1974); the side seam structure is a lock seam.

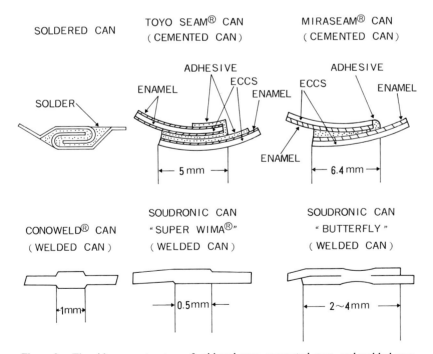

Figure 2. The side-seam structure of soldered cans, cemented cans, and welded cans.

The ends are stamped out from sheet tinplate by automatic presses. Expansion rings are formed in the ends by designing the press to permit ready bulging of the ends during heat processing to relieve strain on the seams of the can. After the rim of the ends has been curled inward by the curling machine, the inside of the curl is lined with a sealing compound. The ends are mechanically joined to the cylinder by a double seaming operation in which two flanges or hooks of the body cylinder and end are mechanically interlocked. When correctly formed, a double seam gives an efficient, air-tight seal. The control of the double seaming operation is essential in the formation of acceptable, safe, hermetic closures. The dimensions of the double seam in terms of its thickness, tightness, and the degree of overlap of the body and end hooks are severely controlled. Figure 3 shows a cross section of a double seam.

Recently lacquered cans with lower tin levels have become dominant for cost reduction. But internally plain (unlacquered) tinplate bodies with higher tin levels are used for packaging of light-colored fruits (apples, peaches, and pears). With these products, the controlled dissolution of tin from the can body is considered to be advantageous in terms of product color and flavor.

2. Cemented Side-Seam Can

The adhesive-bonded side seam can of the ECCS manufacturing process was designed for situations where soldering is not possible, on a commercial basis. In the mid 1960s, the Mira seam can process was developed by the American Can Company in the United States and the Toyo Seam Can

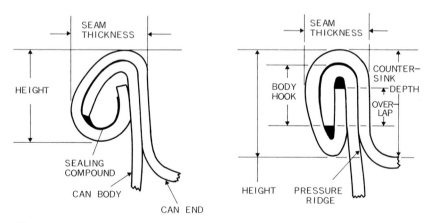

Figure 3. Cross section of double seam. *Left,* first seaming operation; *right,* second seaming operation.

process was developed independently by Toyo Seikan Kaisha Ltd. in Japan. At the early stage, cemented side-seam cans were only used for carbonated beverages and beer, but they are currently being used for thermally processed food and drinks (process condition is at 113–125°C, for 30–90 min.). This has come about through improving the adhesive properties to give resistance to high-temperature water and less deterioration of adhesive strength during long storage. In Japan the rate of shipment of cemented side-seam can has recently increased to about 30% of the total shipment of cans.

The cemented side-seam system is superior to the soldering system in the following respects:

1. The adhesive bonding process is simple and easy.
2. Eliminating the solder bath gives a clean and sanitary can manufacturing process.
3. High speed of production.
4. The adhesive-bonded side seam is stronger than the soldered one.
5. The absence of a solder side-seam margin allows all-around printing except for the minimum side-seam margin, which improves the appearance and gives better display efforts.
6. The absence of solder itself eliminates the risk of solder splash inside the cans and also of tin and lead pick-up from the side-seam areas.

The Toyo seam adhesive bond system is shown in Fig. 4. A nylon adhesive film is used, applied on the preheated primer coated blank

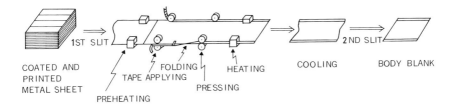

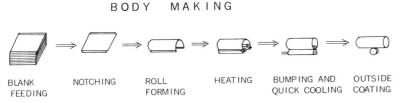

Figure 4. Toyo seam adhesive-bonded system.

by rolling. Cross sections of the cemented side seam cans were shown in Fig. 2. In the Toyo seam can, the internal cut edge of the side seam is encapsulated by adhesive tape completely, and the level of iron pick-up is much lower than that of soldered can. The Toyo seam can is especially suitable for products where flavor, color, and clarity are extremely sensitive to very low levels of dissolved metals.

3. Welded Side-Seam Can

The welding process for container manufacture is essentially based on resistance welding using an alternating current power supply. Two types of welding process have gained commercial acceptance. One is the Conoweld process, using copper roll electrodes developed by Continental Can Co. in 1969 in the United States, and the other is the Soudronic process, using an intermediate copper wire electrode between the roll electrodes and tinplate, developed by Soudronic AG.

The Conoweld process requires mechanical cleaning of the outer 2–3 mm of the can blank edges just prior to welding in order to minimize contamination of the electrode rolls. A solid-phase weld bond is formed over the entire 1.3-mm overlap area. It was used initially for ECCS beverage cans and later for ECCS aerosol cans. Currently it is being used for some food can production.

The original Soudronic welders were used primarily for hand-fed production of tall, large-volume containers, then for tall aerosol cans, large beer cans, and other specialized applications. A can blank edge overlap of 2–4 mm was required, with the actual weld nugget only occupying about quarter of the overlap area. The unwelded portions of the overlap were difficult to protect from possible corrosion and were therefore not suited for most food products. However, because of the use of the wire electrode, this welding process was ideally suited for tinplate cylinders. Recently radical improvements have been made to the basic Soudronic system in order to convert from the butterfly weld nugget structure to a solid-phase weld bond, which extends over the entire 0.5–0.8 mm can blank overlap area, and to permit welding speeds of up to 50 m/min. These new "wire mash" or WIMA welders produce welded tinplate cans suited to beverage and food can applications. The principles of wire mesh welding are shown in Fig. 5.

Inevitably, with a welded seam there are cut edges exposed inside and outside the can. Side stripe coating is a necessity for most food and beverage cans, and this coating quality substantially determines the corrosion resistance of the welded can. Epoxy lacquer, vinyl organosol lacquer, and polyester powder are currently used for side stripe coating, and

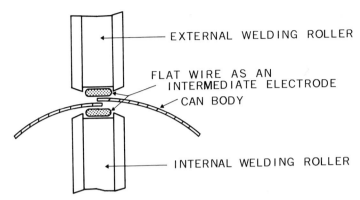

Figure 5. Diagram showing principle of wire mash welding.

they are applied by airless spray, roller coating, or powder coating. The thickness of coating varies according to the corrosivity of food: up to 10 μm for less corrosive food and 20–100 μm for very aggressive food. The construction of the welded portion is connected with the quality of side stripe coating: the thinner weld, smooth surface, and no metal splash are desired. A selection of three-piece food and beverage cans is shown in Fig. 6.

B. Two-Piece Cans

1. DWI Can (Draw and Wall Ironing Can)

There are two main methods commercially used for the two-piece (a seamless body and an end) cans, the DWI and DRD processes.

The DWI can has a thick bottom and relatively thin walls, so it is particularly suitable for beer and carbonated beverages, where the pressurized product structurally supports the containers after filling and closing. For processed food, though currently not as popular, beaded DWI cans are used to avoid buckling from internal vacuum.

Round blanks are stamped from the coiled plate and are drawn once or twice in a cupping press to produce a shallow drawn cup. The wall of the cup is ironed with a punch and ironing dies to about one-third of its original thickness and three times its original height, while the base dimension remains unchanged. By ironing, the can wall is work hardened and it becomes able to withstand the axial load required for double seaming. The external decoration is applied to the can on a printing mandril where the can is rotated between printing rollers. Internal lacquer is applied by spraying. The number of spray coats depends on whether the can is

Figure 6. Photograph of three-piece can: (1) soldered can, (2) adhesive-bonded ECCS processable food can (Toyo seam), (3) adhesive-bonded ECCS beverage can (Toyo seam), and (4) Soudronic welded tinplate beer can.

aluminum or steel and on the type of products. For aluminum beer cans, a single coat is satisfactory. For tinplate soft drink cans, generally a very high-degree of coverage is necessary and the two-coat system is used.

A sequence of operations in manufacturing the DWI can is shown in Fig. 7.

Advantages of DWI cans are as follows:

1. The two-piece can has only one double seam, so integrity can be obtained more easily than for three-piece cans.
2. All-around decoration.
3. Less packaging material is used.
4. The two-piece can stacks well. This is of advantage to the canner, with the can taking up less area and thus reducing storage space and using less packaging material. Another advantage is in the supermarket. The bottom of the two-piece can is designed and formed for better stackability, so it can be stacked vertically without risk of toppling on the shelf.

Chapter 5. Metal Containers

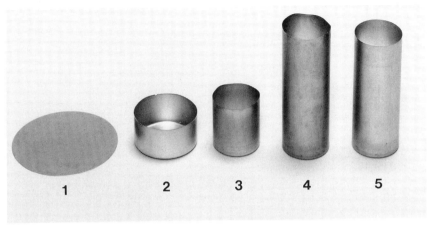

Figure 7. Sequence of operations in manufacturing draw and wall ironing can: (1) body blank, (2 and 3) drawn and redrawn cups, (4) wall ironing and bottom forming, and (5) finished can trimmed to required height.

2. DRD Can (Drawn and Redrawn Can)

The shallow drawn oval cans produced from a single pressing operation have been available for many years. In recent years there has been considerable progress in methods of producing DRD cans.

A draw and redraw process is shown in Fig. 8 (Habenicht, 1976). The

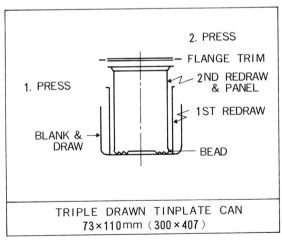

Figure 8. Diagram showing triple-drawn can.

DRD cans are made from preenameled sheets or coils requiring no enamel repair after the can is formed. This may be an advantage since it is cheaper to apply enamel to sheets than to formed cans.

In the can-making operation, the metal is first formed into a cup by the draw method and then further elongated in the redraw operation. During the elongation process, the circumference of the body is reduced. Consequently the final can wall remains substantially equal to the original plate thickness, although some thickening usually occurs at the top of the can. DRD cans are produced from tinplate, ECCS, and aluminum. In recent years the use of ECCS has increased because of its excellent enamel adhesion and low cost.

A selection of two-piece food and beverage cans is shown in Fig. 9.

Modifications of the DWI and draw and redraw process have been developing. They are called draw thin redraw (DTR) and precision sidewall thickness control (PSTC), and they were developed independently. Common features of these cans are in the thickness of the can sidewall. The walls are intentionally thinned in the forming process to maximize material utilization, while those of the conventional drawn can are thicker than the original can stock.

Figure 9. Photograph of two-piece cans: (1) aluminum DWI beer can, (2) tinplate DWI beverage can, and (3) ECCS DRD food can.

In the DTR system, precoated ultra-light-gauge steel (0.18 mm, DR9, ECCS) is used. This system involves stretch forming the steel. Stretch forming helps protect the integrity of the coatings better than ironing, which uses compression to form the steel into cans. Post coating is not required. Figure 10 shows the DTR process. First a large-diameter blank is cut and a cup with flange is formed. Then the can diameter is reduced and height is increased during the first drawing operation, while maintaining the flange. Finally, the diameter is further reduced and height increased to finished can size during the second redraw operation, while maintaining the flange and forming the bottom profile. The sidewalls of the finished can are about 15% thinner than the original metal and thinner than the bottom (Church, 1986).

C. Easy-Open Ends

In the mid 1960s the aluminum ring-pull pour-type easy-open end was developed, and now it is widely used for beer and beverage cans. For high chloride-containing beverages, such as tomato juice and vegetable juice, which may cause pitting corrosion of the aluminum end, a tinplate easy-open end is used. For thermally processed foods, sterilizable full-aperture ring-pull ends made of tinplate or aluminum are used. These easy-open ends are made through a scoring and riveting operation. Metal exposure caused by these operations is repaired with lacquer by an end spray method or body spray method after double seaming for aggressive foods.

Recently, to deal with the enviromental problem, several types of ecology ends have been developed: the non-detachable ring pull end and the tape seal end, prepunched with an aperture that is heat-sealed with organic-coated metal foil or metallized polyester film. Various types of easy-open ends are shown in Fig. 11.

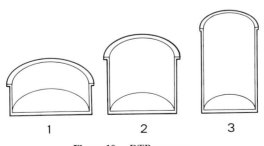

Figure 10. DTR process.

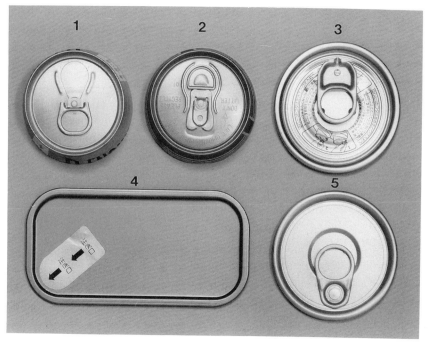

Figure 11. Photograph of easy-open ends: (1) pour aperture aluminum ring-pull end, (2) aluminum nondetachable end, (3) full-aperture tinplate ring-pull end, (4) ECCS tape tab end, and (5) full-aperture aluminum ring-pull end.

D. Internal Can Corrosion

Almost all cans now in use are enameled, except for organic acid-containing fruits. Enameled cans are used for such purposes as preventing sulfide discoloration or the excess dissolution of tin or iron and minimizing the change of color and flavor of products.

Corrosion of enameled cans is briefly explained. The following problems occasionally occur in enameled cans.

1. Pitting corrosion at defects of enamel coating.
2. Anodic undermining corrosion.
3. Cathodic detachment.

One example of pitting corrosion at defects of enamel occurs in the use of enameled tinplate cans with cola drink (phosphate drinks). In this product, iron becomes the anode against tin, forming an electrochemical cell, and is subject to galvanic corrosion. The corrosion becomes localized pitting (Fig. 12a) (Tsurumaru, 1981), which produces an increased risk of perforation and a relatively short shelf life.

Another example of pitting corrosion is found with aluminum easy-open ends at the score line with juices (Horst and English, 1977). Aluminum ends are originally coated, and the corrosion takes place mainly along the pressed-in scoreline if the underside organic coating is torn or damaged by the score operation. This corrosion take place with high-chloride-content products such as tomato and vegetable juice. It is accelerated when the end is used with the steel can body, because the aluminum end is anodic to the steel can body. The perforation resistance of an aluminum end is improved by eliminating aluminum exposure of the end with relacquering or by using a better can body coating to reduce the cathode area.

Anodic undermining corrosion is a most significant failure mechanism for the corrosion of organic-coated metals. It occurs in its simplest form when there is a thin sacrificial metal coating on the metal substrate. For example, where there is a defect in the organic coating on a tinplate can, the tin, being sacrificial to the steel in citrus drinks (Fig. 12b) (Tsurumaru, 1981), will be selectively dissolved out from between the steel and the organic coating, starting from this defect.

One of the examples of cathodic detachment of organic coating is caused by reduction of metal oxide. It is easy to cathodically detach organic coating from steel, which has easily reduced oxide. On the other hand, it is difficult to cathodically detach organic coatings from aluminum and ECCS, which have oxides that are not reduced. Iron oxide is readily reduced to form soluble ferrous ions (Koehler, 1971).

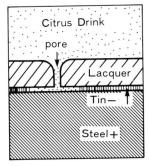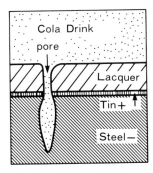

Figure 12. Examples of galvanic corrosion. *Left,* citrus drink: underfilm corrosion → lacquer lifting. *Right,* cola drink: pitting corrosion → perforation.

REFERENCES

American Iron and Steel Institute (1979). "Tin Mill Products," p. 39. AISI, Washington, D.C.

ASTM A623M (1981). "Standard Specification for General Requirements for Tin Mill Product," pp. 662–663. ASTM, Philadelphia.

Church, F. L. (1986). New draw/thin/redraw process makes a super can for Campbell. *Modern Metals* **April,** 28–35.

Habenicht, G. (1976). "Tinplate containers in a changing world of technology." Proc. First Int. Tinplate Conf., pp. 110–121.

Horst, R. L., and English, G. C. (1977). Corrosion evaluation of aluminum easy-open ends on tinplate cans. *Mater. Performance,* **16,** 23–28.

Koehler, E. L. (1971). Corrosion under organic coatings. *Localized Corrosion,* **NACE-3,** 117–133.

Matsubayashi, H. (1982). Progress in TFS for cemented side seam can. *J. Metal Finishing Soc. Jpn.* **33,** 465–473.

Toyo Kohan Co. Ltd. (1974). "Tinplate and Tin free Steel," p. 333. Agne Japan.

Tsurumaru, M. (1981). Corrosion and protection of can. *Corrosion Control Jpn.* **25,** 42–47.

Tsurumaru, M. (1982). Metal can, *JPI Journal* **20,** 23.

CHAPTER 6

Glass Containers

Yoshihiro Yamato
Toyo Glass Co., Ltd.,
Kawasaki, Japan

I. INTRODUCTION

"Soda lime glass," the most popular composition of glass, is used to make ordinary glass containers as well as flat glass. A great number of glass containers are currently manufactured for sealed packagings because glass containers have many advantages for preserving foodstuffs.

Product lives have become shorter due to the diversification of our eating habits and new product variations. The long shelf life of products is not a major concern for food packaging companies because of improved logistics with the cooling chain. Food and beverage companies are looking for convenient, lightweight packagings that are attractive. Therefore the glass container industry has introduced new packagings to cope with these trends.

This chapter describes the forming, properties, and latest trends of glass bottles.

II. FORMING GLASS BOTTLES

Table I shows one example of glass composition, the very popular "soda lime glass." Broken glass or 'cullet" is added to these raw materials for easy melting. This mix is called "batch" and is fed into the furnace.

Figure 1 outlines the glass bottle forming process. The tank furnace is used for melting. The furnace fuel is heavy oil or gas, and in some cases electric boosting is applied as a melting aid. Glass is melted at higher than 1500°C in the melter and goes through a throat, refiner, and forehearth to

Table I Glass Composition

Oxides	Weight (%)	Raw materials
SiO_2	73.0	Sand
Al_2O_3	1.7	Sand, feldspar
Na_2O	13.1	Soda ash, sodium sulfate
K_2O	1.2	Sand
CaO	11.0	Limestone
Fe_2O_3	0.04	Sand

be cooled. At the feeder, glass is at 1000°C and is cut into a glass block or "gob," which will be delivered to the mold on the forming machine. Glass bottle forming is completed through the blank and finish molds. Then glass bottles are conveyed through the annealing lehr in order to eliminate the uneven strain made by the rapid cooling during forming. In this process, bottles are heated again up to 600°C and cooled down slowly over 1 h.

Bottles released from the lehr are sampled for capacity, weight, measurements, heat resistance, and internal pressure strength. All bottles are passed through many sorts of automatic inspection machines, and visual inspection for their appearance is done by human eyes.

After inspection, bottles may be decorated with ceramic colors, prelabeled, or coated with plastics, if necessary.

They are then packed in cases or are bulk palletized. Bulk palletizing is currently growing in popularity.

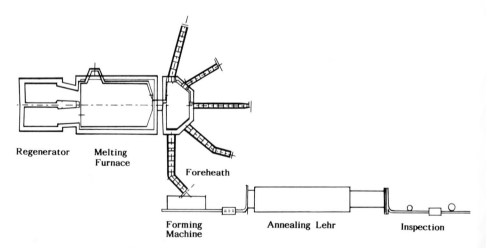

Figure 1. Manufacturing process of glass bottles.

Chapter 6. Glass Containers

III. PROPERTIES OF GLASS BOTTLES

Glass bottles are filled with various products and are delivered to the consumer with much abuse of the glass surface. The quality requirement is to withstand such abuse.

As far as the chemical properties are concerned, glass bottles are safe and stable against most products. This section will discuss the physical properties of glass bottles.

A. Internal Pressure Strength

The more complicated a bottle shape is, the weaker the internal pressure strength is. If the glass wall thickness is the same, the round bottle is the strongest. Therefore, carbonated drinks are bottled in round bottles or similar shaped ones.

The round bottle may be considered as a thin-walled cylinder, and its stress can be calculated from the theoretical formula. The maximum tensile stress of the glass wall is proportional to PD/t where P is the internal pressure, D is the diameter, and t is the wall thickness.

Taking S for the glass strength (normally 300–600 kgf/cm^2), the internal pressure strength P_{max} is determined by the following equation:

$$P_{max} = 2t/DS$$

Therefore, the thicker the glass wall is or the smaller the diameter is, the bigger the internal pressure strength is. In fact, actual figures coincide with the theoretical figures.

The test method of the internal pressure strength is described in ASTM Standards C147-76 and JIS-S-2302.

B. Impact Strength

Most cases of glass bottle breakage are caused directly by impact. It is difficult to theorize the impact strength, which varies considerably depending on the bottle shape, wall thickness, and location and manner of impact. Roughly speaking, the impact strength is proportional to the bottle diameter, and to the third or fourth power of the wall thickness. Method JIS-S-2303 determines a value for the impact strength, but no standard has been established.

C. Thermal Shock Resistance

Glass bottles endure quick changes of temperature at the filling process. Soda lime glass has a comparatively large thermal expansion and shrinkage and small heat conductivity. Under an abrupt temperature change, the expansion or shrinkage rate of the inner wall surface is different from that of the outer wall surface, and this creates a great deal of stress in the glass, which sometimes leads to breakage.

Glass is strong under compressive stress, but is weak under tensile stress. Rapid cooling creates more tensile stress on the glass than rapid heating. Thus glass bottles tend to break under rapid cooling. Figure 2 shows the stress distribution for rapid cooling and heating.

The thermal shock resistance falls off with the inverse root of the glass thickness.

The test method of the thermal shock resistance is shown in ASTM C 149-77 or JIS-S-2304. The test is normally done at a temperature difference of 42°C.

D. Color

Glass can be of almost any color. However, there are only a few colored glasses in the market because of the economy. Figure 3 shows several transmission curves of colored glasses.

The appropriate colored glass is chosen for a particular product, to protect from deterioration by the light, and sometimes to catch the eye.

IV. TRENDS IN GLASS BOTTLES

Two major aspects of the most recent technical developments are described in this section.

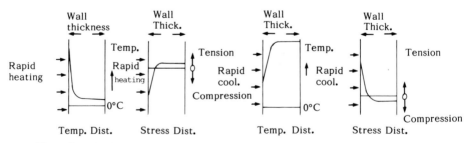

Figure 2. (a) Distributions of temperature and stresses in rapid heating. (b) Distributions of temperature and stresses in rapid cooling.

Chapter 6. Glass Containers

A. Lightweight Bottles

The effects of light weight in glass bottle making are to save the cost of the glass and to reduce the "heavy" image in the market. It is difficult to use light-weight returnable bottles due to the abuse created in filling lines and by consumer handling. However, nonreturnable bottles can be reduced in weight through ongoing efforts.

The correlation between capacity and weight of various glass bottles is shown in Fig. 4. Note that the current light-weight nonreturnables are substantially lighter than the conventional ones.

The technology of light-weight container making involves processes for surface treatment, forming, strengthening, and so on. To assure the quality of light-weight bottles, design and inspection are essential. A combination of these efforts has made acceptable light-weight bottles available.

1. Dual Coating

The strength of glass bottles is substantially reduced when scuffs, scratches, or bruises are made on the surface of the bottles in filling lines or by handling. In order to prevent these surface scuffs, scratches, or bruises,

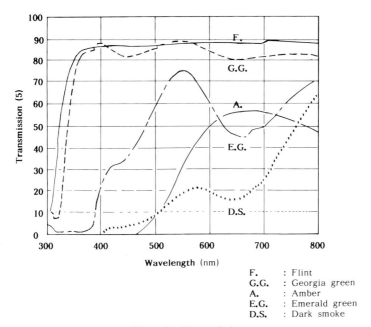

F.   : Flint
G.G. : Georgia green
A.   : Amber
E.G. : Emerald green
D.S. : Dark smoke

Figure 3. Transmission.

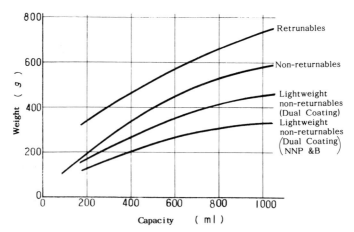

Figure 4. Weight versus capacity ratio.

there are various kinds of surface treatments available. Among such treatments, dual coating is the most effective surface treatment for lightweight bottles.

Dual coating is a surface treatment that utilizes tin oxide or titanium oxide for "hot-end coating" and polyethylene or a surface-active agent for "cold-end coating." The layers made on the surface are very thin and invisible.

The thickness of the hot-end coating is 40–60 Å and the applied polyethylene is ~0.3 mg per 300-ml bottle. These bottles have excellent properties of abrasion resistance and slide angle, and therefore are difficult to damage on the surface. As Fig. 5 shows, bottles with dual coating—the combination of hot-end coating and cold-end coating—have excellent qualities whether the surface is wet or dry. Even if two of these bottles are rubbed against each other, no scratches will appear on the surface of the bottles. Consequently, the chance of reducing the bottle strength is very low.

Figure 6 shows the internal pressure strength of the same type of bottles with dual coating and without such treatments. The average and minimum of coated bottles are approximately twice as high in the figure as for bare bottles. Reducing the weight of nonreturnables can possibly go further.

The dual coating can be applied with simple and compact equipment, and great benefits can be expected at a reasonable operation cost. Consequently, this system is widely accepted and popularly used for nonreturnables. This is seldom used for returnables, since alkali washing gradually reduces its effect.

Chapter 6. Glass Containers

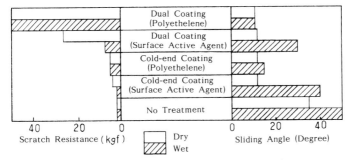

Figure 5. Surface treatments of glass bottles and scratch resistance and lubricity. (Scratch Resistance is the value of load that gives scratches on the surface of the bottles when two bottles are rubbed against each other. Sliding angle is the angle at which a bottle slides off when one bottle is placed on top of another bottle.)

2. NNPB Forming Process

Better glass distribution with improved minimum thickness is necessary for further weight reduction.

In the NNPB (narrow neck press and blow) forming process, a parison of the narrow-neck bottle is formed by pressing, resulting in a more even thickness than by blowing.

Generally, the minimum wall thickness of bottles formed by the NNPB forming process is 20% more than that of the same weight bottles formed by the blow and blow forming process. The NNPB forming process requires an advanced bottle-forming technology and expensive molds.

Therefore it is applied to rather more light-weight bottles, along with the

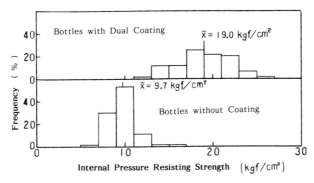

Figure 6. Surface treatment and internal pressure strength. "Bottles with dual coating" are coated with tin oxide and polyethelene.

dual coating. In case of less light-weight bottles, only the dual coating is applied because the necessary glass thickness can be obtained by the blow and blow process.

3. Chemical Tempering

One of the method for further lightweighting of glass bottles is chemical tempering. Chemical tempering is a method that strengthens the bottles by forming a layer of compressive stress on the surface of the glass—by ion exchanging, from sodium to potassium, at about 400°C. Approximately 15 μm of a compressive stress layer exists on both the inside and outside surface of chemical-tempered commercial bottles, and the value of the maximum compressive stress is ~1000 kgf/cm^2 (Fig. 7).

The compresion layer is so thin that even a tiny check made by bottle handling decreases the strengthening effect dramatically.

Moreover, chemical tempering requires a separate processing plant in addition to an existing bottle manufacturing plant, with the resultant additional processing cost. Therefore, chemical tempering is applied to a limited range of products—mainly applied as part of plastic-coated returnables for soft drinks with a capacity of 1 l or more.

4. Strength of Light-Weight Nonreturnables

Currently, the most popularly used light-weight nonreturnables are bottles with a dual coating manufactured by the NNPB forming process.

Light-weight bottles have thinner walls but uniform wall distribution, and coatings to protect from abrasion. Most of their contours are a full shape suitable for light weight, and the ratios of height to body diameter are between 2.0 and 2.5.

Even though light-weight nonreturnables are substantially lighter than

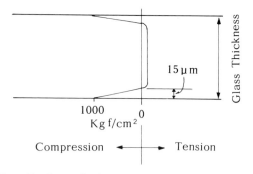

Figure 7. Stress distribution of chemical tempered bottle.

Chapter 6. Glass Containers

Table II Strength of Bottles in the Marketplace

Strength test	Type of bottle	
	Dual-coated bottles (300 ml, 170 g)	Returnables of carbonated drinks (340 ml, 420 g)
Internal pressure $\bar{x}$ (kgf/cm^2)	30.6	25.7
Impact $\bar{x}$ (kgf · cm)	5.2	4.5
Thermal shock $\bar{x}$ (°C)	>70	56.3

conventional nonreturnables, the strength of light-weight nonreturnables is comparatively high. Light-weight nonreturnables in the market are strong enough, as shown in Table II, when compared to collected soft-drink returnables. Thus, excess packing for nonreturnables can be eliminated.

The coating on nonreturnables generates low friction in bottle-to-bottle contact, which enables smooth operations in filling lines. Also, because of the comparatively thin glass wall, faster heating up and cooling down in filling lines can result in high efficiency and productivity.

Light-weight nonreturnables are simple in shape, and this means they are easy to make and easy to handle. Thus the production and distribution costs can be reduced.

B. Prelabeled Bottles

There are two types of prelabeled bottles. One is wrapped with a heat-shrinkable sleeve (~300 μm thick) of foamed polystyrene. The other is wrapped with a solid film, such as polyvinyl chloride (~50 μm thick) or polystyrene. Prelabeled bottles have more graphic area and no need of a labeling machine in filling lines.

The foaming type prelabeled bottles (known as Plasti-Shield) were first developed in 1971 by Owens-Illinois, Inc., in the United States and have

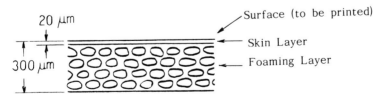

Figure 8. Plasti-Shield bottle.

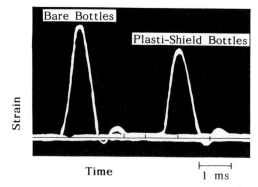

Figure 9. Shape of strain wave at impact.

been accepted in many countries. Bottles of this type are used mainly for "single-service" (smaller than 500 ml) soft drink bottles.

As shown in Fig. 8, the Plasti-Shield label consists of a milk-white foamed layer and a solid layer of approximately 20 μm thickness. Excellent shock resistance is given by the shock-absorbent foamed polystyrene.

Figure 9 shows the strain waves that appear on the oscilloscope when impacts are given to bottles. Under the same conditions, Plasti-Shield bottles show ~80% of the impact waves height compared to bare bottles, and this means that the impact was absorbed by 20%.

When bottles are loaded into vending machines, they receive a certain level of impact. Figure 10 shows the test method for impact strength. The results of the test are shown in Fig. 11. Plasti-Shield bottles are 1.4 times stronger than returnables.

Plasti-Shield film reduces the noise level at the filling station, eliminates partitions in a carton, and protects bottles in a vending machine. The solid layer is good for abrasion and printing.

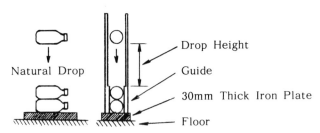

Figure 10. Horizontal drop test method. Fill up three sample bottles with water and lay down two of them on the plate as shown. Then drop the remaining one horizontally at a certain height.

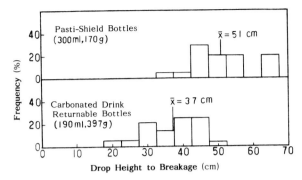

Figure 11. Results of drop test.

Prelabeled bottles with solid film sleeves were introduced into the market in Japan in 1981. The solid film has no cushoning effect, but performs well in fragment retention when bottles are wrapped up the neck with it. It also has excellent heat conduction when compared to foamed labels.

Prelabeled bottles can be standardized for many products simply by changing graphics on the label. Prelabeled bottles may be the best packagings for product diversification. The plastic-coated bottles are much safer when broken and can be possibly further reduced in weight. These bottles are used mainly for large-size soft-drink returnables with a high gas volume.

V. AFTERWORD

Glass bottles are packagings suitable for diversification. Bottles can be formed into a simple shape with lighter weight, a sophisticated contour with rather heavy weight like whisky bottles, or a very specific design like cosmetic bottles. And bottles can be fashionable packagings with a range of glass colors, printing designs, label graphics, and closures.

On the other hand, there is still a very serious problem. With today's changing life styles, many materials are used for packagings, and waste disposal becomes a social issue. From environmental and ecological points of view, glass bottles have benefits as returnable or reusable containers with an effective recycling system.

REFERENCE:

Yamato, Y. (1983). Security and its valuation of "Plasti-Shield Bottle." *J. JPI* **21**(3), 9–13.

CHAPTER 7

Plastic Containers

Koji Kondo
Dainippon Ink & Chemicals, Inc.,
Tokyo, Japan

I. INTRODUCTION

Plastic food container technologies have been developing vigorously in the plastic industry over the past 25 years.

Obviously, these plastic containers are the newcomers to the packaging field, but plastic containers have actually succeeded in replacing metal, glass, and paper containers in many applications.

Various performance advantages and lower prices have been realized, through continued developments in the polymer industry.

Performance advantages of plastics are as follows:

1. Easy coloring.
2. Low cost by mass production.
3. Good productivity for manufacture of complicated shape products (design freedom).
4. Shatterproof plastic containers are rarely, if ever, cracked or dented while in delivery or shipment.
5. No rusting.
6. Good water resistance.

Disadvantages of plastics are as follows:

1. Less heat resistance.
2. Tendency to creep.
3. Cracking and crazing trouble.
4. Static electricity generation.
5. Lower dimensional stability.
6. Lower gas barrier.

Polymers can be clasified into two types, according to behavior during heating. One is thermoplastic polymers and the other is thermosetting polymers. Thermoplastic polymers soften and melt on heating and solidify again on cooling. In the softened state, they can be converted to a paticuliar shape that remains after cooling, like wax. This cycle is a repeatable process without changing its properties. On the other hand, a thermosetting polymer, which is in a liquid or solid state at room temperature, is melted and hardened on heating and converted into an infusible state by some chemical change (cross-linked), like cooking the yolk of an egg. Many polymers for packaging use are thermoplastic polymers, such as polystyrene, polyethylene, and polypropylene.

II. POLYSTYRENE

Polystyrene has been known for many years, and the first investigations into polystyrene polymerization were made in 1839. In 1935, Germany's I.G. Ltd. initiated the first industrial production of polystyrene, while America's Dow and Monsanto followed with production of styrene monomer in 1936, after which Dow began polystyrene production in 1939.

At present, polystyrene is among the four major general-purpose polymers, which include polyethylene, polypropylene, and polyvinyl chloride, and has made up a large volume of the consumption of plastic containers.

Of all polymers used for thermoforming, polystyrene is the largest in terms of volume.

Basically, polystyrene polymer includes general-purpose grade (GPPS), which is polymerized styrene monomer, and a high-impact grade (HIPS), which is polymerized styrene monomer with synthetic rubber to decrease brittleness.

A. GPPS

Styrene monomer is a colorless, transparent liquid with a boiling point of 145.2°C. Styrene monomer is polymerized very easily by heating. As polystyrene is an amorphous polymer, it has a transparent and clear appearance.

Polystyrene is a hard and rigid polymer, but it is brittle like glass. GPPS has a softening temperature of 95°C, refractive index of 1.6, and density of 1.05.

The excellent flowability of GPPS on injection molding machines makes possible production of such things as transparent thin-wall drinking cups by multicavity molds with high cycle, high-speed injection.

Nonoriented (cast) GPPS sheet is not widely used for thermoforming because it is too brittle, so it is difficult to wind up the sheet on the winder.

Instead, GPPS-based biaxially oriented sheet and foamed polystyrene sheet (also called polystyrene paper, PSP) are widely used for thermoforming in large quantities. GPPS is the main raw material used in both of these sheets.

For production of biaxially oriented polystyrene sheet, thick and flat GPPS substrate, which is extruded from slot dies, is oriented in the transverse direction immediately after in-line, longtitudinal orientation at a controlled temperature of 100–140°C. The stretch ratio is about threefold, both longitudinally and traversely.

The orientation relieves the sheets of brittleness and substantially improves their impact strength, stiffness, and low-temperature resistance. The Dart impact strength data of 250 μm thickness of oriented polystyrene sheet show 25 kg · cm and 35 kg · cm at 23°C and −40°C, respectively (ASTM-D1709).

Biaxially oriented polystyrene sheet is not widely used in Europe, despite the large consumption in the United States and Japan. A considerable amount of polyvinyl chloride (PVC) sheets has been displaced by oriented polystyrene sheets in a variety of applications as a result of the problem posed by vinyl chloride monomer migration into foodstuffs from package materials and the problem of hydrogen chloride gas generation during the incineration of PVC material waste. Oriented polystyrene sheets include many grades, such as corona-discharged, antifog, antistatic, easy-release, and slip type.

For thermoforming of oriented polystyrene sheets, the trapped-sheet, contact heat vacuum and pressure thermoforming machine is used. An ordinary vacuum thermoforming machine cannot be used because orientation stress release may occur during infrared preheating and cause sheet to shrink. Accordingly, it is necessary to hold the oriented polystyrene sheet by using a vacuum/blow hot plate installed on the heater base, placing an oriented polystyrene sheet on its top surface, and adding a mold cavity. The hot plate, heated to 130°C, has many small trapping holes. A vacuum is then applied through the hot plate, and the oriented polystyrene sheet is attracted toward the trapping holes and heated upon contact with the hot plate.

The contact duration varies with the thickness of the sheet involved, but it lasts 1 s for a thickness of 150 μm. Then the vacuum is switched to compressed air. Pressure (~3 kg/cm^2) is applied through the hot plate and forced against the sheet surface, while its reverse side is vacuum attracted toward the mold cavity and cooled upon contact with its surface. The cooling water, heated to approximately 70°C, circulates through the mold. It requires 1 s to cool a sheet with a thickness of 150 μm, thereby forming cycle ranges of from 4 to 10 s depending on the thickness of the sheet.

The fabricated products include food contact trays, meat trays, food packs, produce packs, etc. The fabricated products made from oriented polystyrene sheet are transparent and attractive in appearance, while producing tough and rigid trays.

Polystyrene provides a lower moisture and gas barrier than polyolefins, such as polyethylene and polypropylene. However, lower barrier properties can be an advantage, as in the case of vegetable packaging and radish seedling trays.

Polystyrene tends toward environmental stress cracking, and is therefore limited in the amount of contact with fatty or oily food products such as butter, margarine, and other edible oils. Alpha-limonene in orange peel also attacks polystyrene.

Polystyrene also has poor weatherability, so it has a tendency to yellow in color when exposed to direct sunlight.

Since oriented polystyrene sheet containers cannot be used for packaging foods exposed to temperatures in excess of 85°C, it is unsuitable for retortable packaging. Polystyrene is an excellent high-frequency insulator, thereby attracting dust and other foreign substances. To minimize dust collection problems, it is better to use an antistatic type sheet.

Highly brittle GPPS is combined with HIPS for a coextrusion sheet that produces a glossy surface layer. This product is used in large quantities for yogurt containers.

B. HIPS

In order to improve the brittleness of GPPS, synthetic rubber such as polybutadiene or styrene butadiene rubber is added to styrene monomer during polymerization. As a result, high-impact polystyrene (HIPS) is obtained.

Tiny synthetic rubber particles have been dispersed in the matrix of polystyrene. The synthetic rubber is added at a rate of up to 10% by total volume. The value of impact strength varies with the amount of the synthetic rubber added and the polymerizing conditions. When light shines on a polymer, it is scattered by the dispersed particles. As a result, HIPS grades vary from translucent to opaque, depending on impact properties and overall thickness.

HIPS sheets have a draw ratio (depth/diameter) of 2–2.5, compared with 0.5 for oriented GPPS sheet. This is why HIPS is widely used for deep-draw food packaging such as egg trays, soybean curd (*tofu*) trays, ice cream containers, drinking cups, etc.

In addition, many lactobacilli beverages are packed in HIPS small-size injection blow bottles.

C. Expandable Polystyrene

1. In-Mold Bead Forming

Expandable polystyrene beads are produced by incorporating a low-boiling-point hydrocarbon foaming agent, such as pentane, into the styrene monomer during polymerization.

In the first step, a primary foaming process is necessary to obtain a uniform foamed fabricated product. Beads are heated to 90–105°C by steam heating in a vessel while agitating.

The second step is the aging process. The purpose of the aging step is to raise the inner pressure of preformed beads to atmospheric pressure by air penetration into the cell.

The third step is molding or secondary foaming. Aged beads are charged into the mold cavity and heated by steam. The softened beads expand again in the mold, at which point the individual cell walls fuse together at the contact points, due to both expansion pressure and softened walls.

The foaming ratio usually varies from 30 to 60, according to the fabricated product's specifications. A lower foaming ratio provides better mechanical strength. Generally, foamed products are useful for energy conservation and material savings; they are low specific gravity products with economical advantages.

Foamed polystyrene beads possess high rigidity and provide independent foamed cells. Therefore, the product obtained is low in heat conduction, provides insulation against high and low temperature, and is shock-absorbent.

Typical products include insulated drinking cups, instant noodle cups, frozen food carrier boxes, and fish carrier boxes.

2. Foamed Polystyrene Sheet

There are two process for manufacturing foamed polystyrene sheets.

The first process is extruding the expandable polystyrene beads into foam sheets. As this process requires the use of an additional effective foaming agent, it is not generally enployed.

The second process is extruding polystyrene pellets into which Freon 12 or other proper foaming agents such as pentane have previously been injected. Usually, a circular die is used for extrusion in order to obtain uniform foam distribution. The foamed tube is cut off into flat sheets. Following an aging process to eliminate the volatile agents, this sheet is then thermoformed.

The foaming ratio is usually 10. Appearance varies from a china porcelainware appearance to rough surfaces according to the customer's order and cost requirements. The latter process is mostly used to manufacture foamed polystyrene sheets.

Thermoformed products have a wide range of applications, such as meat trays, fish trays, produce trays, egg trays, hot drink cups, lunch boxes, food containers, and dishes. The draw ratio of a foamed polystyrene sheet is up to 1:1.

Laminated foamed polystyrene sheets are made by the in-line thermal lamination process with reverse-printed PS film. Fabricated products made from laminated sheets have a more glossy attractive appearance, better rigidity, and more mechanical strength and stiffness than foamed sheet only. They can thus help save on materials.

Styrene monomer dissolves polystyrene polymer. Accordingly, it is a very difficult to remove the small amount of styrene monomer contained in the polystyrene polymer structure during the polymerization process. Styrene monomer may be produced by alkylation of benzene with ethylene and subsequent dehydrogenation of the ethylbenzene obtained. Accordingly, styrene monomer contains impurities such as toluene, ethylbenzene, isopropylbenzene, etc., as well as by-products of the manufacturing process of the monomer.

The extensive presence in polymer of such impurities as residual volatile matter could possibly constitute a source of claims against the makers of insulated drinking cups etc. because of the obnoxious odors contained therein. Special precautions should be taken on this account.

In Japan, the upper limits of residual volatile matters contained in polystyrene pellets and foamed polystyrene have been stipulated as 5000 and 2000 ppm, respectively.

III. BUTADIENE–STYRENE COPOLYMER

As previously mentioned, general-purpose polystyrene (GPPS) is highly transparent but has the drawback of being extremely brittle. In order to improve this product's brittleness, synthetic rubber is added to polystyrene during the polymerization operation. This process provides high-impact polystyrene (HIPS) with enhanced impact strength. The HIPS is

Chapter 7. Plastic Containers

available in extensive lines of products ranging from translucent to opaque, depending on particle size, quantity and distribution of synthetic rubber particles. A number of manufacturers are conducting research for improving the clarity of HIPS.

As one of the improving process, butadiene–styrene (BS) block copolymers have been developed. According to classification by mechanochemical modifications, a series of copolymers leads to formation of random, alternating, graft, and block copolymer.

If they are binary copolymers of A with B, distributions of individual monomeric building units are given as follows:

1. Random

    $$-A-B-B-A-B-A-B-B-A-A-A-$$

2. Alternating

    $$-A-B-A-B-A-B-A-B-A-B-A-B-$$

3. Graft

    $$\begin{array}{c} B-B-B-B-B-B-B- \\ | \\ -A-A-A-A-A-A-A-A-A-A-A-A- \\ | \\ B-B-B-B-B \end{array}$$

4. Block

    $$-A-A-A-A-A-A-B-B-B-B-B-B-B-$$

A styrene–butadiene block copolymer may be produced by an active polymerization process. Usually, many polymerizations are terminated by side reactions, and the polymers obtained have a wide range of molecular weight distribution. To avoid termination side reactions by impurities such as water, highly purified styrene monomer and organolithium initiator are charged into purified solvent and are polymerized to polystyrene as a first step. After all of the styrene monomer is consumed, the polystyrene obtained still has an activity for further polymerization with styrene or another monomer such as 1,3-butadiene at the end of the polymer chain (active) and has a very narrow range of molecular weight distribution. Therefore, when 1,3-butadiene is added to an active polystyrene block (S –) as a second step, further polymerization occurs to produce an active styrene–butadiene block copolymer (S – B –). After that, styrene monomer is added sequentially to S – B – block copolymer, and consequent S – B – S – active block copolymer is obtained.

$$n-\text{BuLi} + \left[\text{CH}_2=\text{CH}-\underset{\bigcirc}{} \right]_x \longrightarrow \left[n-\text{Bu}-\text{CH}_2-\overset{\ominus\ \oplus}{\underset{\bigcirc}{\text{CHLi}}} \right]_x \longrightarrow$$

$$\left[n-\text{Bu}-(-\text{CH}_2-\underset{\bigcirc}{\text{CH}}-)_{x-1}-\text{CH}_2-\underset{\bigcirc}{\text{CHLi}} \right]_+$$

$$(\text{CH}_2\text{CH}=\text{CHCH}_2)_y \longrightarrow + (\text{CH}_2=\text{CH}-\underset{\bigcirc}{})_z \longrightarrow$$

$$n-\text{Bu}-(-\text{CH}_2-\underset{\bigcirc}{\text{CH}})_x-(-\text{CH}_2-\text{CH}=\text{CH}-\text{CH}_2-)_y-(-\text{CH}_2-\underset{\bigcirc}{\text{CH}}-)_z-$$

K-Resin (Phillips 66 Co.) is a styrene–butadiene block copolymer containing a higher percentage of styrene than 1,3-butadiene. K-Resin is amorphous polymer and provides a good clarity and impact resistance. It is are commercially available in extrusion and injection grades.

In the case of injection molding, resin temperature and mold temperature are set in the range of 190–230°C and 10–60°C, respectively. Production of sheet by extrusion and thermoforming for fabricated products requires the same type of machine and equipment as for HIPS. K-Resin provides transparent and high-impact-strength fabricated products, but it is often blended with other resins such as polystyrene because of its expense. Occasionally it is used as one layer of coextruded multilayer sheets.

ASAFLEX (Asahi-kasei) and CLEAREN (Denki-kagaku) are similar polymers.

IV. ACRYLONITRILE–STYRENE COPOLYMER

The binary copolymers of styrene with 20–30% acrylonitrile (AS, ANS, or SAN) yield amorphous polymer and provide excellent gloss and transparency. They provide more heat, chemical, and oil resistance than GPPS and HIPS, while also showing improved stress cracking.

The binary copolymers of styrene with 70% acrylonitrile are called High-nitrile polymer and provide excellent carbon dioxide and oxygen gas barrier properties and oil and chemical resistance. Therefore, carbonated beverage bottles have been developed from high-nitrile polymer. In 1979, the FDA (U.S. Food and Drug Administration) reported concern about the possibility of carcinogens contained in the acrylonitrile monomer and possible migration of monomer into beverage, and suspended use of the high-nitrile polymers for carbonated beverage bottles except for edible oil bottles. The FDA, however, agreed to their use for beverage bottles in September 1984.

V. ACRYLONITRILE–BUTADIENE–STYRENE COPOLYMER (ABS)

ABS polymer has a heterogeneous structure deriving from the polybutadiene (5–30%) particles dispersed in the acrylonitrile–styrene copolymer matrix. Usually, the ABS polymer is produced an by emulsion polymerization process.

Acrylonitrile provides heat, chemical, and stress-cracking resistance for ABS polymers. Butadiene provides toughnes and improves impact. Styrene provides hardness, rigidity, and good flowability. Properties of ABS polymer vary according to the relative ratios of those three compornents.

ABS polymer has been used as a packaging materials for butter and margarine, but polypropylene containers have replaced them because of ABS's higher price.

VI. POLYVINYL CHLORIDE (PVC)

PVC is a white powder obtained by radical polymerization or chain polymerization from vinyl chloride monomer.

Generally speaking, linear polymers are formed by two basic polymerizations, successive and chain polymerization. In successive polymerization, all of the functional groups of each starting monomer start the reaction simultaneously and polymerize successively. The reactivity of functional groups that remain at both ends of the building polymers is same as for the starting monomers and does not change essentially. Therefore, higher-molecular-weight polymers are formed as time proceeds. The resulting molecular weight distribution curves in the successive polymerization process shows that there is a wider range of distribution but lower polymer weight in each molecular weight fraction as time proceeds.

Successive polymerization can be classified roughly into two types, polycondensation and polyaddition. Polycondensation produces polymers such as polyesters and polyamides from dicarboxylic acids with glycols and diamines, respectively:

$$H_2NO-R-OH + HOOC-R'-COOH \rightleftarrows -(-O-R-OCO-R'-CO-)- + 2H_2O$$
$$H_2N-R-NH_2 + HOOC-R'-COOH \rightleftarrows -(-NH-R-NHCO-R'-CO-)- + 2H_2O$$

Generally speaking, it is required that at least two functional groups be contained in each starting monomer to form a linear structure. If some of the monomers contain more than two functional groups, the polymers obtained have a three-dimensional cross-linked structure. During the polycondensation process, low-molecular-weight by-products are produced simultaneously. Therefore the elements of the polymer structure generated differ from the elements of the chemical structure of the components of the starting monomers. On the other hand, in polyaddition and chain polymerization processes, polymers have the same elements of chemical structure of components as the starting monomers.

Polyaddition produce polymers such as polyurethanes and polyureas form diisocyanates with glycols and diamines, respectively:

$$O=C=N-R - N=C=O + HO-R'-OH \rightleftarrows -(-O-R-OCONH-R'-NHCO-)-$$
$$O=C=N-R-N=C=O + H_2N-R'-NH_2 \rightleftarrows -(-HN-R-NHCONH-R'-NHCO-)-$$

During the polyaddition process, low-molecular-weight by-product does not split off.

In chain polymerization, monomers add on to each other to obtain linear polymer. Therefore, elements of components of the obtained polymer structure do not differ from elements of the starting monomer. For the polymerization to initiate, it is necessary to activate the monomer by adding a small amount of initiator. Activated particles of initiator react with monomer and form activated monomer, which has the same kind of activated particle. Activated monomer reacts with nonactivated monomer, and a new linkage and new activated point are formed simultaneously. It is considered that the activated point for polymerization is located at the end of a polymer chain. If the activated molecule is denoted as a radical $R^{\bullet}$, and the monomer and polymer by the symbols M and P, respectively. the polymerization process can be expressed as follows:

Initiation $R^{\bullet} + M \longrightarrow RM^{\bullet}$
Propagation $RM^{\bullet} + M \longrightarrow RM_2^{\bullet} (=P^{\bullet})$
              $RM_x^{\bullet} + M \longrightarrow RM_{x+1}^{\bullet} (=P^{\bullet})$

Chapter 7. Plastic Containers

Termination $RM_x^{\cdot} + RM_y^{\cdot} \longrightarrow RM_{x+y}R \;(= P^{\cdot} + P^{\cdot} \to P-P)$
(recombination)
$RM_x^{\cdot} + RM_y^{\cdot} \longrightarrow RM_x + RM_y \;[= P(+H) + P(-H)]$
(disproportionation)

For the polymerization of polystyrene and poly(methylmethacrylate), the dominant reactions of termination are recombination and disproportionation, respectively.

During the polymerization reaction a side reaction called transfer may occur, in which the activity of an activated molecule is transferred to a monomer or another substance such as a transfer agent (denoted by the symbol SH):

Transfer $P^{\cdot} + M \longrightarrow P + M^{\cdot}$
$P^{\cdot} + SH \longrightarrow P(+H) + S^{\cdot}$

Chain polymerization can be classified roughly into two types, addition or ring-opening polymerization. Addition polymerization produce various vinyl polymers such as polyvinyl chloride, polystyrene, polyethyrene, polyvinylidene chloride, polymethacrylate, etc. Ring-opening polymerization produces polymers such as poly(ethylene imine) and nylon-6.

According to the activated particles (chain carrier) structure of initiation, addition polymerization is classified as radical or ionic polymerization. As stated previously, in the active polymerization process of styrene–butadiene copolymer, this reaction proceeds by the anion chain carrier, so it is called anionic polymerization. The radical chain carrier for polymerization of PVC can be induced by heating of peroxides as a catalyzer. As methods of polymerization of PVC, heterogeneous phases such as emulsion or suspension are often used.

PVC has been studied for many years. Vinyl chloride monomer was first discovered in 1835. Germany's I.G. Ltd. initiated its emulsion polymerization process in 1931 and Union Carbide (USA) followed with its solution polymerization process in 1933. In 1941, Goodrich (USA) added its suspension polymerization process, which is the dominant process in the industries today.

A large volume of PVC is currently consumed due to relatively cheap production of vinyl chloride monomer and useful properties of the polymer. PVC is a versatile polymer, and provides good light, oil, and chemical resistance. PVC offers good clarity with high impact resistance. Unplasticized PVC is highly impermeable to oxygen and carbon dioxide. However, the gas barrier properties of PVC are strongly impaired by plasticization.

PVC grades vary from rigid to soft, which contains plasticizer, depending on the application and performance. In addition to food packaging, PVC is also widely used for industrial uses such as blisterpacks, press-through packs for pill-form medicines, agricultural films, electric wire insulations, pipes, etc.

Until the FDA reported concern about the possibility of carcinogens contained in the vinyl chloride monomer and possible migration of the monomer into food products, PVC was also an important thermoforming material for the transparent packaging used for many food products.

In addition, combustion of used packaging materials made from PVC releases hydrochloride gas, which attacks furnaces and pollutes the air.

In thermoforming for food packaging, the PVC sheet has been replaced by biaxially oriented polystyrene, high-impact polystyrene, polypropylene, and coextrusion sheets. In the container fields, PVC bottles have been replaced by polyethylene terephthalate stretch-blow and multilayer direct-blow botles.

The limitation on vinyl chloride monomer maximum content in fabricated products is set at 2 ppm in the United States (1 ppm in Japan) at present. It is considered that there is almost no problem with monomer migration into food products at such a monomer content level, but consumers remain hesitant to begin using PVC food containers again. Therefore, the consumption of PVC packaging materials has not recovered to previous levels.

Whereas thermoplastic sheets such as polyethylene and polypropylene are made by the T-die extrusion process, PVC sheet is generally made on calendar machines. PVC powder is mixed with a high-molecular-weight non-migration-grade plasticizer, heat stabilizer, and other components in a blender. Once it reaches an adequate plasticity on a heated roll, a hot strip is then made on the calender roll unit on which hot rollers are arranged in L or Z letter shapes. These are then cooled to produce the finished PVC sheet.

PVC sheet has better thermoformability than most other sheets. Thermoformed cups and trays have been used for many food products such as water, soft drinks, milk, jam, ice cream, egg, sliced ham, Vienna sausage, frankfurters, luncheon meat, snack food, candy, and fruit.

PVC bottles are made by direct blow molding from an extruded hot parison.

PVC bottles for sauce, tomato ketchup, and mayonnaise containers have generally been replaced with other products.

Chapter 7. Plastic Containers

VII. POLYETHYLENE

Polyethylene (PE) was discovered by ICI Ltd. during the fundamental research program experimentation on chemical reactions of gases under super high pressure.

One day, the apparatus happened to have a leak, and the air comprised of nitrogen and oxygen that acts as a catalyst for radical polymerization of polyethylene was discharged into ethylene gas through part of apparatus accidentally. When the apparatus was opened after the work, the workers found polyethylene in it. ICI continued the work and initiated commercial production of polyethylene in 1939. Polyethylene served as an electrical insulating material for radar and for cables at the bottom of the sea. The performance of radar has increased dramatically, and it is said that about 100 submarines were sunk within the first week after being equipped during World War II.

The polymerization technology of polyethylene was transferred from ICI to DuPont and Union Carbide and Carbon (UCC) in the United States. They started the production of polyethylene on a commercial scale in 1942. After World War II, polyethylene packaging films and extrusion coating technology were developed.

Production conditions of polyethylene under high pressure are as follows:

Ethylene monomer pressure:	1500–2000 atm
Oxygen concentration:	0.06%
Temperature:	190–210°C
Molecular weight:	~6000–12,000, max. 65,000

Oxygen catalyst is changing to peroxide catalyst in order to control the process of the reaction at present.

Later, the reaction's application was rapidly diversified and output was upgraded to one of the most favored products of the petrochemical industry. The monopolistic production of polyethylene in the United States came to an end when the 1952 Anti-Trust Act was invoked. The number of polyethylene manufacturers increased to six in 1956 and 10 in 1965.

Dr. K. Ziegler, head of Max Plank Coal Research Institute, discovered that polyethylene could be produced even by low-pressure polymerization of ethylene, if a catalyst comprised of triethylaluminum combined with titanium tetrachloride was used.

Since until then everyone had taken it for granted that super high pressure had to be applied in the polymerization of ethylene monomer, the discovery of the low-pressure process took technological circles by surprise. Moreover, polyethylene polymerized at low pressure has a high crystallinity, a linear molecular structure, and a high density, as compared to high-pressure polymerized polyethylene.

Commercial production of low-pressure polymerized polyethylene was started by Montecatini (currently known as Monte-Edison) of Italy in 1954, followed by Hoechst (FRG) in 1955.

After development of the low-pressure process, the conventional process was termed the high-pressure process. Therefore, low-density polymer obtained from this process is called high-pressure polyethylene or LDPE (low-density polyethylene).

Recent progress in catalyst technology facilitates production of low-density polyethylene even under low pressure. Since polyethylene molecules become linear when polymerized in a low-pressure process, such a product is expressed by the code LLDPE.

Polyethylene is a long-chain, partially crystallizable polymer produced by polymerizing ethylene gas monomer. It is tasteless, nontoxic, lighter than water, and somewhat whitish in appearance. The film obtained is translucent. It is highly water resistant, water-vapor resistant, chemical resistant, low-temperature resistant, and provides good electrical insulation.

The properties of polyethylene vary with the pressures and catalysts involved in the polymerization process and with its density. Depending on pressure applied during polymerization, polyethylene is classified as low-pressure, intermediate, and high-pressure polyethylene. Depending on own density, polyethylene is classified as very-low, low-, and medium-, and high-density polyethylene. In general, high-pressure processes produce branched side chains on the polyethylene main chain, giving bulky and low-density polyethylene.

The ranges of density are as follows:

Very-low-density polyethylene (VLDPE): max. 0.909
Low-density polyethylene (LDPE): 0.910–0.925
Medium-density polyethylene (MDPE): 0.926–0.939
High-density polyethylene (HDPE): 0.940–0.965

A. Low-Density Polyethylene (LDPE, LLDPE, and VLDPE)

Low-density polyethylene is classified into two types. One is high-pressure-process polymerized bulky polymer with long, highly branched side chains, and the other is linear-structured low-pressure-process polymerized polymer. The former is called LDPE and the latter, LLDPE. VLDPE is mainly polymerized by a low-pressure process (it is reported that LLDPE and VLDPE are even polymerized by the high-pressure process recently) and is somewhat bulky with short, branched side chains on the polyethylene main chain. LLDPE and VLDPE are produced by copolymerization of alpha-olefins such as butene-1, hexene-1, 4-methylpentene-1, or octene-1 as comonomers with ethylene gas monomer.

Their physical properties have been upgraded over LDPE in tensile strength, puncture resistance, and tear resistance, especially in heat seal strength and hot tack seal strength.

LDPE softens at 100–105°C,. but it is not thermally decomposed until the temperature exceeds 80°C. Crystallinity ranges from 60 to 70%. It shows excellent cold resistance, withstanding extremes of −70°C. It provides a good (low numerical value) water vapor transmission rate but at the same time poor oxygen and carbon dioxide gas barriers, enabling easy permeation. LDPE excels in heat sealability and is extremely useful as a packaging material because of its low cost.

Half the output of LDPE is processed into films as heat-sealable substrates for laminated packaging materials, such as a pouches and bags for hardware, groceries, frozen foods, milk, other foods, and trash. Electronic part and integrated circuit bags are coated or embedded with an electrostatic agent.

In 1954, a technology for applying extrusion coating to cellophane and paper was developed in the United States. This innovative technology was shockingly new to many people in the industry as it endowed paper with entirely new properties, such as heat sealability and water resistance. This technology spread rapidly in the flexible packaging industry. Moreover, it was diffused to cover not only paper and cellophane, but also aluminum foil, biaxially oriented polypropylene film, and polyester film.

Developed next was the LDPE sandwich laminating process, wherby a thin membrane of molten polyethylene extruded from a T die is inserted as a kind of adhesive between cellophane, oriented polypropylene film, aluminum foil, etc. and polyethylene or polypropylene copolymer sealant film to laminate them. Packaging material produced by sandwich laminating a trap-printed substrate on the reverse side of a cellophane, or oriented polypropylene film and cast polypropylene copolymer film by means of LDPE, is called trilayer lamination (cellophane/printing/LDPE/CPP) and is used in large quantities for snack foods. As CPP is highly scratch resistant, it leaves no scars on its surface if struck by a sharp or pointed edge of the package contents. If scarred, its surface turns slightly whitish, but its transparency remains unimpaired.

As neither nylon nor polyester films are heat sealable, they are dry (adhesive) laminated with LDPE or LLDPE film of over 60 μm thickness as a sealant. In this case, the LDPE film surface is corona discharged to 42 dyn/cm of surface tension (check by inspection kit) to assure tight adhesion, often by a urethane-based adhesive.

B. High-Density Polyethylene (HDPE)

HDPE is a translucent polymer, with its softening point ranging from 100°C to 125°C and crystallinity from 75 to 95%. It is highly resistant to chemicals and provides better performance as a gas and water vapor barrier than does LDPE. HDPE-based blow-molded bottles are used for detergent, hair shampoo, milk, kerosene, and other liquids. Using HDPE or polypropylene, a nylon and/or saponificated ethylene–vinyl acetate copolymer (EVOH) multilayer squeezable barrier bottle can be produced (PE/tie/barrier/tie/PE). These bottles have also been used for mayonaise and tomato ketchup. They are manufactured by the so-called direct blow process whereby compressed air is blown into a multilayered tube extruded from a multilayered circular dies. Although they provide high barriers and squeezable bottles, these HDPE-made bottles have a drawback in that they have a translucent appearance. For this reason, some of those bottles were replaced by PET bottles or multilayered barrier PET bottles for soy sauce, cooking oil, carbonated beverages, soft drinks, wine, etc.

It was later found that after blending HDPE with amorphous nylon, blowing would provides a gas barrier. In other case, fluorine gas amply serves as the gas barrier. Those two processes, though patented, are also beeing applied to agricultural chemical containers and gasoline tanks.

HDPE injection-molded containers have been widely used, especially as crates for glass bottles of beer, juice, carbonated beverages, soft drinks, soy sauce, etc. They are also used for baskets of marine and agricultural products such as fish, seafood, vegetable, and fruits, and as containers for bread, other foods, machine parts and components, office supplies, and for making pails of water-base paint. Frequently, they are also used by the merchandise distribution industry, as pallets for food processing plants, etc.

High-molecular-weight HDPE is processed into film, using circular dies. Compared to conventional HDPE film, it is a resource-intensive product because it has high strength and can be thinned. It is sold in supermarkets in large quantities in the form of shopping bags.

VIII. POLYPROPYLENE

Polypropylene (PP) is obtained by polymerizing pure propylene gas with Ziegler–Natta catalyst. As previously mentioned, Dr. K. Ziegler discovered in 1953 that linear polyethylene could be produced by the low-pressure process, using a catalyst—a compound of triethyl aluminum and titanium tetrachloride (the Ziegler catalyst).

Dr. G. Natta, based on that discovery, pushed forward with more advanced research into the possibility of polymerizing propylene gas by means of the Ziegler catalyst. As the result of his efforts, Dr. Natta succeeded in developing polypropylene, in 1955, by using titanium trichloride instead of titanium tetrachloride in the Zielger catalyst (the Ziegler–Natta catalyst). He furthermore discovered that the polypropylene obtained was a stereospecific polymer, and that polypropylene's methyl radicals were in a regular configuration on the same side of the main chain. He pointed out that the high molecular weight, high melting point, and high crystallinity of the polypropylene polymerized with the Ziegler-Natta catalyst are derived from the three-dimensional regular configuration of their molecular structure. Dr. Natta described such three-dimensional regularity of configuration as "tacticity." He called "isotactic" a configuration wherein methyl radicals were arranged on the same side of the main chain. Where methyl radicals were arranged in an alternate position with respect to the main chain, he called such a configuration "syndiotactic." When methyl radicals were arranged at random toward the main chain, he called such a configuration "atactic."

Montecatini (Italy) entered into commercial production of polypropylene in 1959. During polymerization of polypropylene, small amounts (about 5%) of atactic polypropylene are produced as by-products. Since this material dissolves in solvent, solvent is added to increase solvent-insoluble isotactic polypropylene, and the admixtures are washed out. During the washout process, used catalyst is also removed after killing the activity of the catalyst by addition of an agent. Montecatini-Edison set up a joint venture with Hercules (USA), and polypropylene is currently being produced by Hymonte, Inc., with process know-how licensed by Mitsui Petro Chemical of Japan. This technique is known as the "fourth generation gaseous process." It provides high catalytic activities and high yield rate, and dispenses with solvent washout for used catalyst and atactic polymer.

Polypropylene is the lightest weight polymer among the general-use polymers, with a density of 0.90–0.91. It provides good rigidity and surface hardness and shows a brilliant surface and a translucent appearance. It has a very sharp crystaline melting point of 162°C. This means that it provides good heat resistance and can be used in sterilization by boiling water in retortable pouches. Polypropylene provides a better stress crack resistance than polyethylene, and also has excellent hinge properties, which permit the fabrication of integral hinge products.

Although high in heat resistance, polypropylene is poor in heat sealability due to its high and sharp melting point. To remedy such a defect, ethylene is added at 3–5%, to produce ethylene-propylene random copolymer.

Polypropylene is much inferior to polyethylene in its low-temperature impact strength. To remedy such a drawback, ethylene–propylene block copolymer was developed, with a maximum ethylene content of 20%. The greater the ethylene content, the less rigid the product. Usually, to produce the ethylene–propylene block copolymer (EPR), the ethylene gas and propylene gas mixture is charged while activated catalyst is still alive in the final step of the polymerization process of polypropylene homopolymer. Polypropylene block copolymer has been improved to exceed PP in low-temperature impact strength. Its shortcoming is its opacity. A sizable quantity of PP block copolymer has been applied to crates and containers for foods including beer, soy sauce, carbonated beverages, soft drinks, bread, etc. Processed agricultural and marine products are additional applications.

In the case of PP block copolymer, it seems that the polyethylene particles exist in the polypropylene matrix, and the interface of each particle is covered with ethylene–propylene copolymer (EPR). In other words, a trilayered structure has been formed of PE–EPR–PP. Impact-triggered microcracks continue to grow in a PP phase until they reach a boundary between the PP phase and the EPR phase. It was found that impact strength is proportionate to the heat-seal strength of EPR with PP and PE homopolymers, and that the greater the heat-seal strength, the higher the impact strength.

The PP homopolymer tray is poor in low-temperature impact strength and cracks when it is dropped at 10–15°C, whereas a PP block copolymer tray dose not crack unless it is dropped at 0–5°C.

Nonoriented polypropylene film is produced by both a casting process (CPP film) using T dies and an inflation process (IPP film) using circuler dies. The casting process is more productive and yields transparent film. On the other hand, the inflation process is suited to producing thick film with well-balanced film strength in the longitudinal and transverse direction. Investments in inflation equipment are less costly.

Although inferior to PE film in low-temperature resistance, CPP film excels in heat resistance, transparency, luster, and scratch resistance.

CPP film has been widely used as a sealant layer, beside as a general packaging film material, in packaging materials for snacks and retortable foods. When used as a sealant layer, PP random copolymer is, of course, employed in a dry lamination process.

The biaxially oriented PP film (BOPP film) is produced either by the tenter process or the double bubble inflation stretch process. Films oriented longitudinally and in the traverse direction by sevenfold each under the tenter process have been widely used as flexible packaging materials. Orientation contributes to improving strength, stiffness, transparency, and low-temperature impact resistance.

Cellophane, which once enjoyed extensive application for transparent packaging, has been largely replaced with BOPP film in many applications, the latter being superor in respect of water and humidity resistance. BOPP film surface has a lustrous, brilliant appearance. Since it is based on high-melting-point polymer, it is more heat resistant than PE. The cold impact resistance is improved dramatically by orientation, so it can be used in packaging material for refrigerated foods.

Some of the BOPP films are coated with PVDC (polyvinylidene chloride) to provide a better gas barrier or with an acrylic heat-seal coating agent; some are corona discharged to improve adhesion of urethane-type laminate adhesive; some undergo antistatic treatment; and some undergo multilayered coextrusion coating (PP homopolymer/random PP copolymer) to improve heat sealability.

Polypropylene is a partially crystallizable polymer. Accordingly, when polypropylene spherical crystals gradually grow into sheets and films up to the size of light wavelengths, and after extrusion of sheets and films, the spherical crystals begin to reflect light and take on a translucent to whitish opaque appearance. Since grains so tiny that they do not reflect light do not affect the appearance, crystallizing agents are added to polymer upon production of sheets and films in order to minimize the size of crystal grains. Then the extruded products are quenched on the cooling roll. In some cases, products extruded from the dies are rolled up after water-dip quenching. Crystal structure is destroyed by orientation. For this reason, BOPP film features high transparency and provides a smooth, lustrous surface when oriented.

Crystallinity causes problems in thermoforming. Below the melting point PP is difficult to form because it does not flow well in vacuum forming; above the melting point, the sheet has little melt strength, sags, and is very difficult to make into uniform-thickness fabricated products.

A high melting point requires a high forming temperature and extra energy, especially since longer thermoforming cycle time has to be controlled adapt to the high thermoforming temperature and a low temperature in the cooling process. Slow cooling allows recrystallized spherical crystals to grow in formed products, thereby turning the surface opaque. As mentioned above, compared with PVC and PS, PP is poor in thermoformability, but better in resistance to heat and chemicals. Since PP contains no plasticizer, it poses no problem of migration, safety, residual monomer, cancer hazard, or air pollution by hydrogen chloride gas.

In recognition of polypropylene's merits, many polymer producers are carrying out research on the development of a new thermoforming process for polypropylene. One such method, a vacuum pressure thermoforming process, has been applied to oriented polystyrene sheets in place of the vacuum forming process. For this type of sheet, the contact heating system is occasionally employed, using the same forming machine as for a oriented polystyrene sheet.

In the case of PP, however, a direct heating system using an infrared heater is employed, since this sheet is commonly thin (about 0.5 mm). The PP sheet is clamped and heated to 170°C. When heated, a sheet—both sides of which are clamped—is expanded and wrinkles are generated in the center part of sheet. In the next step, generated wrinkles become a smooth plane again because the sheet shrinks to its original form by "orientation release," which activates a contracting force to act on the sheet. At this very moment, a PP sheet is vacuum-drawn from the reverse side, and pushed tightly against the mold by application of air pressure (compressed to 3 kg/cm^2) on its surface in the forming and cooling process.

Polypropylene is low in melt strength and prone to sag, so PE is customarily blended to minimize sag produced during vacuum formation.

The solid-phase pressure forming (SPPF) process developed by Shell, Inc., is another new thermoforming method for PP sheet. The SPPF process employs thermoforming machines by Illig, Inc. The PP sheets are heated to 150°C—lower than their melting point—and are allowed to stand at this temperature. Therefore, PP sheets are kept in a solid phase. While in a solid phase, sheets are fed to a forming tool equipped with "plug assist" and formed. Since products obtained by the SPPF process have been oriented, crystals in the sheet are destroyed and these sheets turn more transparent. In addition, the products have improved sparkle surface appearance and rigidity. They show better water vapor and gas barrier properties than vacuum-formed products.

As the SPPF process provides low-temperature thermoforming, it is energy saving because it requires less energy input than vacuum-forming process. On the other hand, it has a shortcoming in terms of longer cycle time, higher operating costs per hour, and lower dimensional stability at elevated temperature.

Hercules, Inc., developed a new rotary thermoforming (RTF) process and started commercial production. The RTF process extrudes hot PP sheet from a T die and feeds it to a specially designed mold for vacuum forming. The RTF process provides an in-line process in which several molds are mounted on the forming machine and the rotary thermoformer that rotates beneath the extruder. It performs continuous forming and cooling of products.

The RTF process has high productivity, yielding a high rate of output, and enables low-cost operation. Since molds of different sizes and contours can be used, efficient, small-lot-size production is available. As PP sheets are heated above their melting point, they are highly fluid, and deep-drawn products can be produced easily, but they are translucent in appearance and are not oriented because of the higher processing temperature.

Since thermoformed PP products have high heat and oil resistance, they have been widely used for milk portion cups, juice, yogurt and butter containers, medical device trays, hospital supply trays, etc. Because PP has good contact clarity, it looks as if it were transparent when juice is poured in a cup and the surface haze disappears, although the PP cup itself is slightly translucent.

Hercules has also developed a transparent polypropylene plastic can. In cooperation with machinery makers, they have also developed an oriented blow molding system. The oriented polypropylene (OPP) can produced by the Hercules process is fully oriented, including the neck and bottom seam. This means the OPP can has good impact resistance and drop test results when filled with product. OPP cans are lighter in weight than cans made from other materials and are much lighter than glass jars. The OPP can provides a very good moisture barrier and a maximum shelf life for products susceptible to moisture gain or drying.

Oriented polypropylene bottles can be made from both extruded pipe and injection parison. OPP bottles provide better heat resistance and moisture barrier but less gas barrier than oriented polyethylene terephthalate bottles. Therefore, OPP bottles are used for medical Ringer's solutions but not in the soft drink field.

IX. NYLON

Nylon is a polyamide, and has an amide structure (–CO–NH–) in its main chain. The most widely used nylon polymer are nylon-6, –66, –11, –12, –6/66, and MXD6. They are produced mainly by two types of reaction:

1. Polymerized by polycondensation of aliphatic diamines with dicarboxylic acids: for instance, hexamethylenediamine (number of carbon

atoms is six) with adipic acid (six carbons) gives nylon-66, and hexamethylenediamine with sebacic acid (10 carbons) or dodecanedioic acid (12 carbon) gives nylon-610 or nylon -612, respectively. When an aromatic diamine or aromatic diacid such as metaxylylenediamine (MXD) or isophthalic acid (IPA or I) is used, the polymer is called by their abbreviation and the carbon number of the partner—for example, metaxylylenediamine with adipic acid (six carbon) gives nylon-MXD6.

2. Polymerized by addition reaction of ring compounds that contain both acid and amine groups on the monomer, such as amino acids, or of their cyclic amides (lactams): For example polycaprolactam (six carbons) and poly (omega-aminoundecanoic acid) (11 carbons) give nylon 6 and nylon-11. The polymer of the lactam of 12-aminododecanoic acid (12 carbons) gives nylon-12.

Gaebriel succeeded in polymerization of nylon-6 by heating epsilon-aminocaproic acid in 1899. Paul Schlack discovered that nylon-6 is obtained by heating water and a caprolactam mixture in 1938, and this process was commercialized by I. G. Ltd. in 1943. In 1934 W. H. Carothers invented nylon-66, and DuPont pioneered its commercial production in 1941.

Of the many different types of polymers, nylon-6 has been used most widely as a packaging material due to its flexible orientability and low cost, as compared with nylon-66.

Biaxially oriented (BO) nylon-6 film is produced either by a flat or circular die process. When a flat die is used, there are two ways of orientation: simultaneously longitudinal/transverse direction orientation, and sequential orientation. If the nylon-6 substrate for film is first longitudinally streched, the polymer molecules are oriented in an extended direction. Then, as it is streched in the transverse direction, longitudinal tear occurs by intermolecular force in the substrate and further orientation is disabled. Under the circumstances, the simultaneously biaxial orientation process was developed.

Developed later was the successively biaxial orientation process, whereby nylon-6 is blended with nylon-MXD6 and crystallization of nylon-6 is controlled so as to provide transverse orientation subsequent to longitudinal orientation.

It was found that BO nylon film is superior to nonoriented (NO) nylon film in providing a high gas barrier and impact strength at low temperature and dart thrust strength. Some of the properties of this film have been improved beyond predevelopment expectation. Furthermore, BO nylon film is rigid and highly resistant to pinhole and scratches. Being highly heat resistant, BO nylon film has a wide range of application as the outermost

Chapter 7. Plastic Containers

layer of flexible packaging material contacting a heat seal bar. These applications include processed agricultural foods, processed marine foods, processed meat, cheese, soup, pickles, and frozen foods.

As nylon film provides a good gas barrier and is resistant to heat and chemicals, it is useful for production of retortable and boilable pouches. Being oriented, BO nylon film is prone to thermal shrinkage. Accordingly, it should be used at a temperature not higher than 121°C. On the other hand, NO cast nylon film can be used at a higher temperature than 121°C also. Since NO cast nylon film excels in thermoformability, it is used for deep-drawn packaging materials, such as for processed meat, including Vienna sausage, frankfurters, and ham. NO cast nylon film does not exhibit heat sealability, and various resins, such as PE, ionomer, or PP random copolymer, are applied by a variety of processes as a sealant layer.

In such applications, diverse processing techniques have been developed and implemented, including, for instance, an extrusion coating process by which PE is extruded from the T dies onto the NO cast nylon film surface and coated for production of NO nylon film/PE construction. A dry lamination process whereby urethane-type adhesive is coated onto corona-discharge NO cast nylon film and laminated with PE film is applied to make NO nylon film/adhesive/PE film construction. Besides using the coextrusion process, a multilayer film is produced by construction of nylon-6/tie layer/PE. An ionomer/PE construction is coextruded onto NO cast nylon film for multilayer coextrusion coating. In this case, the role of the ionomer layer is to make sure of the adhesion to nylon with PE.

Which method to employ depends on customer needs, cost, delivery time, and converting machine installed on the converter. The demand for coextruded multilayer film is rapidly increasing since it is more cost effective.

Depending on usage, increased barrier properties may be required for nylon construction. In such a case, saponified ethylene–vinyl acetate copolymer (EVOH) is used together with nylon in the coextrusion process. As EVOH and nylon-6 each have good adhesive properties, no tie layer is necessary between them. Nylon/EVOH/X/ionomer construction is used in many applications.

Since epsilon-caprolactam does not react completely during polymerization, low-molecular-weight polymer and monomer are left in the polymer. Although not toxic, residual monomer should be thoroughly water washed to reduce its level below 15 ppm. As the surface of nylon film is not heat sealable, it is not brought into direct contact with foods, but close attention should be paid to monomer when dealing with meticulous clients and customers.

Nylon-6 is a useful polymer for producing high-barrier coextruded multilayer squeezeable bottles with polyethlene. Many of these bottles are used as mayonnaise and tomato ketchup containers.

Nylon-11 inflation film is used as a shrinkable casing material in Europe.

Nylon-6/11 copolymer is used in the area of coextrusion coatings and films.

Nylon-MXD6, as the PET/MXD6/PET coextruded container, has been used for beer and wine bottles, to improve barrier properties. They appear transparent in multilayer lamination, but translucent if blended.

X. SAPONIFIED ETHYLENE–VINYL ACETATE COPOLYMER (EVOH)

EVOH is obtained by hydrolysis of the acetyl group of ethylene–vinyl acetate copolymer (EVA). Kuraray (Japan) commercialized EVOH in 1972, followed by Nippon Gousei (Japan) in 1977, and later on, by the Eval Company of America (EVALCA, USA), Solvay & Cie (Belgium), and DuPont (USA).

EVOH is a crystalline polymer and provides an extremely high oxygen and carbon dioxide gas barrier. It is generated from polyvinyl alcohol. EVOH is considered as a polymer containing PVA plus an ethylene unit. In other words, EVOH may be considered as a ethylene–vinyl alcohol random copolymer. Its qualities of water resistance and extrudability increase as the ethylene units increase, but at the same time the qualities of gas barrier and crystallinity decrease. The ethylene units and PVA content in EVOH range from 29 to 48 mol% and from 71 to 52%, respectively.

EVOH is very stiff and transparent, and at first sight looks like cellophane. It is resistant to all solvents except alcohol. EVOH film gives an excellent gas barrier in dry condition, but it is considerably affected by ambient humidity. It is preferable, therefore, to laminate the film on both sides with polyolefins such as water-resistant PP or PE.

As already mentioned, since EVOH and nylon-6 adhere to one another, they can be laminated by means of the coextrusion process. If these materials are blended, however, the terminal radicals of both polymers react with each other, and the compound take on a yellow hue. Recently, a new version of nylon polymer was developed, free of such coloration, and it was found that a mixture of the newly developed nylon polymer and EVOH could be produced in a pinhole-free layer with slightly diminished barrier property.

The coextruded films such as nylon-6/EVOH/ionomer or nylon-6/nylon–EVOH mixture/ionomer construction are extrmely important as a deep-drawn packaging material. They have been used in large quantities for Vienna sausage, frankfurters, sliced ham, and meat, especially for the body material of containers.

Coextruded direct-blow squeezable containers are used for mayonnaise, tomato ketchup, and edible oil. Research has been done on PET/EVOH/PET strech-blow bottles for soft drinks and carbonated beverages.

A plastic can of unique design was recently placed on the market: "omni can" developed several years ago by American Can Co. "Omni can" is made from basically PP/tie/EVOH/tie/PP construction, whereby an EVOH layer is sandwiched between a pair of tie layers containing a dessicant. This absorbs the internal moisture of the EVOH layer and maintains its dry state, as EVOH loses barrier property in a wetter condition. These plastic cans have been placed on the market after retort sterilization at 120°C. Even after a year or more, these containers retain shelf life. This container is formed by an injection-blow process, wherein five layers are formed by extruding, with precise timing, the respective polymers from three extruders.

XI. POLYCARBONATE

Polycarbonate (PC) is one of the engineering polymers and an amorphous polymer. It provides excellent impact strength, heat resistance, and transparency. Although it passes more than 90% of visible rays, it absorbs a sizable quantity of ultraviolet rays. Its gas barrier property is moderate or intermediate, but it provides a very good barrier against flavor and aroma.

Polycarbonate sheets are pressure formed at a compressed air pressure of 3–4 kg/cm^2 and at a forming temperature of 170–200°C. Products obtained are as transparent as thermoformed PVC products, but are incomparably superior in heat resistance.

Polycarbonate has a melting point of 230°C and brittle temperature of −100°C. This means that PC has significantly better heat and cold resistance than ordinary polymers. It is possible to put pasteurized hot food products, directly into polycarbonate containers on the packaging line. In addition, these containers can withstand rapid temperature increase and quick freezing of frozen foods.

Polycarbonate containers are used for hot filling or for pasteurized foods such as puddings, jellies, and jams, and rice cakes.

Coextruded polycarbonate with polyetherimide sheet, PEI/PC/recycle/PC/PEI, was placed on the market for the dual ovenproof tray. Polycarbonate and polyetherimide are both engineering thermoplastics, amorphous polymers with a glass transition temperature (T_g) of 149°C and

219°C respectively. The PEI surface skin layer of the tray provides good thermal and mechanical properties and is also an excellent choice for food contact due to its easy release and lack of taste or odor transfer. The second PC layer provides high heat resistance, excellent low-temperature impact strength, and outstanding toughness. The recycle center layer contains the recycle generated during the coextrusion and forming operation. The recycle is a homogeneous, stable polymer alloy of two starting polymers. This tray and the CPET (crystallized polyester) tray have approximately the same stiffness at room temperature. Above the glass transition temperature of CPET (77°C), this tray exhibits more than three times the stiffness. This tray absorbs significantly less infrared energy than the CPET tray during cooking. Furthermore, it provides a better drop test result at low temperature than CPET.

XII. POLYETHYLENE TEREPHTHALATE

ICI Ltd. (U.K.) developed polyethyleneterephthalate (PET) in 1940 and launched its commercial production in 1944, followed by DuPont in 1945. Presently, America's Eastman Chemicals and Goodyear are also producing PET in large quantities for polyester fiber, film, and PET bottles.

PET is a partially crystallizable polymer that melts at 264°C and is obtained by polymerizing pure terephthalic acid with ethylene glycol. PET polymer used for producing synthetic fiber and film is polymerized by a fusion process, but for production of bottles and pipes, polymer of higher molecular weight is required. For this reason, after polymerization in the fusion process, solid-phase polymerization is conducted as the second step. Germanium, antimony, titanium compounds are used as catalysts. In the United States, antimony-based catalysts are used for PET bottles, but in Japan, more expensive germanium-based catalysts have been used at customers' persistent demands.

In the second step, PET is heated for solid-phase polymerization to a temperature 20–30°C lower than its melting point.

PET is a stable polymer with resistance to light, heat, and chemicals. PET has almost no additive or stabilizer compared with polyethylene, polypropylene, and polyvinyl chloride. It is thus reported that PET is the most suitable polymer for packaging materials that come in contact with foods.

PET film is produced by biaxial orientation about 3×3 times for the longitudinal and transverse directions, respectively, to provide good stiffness and clarity. With their good dimensional stability and heat resistance, PET films are used as the outer layer in flexible packaging construction and are particularly useful as retortable food packaging material.

Although PET films themselves are good gas barriers, if a better barrier is required, polyvinylidene chloride-coated film is available.

For the thermoformed containers, nonoriented cast PET sheet is used. Ordinarily, nonoriented sheet is poor in heat resistance and its surface becomes sticky when heated to 100°C or slightly less. Although it is transparent and pliantly formable, no extensive application has been made, other than for Japanese sake cups.

By blending a small amount of polymer such as PP or LLDPE as the crystal nucleus, nonoriented sheets with good crystallinity are produced. During thermoforming, they are activated to crystallization by heating. Then they lose transparency and take on a whitish coloration. Crystallization embrittles them but improves their heat resistance to 230°C. Containers produced with such a process are called CPET containers. Demand for this type of tray and container is fast increasing as a dual oven-proof container.

PET beverage bottles produced by the stretch-blow process were placed on the market in 1977. The rapid growth of the PET bottle business around the world over a short time has been unequaled by other products in the recent packaging business. The bottle market in Japan began with the soy sauce bottle and the 1.25-l Pepsi container and has expanded to detergent, beer, wine, liquor, vegetable oil, hot-filled juice, and other items.

Two basic process exist for stretch blowing. One is called the single-stage or hot-parison process, in which a parison is made by an injection unit and blown on the same machine. The second process is called the two-stage or cold-parison process, in which a parison is made by an injection machine and blown on another separate machine at a later stage. The single-stage machine is useful for small lot sizes and modified bottle shape. The two stage machine is useful for mass production and regular bottle shapes at higher production rates.

The rapid diffusion of PET bottles is attributed to a number of factors, such as high impact strength, transparency, shatterproof quality, light weight, ease of handling, and low cost per unit volume in the case of large bottles. Compared with glass bottles, PET bottles provide greater inner volume, one of the merits that makes sales promotion easier.

Many manufacturers are involved in development of hot-fillable, high-temperature-sterilizable, high-speed-fillable, heat-resistant, and dimensionally stable bottles. Bottles formed by conventional processes are prone to contraction and tend to revert to preoriented parison contour dimension when reheated above PET's glass transition temperature. In the case of hot filling, PET bottles should be heated above the filling temperature with inner compressed air pressure to make them heat set. Unless a heat set is conducted, bottles are subject to contraction during hot filling.

For efficient production of heat-set bottles, many manufacturers are tackling the problem of how to effectively cool blow molds that have been heated to the heat-set temperature, and how to take bottles off hot molds and reheat them to the heat-set temperature. If this cycle time is lengthy, it will entail substantial loss of production efficiency and waste of capital investment.

It is important to set the heat-set temperature and bottle release temperature at optimum levels. Temperature variation would not only prolong the cycle time, but also increase energy loss in plant operation and production cost. If the release temperature is set too high, it deforms bottles after release. At any rate, how best to eliminate residual stress from oriented bottles is a matter of primary concern.

If the body of a bottle lacks dimensional stability, bottles filled with contents of the same volume show variance in filling level when inspected at the neck. Such unevenness is unsightly when the bottles are placed on the shelf for sale. Since the bottle neck finish has not been oriented, it is less heat resistant and is liable to deform during hot filling. To correct such a drawback, some bottles have their neck finish crystallized. Any neck finish deformation would cause defective capping and consequent leakage of carbon dioxide gas or other contents.

In wide-mouth containers, neck finish parts are also completely oriented. The unoriented portion is cut off to ensure tight coupling with the metal lid.

Many manufacturers are presently carrying out research on improving the PET bottle's oxygen and carbon dioxide gas barrier properties. This is for minimizing changes in taste and loss of gas due to gas permeation through the polymer of the bottle wall. As a relatively easy method, it is a good idea to coat the bottle surface with high-barrier polyvinylidene copolymer emulsion. This method is suited for beer bottles, for instance, where carbon dioxide gas is filled in two volumes, but it is not good for soft drink bottles, where such gas is added in three volumes. Thus, it is not recommended for the latter application, because blisters and peelings occur in these containers when carbon dioxide gas transmission through the container wall detaches the surface-coating film. This is attributable to the weak and insufficient adhesive of PVDC coating on PET bottle surface. Accordingly, research is underway on means of improving PVDC coating material, and on application of primer and surface treatment.

As a means of improving barrier property, it is recommended that nylon-MXD6 be blended with PET at a rate of about 10%. A parison should be prepared in accordance with the aforementioned procedure before blow molding is conducted. This bottle appears white and opaque in color and is used for beer.

Research has been recently conducted on multilayer barrier stretch bottles, with promise of an even newer method. Coextruded pipe or a coinjected parison is streched. As barrier layers, EVOH and nylon-MXD6 are used. They look transparent, but have some problems with interlayer adhesion. It is reported that they withstand practical use. Shelf life is extended to more than a year, but in view of the fact that production costs must be raised substantially to extend the shelf life, some argue against the need for extending the shelf life of soft drink containers at the sacrifice of production economy. For wine, this container enjoys a high reputation.

Research has also been made to modify PET itself, and fairly successful results have been achieved. There remain some problems, however, with the difficulty of drying the polymer in commercial production.

PART III

Food Packaging and Energy

CHAPTER 8

Food Packaging and Energy in Japan—Energy Analysis of Consumer Beverage Containers

Takashi Kadoya[1]
Japan Packaging Research Association,
Tokyo, Japan

I. INTRODUCTION

Since the oil crises of 1972 and 1978, people all over the world have been strongly interesting in saving energy. Meanwhile, saving packaging materials and consumption of energy during transportation or storage has been an important subject in the field of packaging.

This chapter examines the total amount of energy consumption necessary for the packaging of several beverages such as milk and juice. Energy costs include from the cost of the raw materials used in packaging as well as shipping, manufacturing, and eventual transportation to the supermarket. These energy costs are discussed here mainly for the conditions of Japan. The amount of energy consumption of packaging is very different among other countries. For example, the amount of energy consumption of packaging materials depends on the manufacturing treatment of the raw material, converting conditions of packaging materials, packing systems at the filling line, and physical distribution of packaged goods.

The results of calculating the energy involved in manufacturing packaging materials and containers all the way to end uses have been published recently concerning oil resources from overseas countries. However, as these reports are estimated from the uncertain information and many

---
[1] Present address: Kanagawa University, Hiratsuka, Japan.

different preconditions in various countries, they are not directly applicable to Japan.

Accordingly, this chapter describes case studies to estimate the amount of energy consumption of packaging materials and containers from an independent point of view.

In order to calculate the total amount of energy consumption from manufacturing of packaging materials and filling of containers to transportation to the supermarket, glass bottles, metal cans, cartons, and retort pouches were selected as reference containers, and total energy consumption of each container all the way to the supermarket was estimated in this chapter. In the case of glass bottles, the returnability and washing energy are of course included.

Table I shows the representative mean calorie values of various kinds of resources in Japan (Food Production Center, 1983) for calculation of packaging energy.

The contained volume of the consumer beverage container was selected as 200 ml for steel cans and glass bottles as a standard that is commonly used in the Japanese consumer market, as well as 500 and 1000 ml for paperboard milk cartons which are now being distributed instead of glass bottles. Figure 1 shows the distribution of milk cartons and glass bottles from 1979 to 1985 in Japan.

Table I Calories for Various Kinds of Energy Resources

Energy resources	Unit	Mean value of energy (kcal)
Electricity	kWh	2450
Volatile oil–naphtha	l	8600
Paraffin oil–jet fuel	l	8900
Light oil	l	9200
Fuels	l	9400
Heavy fuel oil	l	9900
Other petroleum materials	l	9400
LPG	kg	12,000
LNG	kg	13,300
Blast fuel gas	m^3	800
Coke burned gas	m^3	4800
Town gas	m^3	10,000
Refined gas	m^3	20,000
Coke	kg	7200

[a] From the Food Production Center (1983), with permission.

II. CALCULATION OF THE ENERGY OF THE TRANSPORTATION FOR REFERENCE CONTAINERS

Because the transportation process and distance of packaged goods from warehouses through filling plants to supermarket are very different for various kinds of packaging container, the transportation energy is introduced as results of a transportation process that are applied to each container type individually, instead of estimating from a transportation model in order to compare containers.

The transportation of loads is assumed to occur only by truck, and its fuel costs are 10 km/l for a 2-ton truck and 5 km/l for a 5-ton truck. The ratio of load to truck was estimated from actual results of reference models. Furthermore, transportation for mixed goods was treated as using trucks exclusively.

There are two situations in transportation processes. One is when products are carried by returnable container or exclusive plastic crate container. In the other, products are shipped in by corrugated board containers that are nonreturnable. The life of plastic crate containers was estimated to have a mean value of 100 returns because the life of the plastic crate container is uncertain.

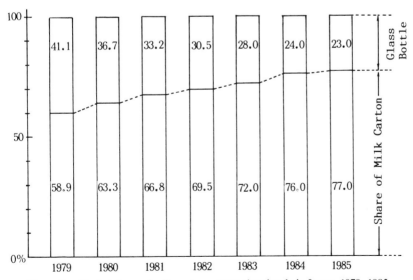

Figure 1. Market share of milk carton versus glass bottle in Japan, 1979–1985.

Table II shows the types of container and capacities as well as the difference between returnable and nonreturnable shipping containers.

Table III shows the fundamental conditions for calculating the transportation energy from filling plant to supermarket for various kinds of packaging materials.

III. CALCULATION OF MANUFACTURING ENERGY OF BASIC PACKAGING MATERIALS

Tinplate, aluminum plate, glass, paperboard, and plastics were taken as typical packaging materials for drinking containers. The manufacturing energies of these packaging materials were calculated from units of typical Japanese makers that were manufacturing these materials. Accordingly, the mining energy from oil fields to raw petroleum or exploration of oil fields are not contained, but raw petroleum, iron ore, bauxite, wood chips, and coal were treated as the starting materials.

As a result, the energy of retentions during the manufacturing process, losses in the packaging process, and carrying or handling are included in energies of transportation and the manufacturing process. Electric power is different in efficiency in terms power plant and oil dependency, but in this calculation 2480 kcal is estimated. The energy units introduced for calculation is shown in Table IV. This report does not contains raw material energies for tinplate and aluminum as do some other existing reports on energy application.

Energies for paper and plastic film such as polyethylene are calculated from the manufacturing energy of raw materials because both materials are able to transferred as resource energy. And the energy of glass bottles is

Table II Various Kinds of Packaging Containers

Number	Type of container	Capacity (ml)	Use	Nonreturnable/Returnable
1	Glass bottle	200	Milk	Returnable
2	Glass bottle	200	Juice	Returnable
3	Glass bottle	200	Juice	Nonreturnable
4	Carton	1000	Milk	Nonreturnable
5	Carton	500	Juice	Nonreturnable
6	Metal can	250	Juice	Nonreturnable
7	Retort pouch	200	Juice	Nonreturnable

Chapter 8. Food Packaging and Energy in Japan

Table III   Transportation Energy of Packaged Goods

Case 1. Glass bottle (milk, 200 ml): 40 bottles/crate, bottle weight with crate 19.3 kg, 90 crates carried by 2-ton truck, tranportation and return 50 km, fuel cost 10 km/l   13.47 kcal/bottle

Case 2. Glass (juice, returnable): 24 bottles/crate, 170 crates 4080 bottles carried by 4-ton truck, transportation and return 200 km, fuel cost 5 km/l   95.10 kcal/bottle

Case 3. Glass bottle (juice, nonreturnable): 30 bottles/case, 470 cases 14,100 bottles carried by 4-ton truck, transportation and return 200 km, fuel cost 5 km/l   16.40 kcal/bottle

Case 4. Paperboard carton (milk, 1000 ml): 12 cartons/crate, carton weight with crate 13.78 kg, 120 crates carried by 2-ton truck, transportation 50 km, fuel cost 10 km/l   33.68 kcal/bottle

Case 5. Carton (juice, 500 ml): 20 cartons/crate, carton weight with crate 11.8 kg, 145 crates carried by 2-ton truck, transportation 200 km, fuel cost 10 km/l   64.67 kcal/carton

Case 6. Metal can: 30 cans/case, 400 cases carried by 4-ton truck, transportation 200 km, fuel cost 5 km/l   32.33 kcal/can

Case 7. Retort pouch: 56 bags/case, bag weight with case 11.9 kg, 200 cases carried by 4-ton truck, transportation with return 200 km, fuel cost 5 km/l   26.65 kcal/bag

estimated as a continuous process because the glass bottle is converted directly from raw materials. In the case of aluminum plate, its refining energy has been calculated in other reports, but in this paper total process energy was calculated by considering the energy of the manufacturing process of an alloy in a metal can.

Table IV   Calorie Equivalent of Various Kinds of Energy Resources

Energy Resource	Calorie Equivalent
Light oil, gasoline	9700 kcal/l
LPG	11,000 kcal/kg
Electric power	2480 kcal/kWh
Tinplate	1 0.00 kcal/g
Aluminum plate	6 5.00 kcal/g
Liner board	1 6.00 kcal/g
Polyethylene (LDPE)[a]	2 5.00 kcal/g
Glass bottle[b]	4.13 kcal/g

[a] Includes 17,185 kcal/gal as resource energy.
[b] Glass energy included in bottle manufacturing process.

IV. ENERGY TO MANUFACTURE THE CONTAINER

The amounts of raw material to make up each container and the retention losses in the container manufacturing process were calculated, and then energies of to finish the body of the containers, such as printing, inner coating, label, adhesives, and caps, were also summed. In addition, actual consumption energy in the container manufacturing process was summarized as total amounts of consumed electric power, civil gas, LPG, heavy fuel oil, and paraffin oil. From these data, the energy of reference containers in model plants was calculated as basic amounts of energy. The energy of surface printing ink, inner coating materials, other coating, glazing and drying, sealing compound, and converting energy during manufacturing process of metal can were included in the energy of metal cans.

For milk cartons, energy consumption was calculated as polyethylene lamination on one side of the paper and surface printing are converted during the papermaking process with inplant forming and filling processes. It was assumed that there are no losses or scrap during the forming and filling process.

The retort pouch is manufactured at a converting plant for packaging materials and supplied to beverage makers by roll. The model composition of retort pouch used was 30 μm PE / 9 μm aluminum foil / 70 μm PE.

V. TOTAL ENERGY CONSUMPTION

Table V shows the results of summing energy for basic materials, each manufacturing process, packaging accessories, transportation, recycling, and other necessary operations as described above. For the calculation of recycling energy, the return rate of metal cans is estimated at about 20% for aluminum and 40% for steel cans under current scavenger conditions in the case of recovery as scrap iron at the final incinerator plant.

However, in Japan, a mean level of about 30% incineration is estimated, and the recovery ratio as heat energy from burning plastic or paper is estimated as 30%.

The recovery rate of aluminum as a composite material was chosen as zero because it is impossible to recover from burning.

The recovery rate of steel was considered as 30% and the energy of recycled scrap iron was calculated to be the same as that of pig iron, but this energy does not account for recovery or manpower energy.

When the transportation energy compares to the total energy, which consists of manufacturing and transportation energies, returnability of glass bottles is about 50% as shown in Fig. 2.

Table V Energy Consumption (kcal/unit) of Various Kinds of Containers

Container	Glass bottle			Paperboard carton		Metal can, juice	Pouch, juice
	Milk	Juice	Juice	Milk	Juice		
Capacity (ml)	200	200	200	1,000	500	250	200
Substance (g)	244	355	122	286	17.7	26.8	1.14
Closure	Paper	Tinplate	Tinplate	Polyethylene (PE)		Aluminum (AL)	
Cap (g)	0.8	2.7	2.7	3.4	(PE) 2.3	3.6	(PE) 4.68
Manufacturing energy							
Materials	1,007.72	1,466.15	503.86	457.60	283.20	268.00	190.90
Cap	12.80	27.00	27.00	(PE) 85.00	(PE) 57.50	(AL) 234.00	
Total	1,020.52	1,493.15	530.86	542.60	340.70	502.00	190.90
Packaging materials energy for transportation	10.00	14.08	11.27	48.37	17.50	6.16	4.53
Transportation energy (to filling plant)	19.60	41.66	30.00	3.88	12.19	9.48	0.37
Filling energy (filling, capping)	0.30	0.30	0.30	1.42	1.42	1.32	0.30

(*continued*)

Table V (Continued)

Container	Glass bottle			Paperboard carton		Metal can, juice	Pouch, juice
	Milk	Juice	Juice	Milk	Juice		
Filling energy (heating, cooling)	56.60	56.60	56.60	113.20	113.20	56.60	56.60
Container manufacture energy	—	—	—	37.22	27.42	198.47	36.00
Transportation energy of packaging goods	13.47	95.10	26.40	33.68	64.67	32.33	26.65
Packaging subtotal	99.97	207.74	124.57	237.77	236.40	304.36	124.45
Materials and packaging total	1,120.49	1,700.89	655.43	780.37	577.10	806.36	315.35
Mannufacturing energy of materials (30 times returnable)	34.02	48.87	—	—	—	—	—
Material energy of breakage loss	12.80	27.00	—	—	—	—	—
Bottle washing energy	34.02	48.87	—	—	—	—	—
Recycling energy of bottle	28.70	28.70	—	—	—	—	—
Manufacturing total	13.17	95.10	—	—	—	—	—
	122.70	248.54	—	—	—	—	—
Recycling energy (recycling rate 30%)	—	—	—	-59.82	-38.49	-80.40	-35.10
Linerboard for shipping container	—	240.00	212.64	—	—	177.67	128.57
Recycling of crate, etc. (plastic container)	10.00	14.08	11.27	29.05	17.50	—	—
Recycling energy of outer packaging	—	-18.00	-15.95	—	—	-13.33	-9.64
Return and recycling total	—	470.54	852.12	—	—	890.34	399.18
Corrugating board total	132.70	202.62	666.70	749.59	556.10	—	—

Chapter 8. Food Packaging and Energy in Japan

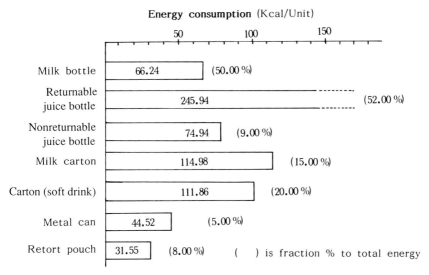

Figure 2. Transportation energy of various kinds of beverage containers.

For nonreturnable containers, the paperboard container had an unexpectedly large value of 15–20% and the metal can, flexible pouch, and nonreturnable glass bottle were 5.8 and 9%, respectively.

The ratio of material energy to total energy consumption for several kinds of containers is shown in Fig. 3. The values given in Fig. 3 for the returnable milk bottle is 91%, returnable glass bottle 88%, and nonreturnable bottle 81%. Furthermore, the values for that milk cartons are 59 and 70%, retort pouches 61%, and metal cans 62%, respectively.

In addition, when the manufacturing energies of containers are compared with each other, the manufacturing energy of nonreturnable glass bottles decreases under 1/2.8 because of progress in promotion of lightweight containers.

It is assumed that the retort pouch shows the lowest value in the total illustration because of its light weight; on the other hand, the energy consumption is large from the viewpoint of total energy in spite of its light weight, because of its lengthy manufacturing process.

As it is meaningless to compare directly the returnable bottle with nonreturnable containers, the number of times recycled was estimated as 8, 20, and 30 times, and as a result, amounts of total consumption energy were summed for the bottle washing energy and recovery energy of the empty bottle. The calculated data are shown in Table VI.

In some kinds of packaged goods, it may be unreasonable to set the rate of recycling at 20 or 30 times, while in some other packaged goods, a higher recycling rate is carried out.

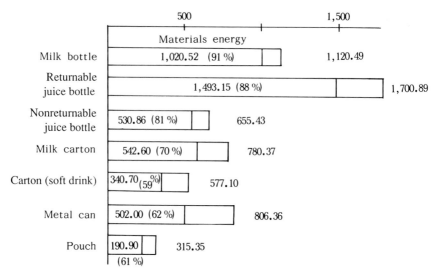

Figure 3. Total energy consumption of various kinds of beverage containers.

The problem of leakage and sealing of bottle caps or crowns has been decreasing due to improving seal accuracy; however, the rejection rate of defect bottles during the returning processes is going to increase now. Therefore, the returnable rate should be decided with consideration of the correct physical distributions.

Table VI Energy (kcal) of Returnable Bottles[a] as a Function of Returnable Rate

	Recycled 30 times	Recycled 20 times	Recycled 8 times
Materials energy	1466.15	1465.15	1465.15
Cap energy	27.00	27.00	27.00
Materials energy per cycle	49.77	74.66	186.64
Material loss rate	1/30	1/20	1/8
Energy of material	49.77	74.66	186.64
Bottle-washing energy	28.70	28.70	28.70
Recycling energy of empty bottle	95.10	95.10	95.10
Filling energy	56.90	56.90	56.90
Transportation energy (Including packaging materials for transporation)	150.84	150.84	150.84
Recycling energy of crate	14.08	14.08	14.08
Total	445.16	494.94	718.90

[a] Bottle capacity 200 ml (355 g).

One example is illustrated here. Suppose the return rate is 20 times: then the amount of energy consumption may approach the total energy of the retort pouch. If the recycle rate is instead about 5 times, the energy may be almost same as the metal can.

Energy consumption of cartons could not be compared to others directly because of the different physical distribution found. The transportation distance for metal cans or glass bottles is taken as 200 km (one way), while in the case of milk or dairy products, the transportation distance is 20 km or one-tenth of the former.

As described above, it is not easy to compare the total consumption of energy mutually because each kind of packaged goods has different physical distributions. We should understand sufficiently that there are different distribution processes in packaged goods, such as apply to the returnable plastic crate and the nonreturnable corrugated fiberboard shipping box for packaged goods.

In order to calculate the energy in each packaging process, the industrial water necessary in manufacturing the returnable glass bottles and boxboard containers, essentially flowing out as effluent wastes depending on its industrial scale, must also be considered. Expenses of treatment for effluent waste water should be accounted for as follows:

Service water	2900 kcal/m
Effluent water	280 kcal/m (degree of pollution was estimated as one-tenth of effluent water from the pulping process)
Drainage load	100 kcal/m in town
Total	670 kcal/m

As a reference, the mean value of expenses of water treatment in a bottling factory in the United States was reported to be 699 kcal/m.

Considering the amounts of energy for consumer packaging containers in this calculation is significant in understanding which process is spending the most energy and which is the most efficient package for a given volume of contents. From this point of view, Table VII is introduced from the results of Table IV. In comparing amount of energy to volume of contents, note that the energy almost stays in the range of 2–3 kcal/ml.

VI. CONCLUSION

There are various difficult problems in the process of calculating the total energy for packaging containers. In the special field of packaging materials, it should be understood that the calculation of packaging energy has to be carried out by methods different considerably from ordinary energy calculations. If energy saving is investigated in detail, this investigation

Table VII Comparison of Total Energy to Weights of Container and Contents

	Glass Bottle			Metal can, 250 ml	Carton, 1000 ml	Carton, 500 ml	Retort pouch, 200 ml
	Nonreturnable bottle, 200 ml	Returnable bottle, 200 ml					
Times recycled	1	8	20	30	1	1	1	1
Total energy per weight of container (kcal/g)	6.83[b] 5.35[a]	2.00[a]	1.38[a]	1.24[a]	28.53[b]	23.42[a]	27.81[a]	68.58[b]
Total energy per weight of contents (kcal/ml)	4.26[b] 3.33[a]	3.59	2.47	2.22	3.56	0.75	1.11	1.99

[a] Plastic crate.
[b] Outer packing box of corrugated cardboard.

represents only one suggestion in order to select concretely one model of packaging materials.

Accordingly, care should be taken in accepting the value of results that ignored the preconditions of data proposed in these case studies. This data may be very useful from the viewpoint of energy saving and in its effective utilization concerning theoretical analysis of energy for packaging. However, when the cost of production is looked over in terms of energy cost, it is necessary to understand in advance that the results of data may depends on conditions that are modified each time by the selection of packaging materials for the container and its reference conditions.

The construction of a model of energy consumption should be carried out by assuring a reasonable basis for comparisons.

In addition, it is most important to compare from viewpoints of total cost, such as the problems of water pollution in the environment and the expenses of treatment for air pollution or solid wastes. Such situations may influence the total cost.

This report is only one trial as a case study in order to analyze on the energy of containers by construction of a model from the viewpoint that the various variable factors result in different energy consumption for packaging.

ACKNOWLEDGMENT

This paper was based on the "Abstract of a Study from the Consideration by Energy for Packaging" in *JPI Journal* (T. Kadoya, 1981, JPI Technical Committee), and the main data were picked up from the report of chapter 4 (Y. Oki, 1981) in the journal. The author thanks *JPI Journal* for permission to use this material. Deep gratitute and sincere thanks are also offered to Dr. Y. Oki, a member of the JPI Technical Committee, Toyo Seikan Kaishya, Ltd., for his valuable guidance and kind advice on translation of technical terms on various kinds of packaging materials and peripheral technologies.

REFERENCES

Food Production Center (1983). Packaging Materials for Foods, Report of Food Production Center, March.

Kadoya, T. (1981). Abstract of "Study from the consideration by energy for packaging," *JPI J.* **19**(4):3–4.

Oki, Y. (1981). Comparison of energy for various kinds packaging containers. *JPI J.* **19**(4):18–24.

PART IV

Packaging Systems and Technology of Food Materials

CHAPTER 9

Recent Development of Packaging Machinery in Japan

Hidekazu Nakai
Toshima-ku,
Tokyo, Japan

I. COMPUTER-CONTROLLED PACKAGING MACHINES

Packaging machinery in Japan has made remarkable progress in the past 40 years, along with the rapid growth of the Japanese economy. Case packaging machines and case forming machines were developed in the industrial packaging field beginning in 1966. Further, package lines consisting of many machines connected in process order became popular in the packaging machinery field. The technology for operation control for the line, assembling stackers, and buffer stock for individually packaged products has gradually become highly developed in this field. In addition, electronic technology has also been actively applied to packaging machinery.

However, at the end of this 40-year era came the so-called "oil shock," which caused a recession in the economy. The storm of diversification that had been approaching before this time arrived in a whirl. In the general merchandise field, new products appeared and disappeared in rapid succession, because of the constant changing of product appearance intended to increase sales volume.

For this reason, packaging plants have been forced to handle many types of packaging. Also, due to the brevity of the product life cycle, we must give packaging machinery the capability of handling various sizes of products easily. It has become necessary for machines to be able to handle random flows in order to deal with varied products on one line at the same time.

This led to development of the computer-controlled packaging machine. Originally, packaging machines were typical preprogram types mainly constructed of mechanical devices, such as linkages and cam mechanisms. Therefore, packaging materials as well as products that the machines handle should be unified and should provided at regular intervals, and then the machines are operated at high productivity through comparably small investments. Packaging machines thus responded to community needs, and their industry has shown rapid expansion.

In this period, packaging machines required automatic control only for accessory functions such as overload release and "no product–no wrapper" interlock. However, due to the diversification of products, the same machine was required to handle several kinds of products in a parallel manner.

On the other hand, in a packaging line consisting of many machines operating simultaneously, automatic controls in the integrated manner of those machines were necessary. In Japan, thanks to the development of electronics, we could overcome these problems considerably.

In 1972, the Japan Packaging Machinery Manufacturers' Association (JPMMA) began studies of the automatic controlled packaging line, and in 1976 it began research work in the flexible packaging line through computer application. Packaging machinery using computers had been developed already as far as weighing machines were concerned. But they didn't cover actual packaging process such as filling, sealing, and wrapping. Then microcomputers came into existence, and we hoped it would be possible to incorporate the computer into packaging machines at reasonable costs.

However, packaging machinery was of such a nature, as already mentioned, that it would not be compatible to use computers for control, so how to apply computers to the control of packaging machinery was our biggest problem.

One of the applications of computers that we thought of was group control of packaging machines. We developed computer controls in some packaging lines of the large companies. But in our opinion user companies should take the initiative in this kind of work, and packaging machinery itself should have a constitution controllable by a central computer system.

The next step was the direct control of machine sequence making use of microcomputers. Around 1975, certain machines already were provided with complex wired logic controllers, and they were compatible with computer control from the standpoint of cost. However, in such a case, the engineers of machine manufacturers themselves had to be educated for computer application and at the same time user companies had to familiarize themselves with the maintenance aspect of the control equipments.

Further, output of 100–1000 packagings per minute should depend on the mechanical driving, so auxiliary controls such as checking the feeding of packaging materials would be possible.

Chapter 9. Recent Development of Packaging Machinery in Japan

However, the main primary control is difficult, if you want to make use of the computers. So in the case of the outer packaging machines with the low speed having such components as pneumatic cylinders or individual motors, the systems already made use of this computer control under the name of sequence controller. The most effective use of the microcomputer in packaging machines, in our opinion, is for size changeover in multipurpose machines with medium-sized production capability for various kinds of products. For the first target we chose in 1978, under the guidance of MITI, we began the study work on CAC-I. We made a pilot machine. This machine was designed to accumulate m lines by n pieces of one-sized products, and wrap the products with a blank sheet. This m and n can be freely varied. The function of the computer was first to control the sequence. Second, it was to control the parameter changes, for instance, if $m \times n$ changes, the change is also controlled by computer. Third, it was a diagnostic one, as already discussed. This diagnostic function is designed to utilize the sensor information by computer. Formerly, displays of failures or accidents were indicated by lamps in the traditional mode. When we use the computer to control the machine, we can display the information using a cathode ray tube (CRT). From the standpoint of the machine, the second control, which is the size changeover control, is the main focus

Figure 1. Computer-controlled packaging machine.

of our attention. Due to great diversification within the market, all the manufacturers are trying to direct themselves toward production of numerous types of products. Of course, manufacturers will not decrease the volume of production, so the real mission of the packaging machine is to take care of the packaging of numerous types of products in great quantity and at low cost. We are speaking of diversified product lines, but of course if we limit ourselves to a certain time frame, there should be a certain amount of mass production of single items. However, the time frame is shorter than before and there are frequent size changeovers. This means we have to adjust the size set of the machines more frequently. Therefore we are required to program and give input to the computer that will cover the size changeover for the machines. In other words, by giving a new code for the new product type, the machine will take care of the size changeover automatically.

This main drive system controls by computer the rotation of a screw, which receives the electric pulse by virtue of the rotation. The female screw moves toward the axis to adjust the position required. By adjusting this way, the computer can control the required portion of the machine within a few seconds, which used to take 15–30 min manually. In this manner, we can eliminate the stoppage of the equipment for the sake of size changeover. There was a trial machine, but this development triggered actual production of such a system by machine manufacturers. Some of the machines have been already used by some companies.

The first concern that we had for this machine was as follows: We worried about the accuracy of the adjustment by computer control in actual working. In other words, packaging machines require detailed adjustment. Therefore, we worried about the reproduceability by computer control. However, based upon our past experience, our system never had any problems with corrugated board, cartons, or paper. After developing this machine, we have found one interesting fact: That is, the computer looks as if it is an all-around player. But as a matter of fact, the computer is a restricted machine. Therefore we have to make machines suitable for the computer. After completing such a machine, we tried to do a size changeover with our human hands without a computer, and to our great surprise we completed the task within only a few minutes. And we joked that we really don't need computers at all! We once received many requests from customers to make an easy size-changeable machine. However, we couldn't meet their request. We were forced to simplify the mechanism of the packaging machine itself, in order to make use of the computer. As a result, we have improved the machine mechanism so as to easily do a size changeover by human hands. Another important thing is that numerical control of packaging machines is now possible. Previously we believed that the adjustment should rely on human experience. When it comes to fine critical adjustment, we have to have a oval hole for screws or we have to use turn-buckle. In other words, we left the fine adjustment work to the assembly experts. Experience was the only thing when it came to fine adjustment. We thought that numerical control was just impossible. However, at this time, we have discovered that fine adjustment is also done through the manipulation of figures. Thus the shape and size of the product can be matched through the manipulation of figures, and instructions can be shown on the assembly drawings. By so doing, we can shorten the time required for the assembly of the machines. Not only that, we can improve the reliability and quality of the packaging machines themselves.

II. COMPUTER-CONTROLLED OPTIMUM ACCUMULATING AND PACKAGING SYSTEM FOR MULTISIZED PRODUCTS

I would like to move on to the topic of computer-controlled optimum accumulating and packaging system for multisized products. We studied this sytem in 1980, and tried to minimize the waste of space among different sized products. For example, there is one operation at a packaging and shipping place for electric parts that requires packing different-sized products into a corrugated box according to requirements from customers.

The products, which are picked up by referring to the order slip, are fed on a cart to the packing place. The operator chooses the most suitable corrugated box from among three sizes of boxes—big, middle, and small—and packs the products into the box. Based on his long experience at this work, the operator can do the most efficient and wasteless packing work. Therefore, we can't reduce the number of the operators for this work even if we can mechanize the latter processes such as flap folding and taping. However, we attempted the use of the computer to do the operator's work. In other words, the computer chooses the most effective pattern and the most suitable box according to the sized of the different products in effective way. As the studying time was limited, we avoided challenging the study of three-dimensional space, and we tried to study the accumulating and packing of different-sized products on a plane (Fig. 2).

In our study, we limited the sizes of the products to six kinds. We specified the length and width of the product. The rectangular board that is regarded as the box comes in two sizes, which are also specified. When data concerning the kinds and number of the products to be packed are entered, the computer program tries various packaging patterns sequentially according to a certain rule, and then chooses the best pattern of packing to maximize use of space. If we cannot pack the products in one board, we have to repeat the same operation with the remaining products.

Figure 2. Computer-controlled optimum accumulating and packaging system for multi-sized products.

If the remaining products can be packed in an area less than half that of a board, the machine chooses a half-sized board. After completing the calculation, the system actually picks up the products from a magazine or storage conveyor one by one according to the selected pattern, and loads them on the board in accordance with x and y coordinates. This idea can be applied for loading different-sized products on a pallet or lorry.

In general we can handle a large number of the same products continuously in the case of a packaging operation connecting to a production line. However, we must pack and ship the various kinds of products from the warehouse according to orders. Therefore, this is a future theme that we will develop in a highly flexible automation system for that purpose.

III. HEAT-SEALING DEVICE USING A HEAT PIPE

The third topic I would like to talk about is the heat-sealing device using a heat pipe.

In packaging, it is quite popular to seal plastic film by heat sealing. The major mechanisms used for heat sealing are melting with heat, pressing, and solidification by cooling. The most popular way is to use the hot panel. This means heating the block to a certain temperature and pressing it to the film that we want to seal. In order to have a better seal, we have to control the temperature of the hot panel, the pressure, and the pressing time. Each different film has its own melting point. Therefore, it is ideal to press the hot panel with homogeneous pressure at the ideal temperature. However, due to the speed of the machine, the time available is limited only to 0.1–0.5 s. Therefore, in most of the cases, we heat up the hot panel far beyond the melting point of the film and try to give the required heat in a shorter period of time. Thus, this is a heat conduction with transient states, so the temperature of the hot panel must be controlled in a real, acurate sense. Normally we place the heat element close to the sealing portion of the hot panel block, and the temperature sensor is placed in between, so that we can maintain the temperature of the sealing surface at a certain level by controlling the electric current of the heater.

If we have a relatively large sealing area, we tend to have a difference in temperature in the sealing area. Even though we may try to control the heating element, heat conduction in the block differs over the area. Film width and film thickness will give an impact on homogeneous surface temperature of the sealing area. In the past, we tried to vary the density of the heating element in order to achieve this goal.

There are also timing problems. Even though we control the temperature by the sensor, there is a certain physical distance between the sealing surface, sensor, and heater. There will be some time lag between the heater and the heating surface in terms of heat conduction. This might cause overshoot, so the temperature difference on the surface is going to be larger than the temperature difference of the sensor. Therefore, it was thought inevitable to have a 20 or 30°C difference in temperature of the sealing surface.

Heat pipe has an activating liquid and wick in the vacuum pipe. The wick is porous and distributes the activating liquid. When a certain portion of the pipe is heated, the activating liquid near this portion will evaporate. The vapor pressure will spread instantly throughout the pipe, and the vapor will liquefy at the lower-temperature portion. This will bring a lot of heat. Therefore, this gives better heat conduction than copper or other metals. This system heat conductivity is several hundred times better per weight than the metals.

The heat pipe is thinner than the cartridge heater. We can built the heat pipe into the area close to the sealing surface, even in the case of a narrow-edged heat block. This is quite effective in a situation when we lose a lot of heat on a certain portion due to the sealing process. By having a heat pipe between the sealing surface and heater, we can substantially minimize the distance between them in terms of physical action. As a result, we can decrease the lack of heat conduction.

In 1979, we built a horizontal pillow-type packaging machine that uses a heat sealer with a heat pipe as an experimental model. We could obtain better results than we expected. In Fig. 3, the top shows the upper side of the heat block and bottom shows its lower side. These figures show the point where we measured the temperature. We designated 10 positions from 00 to 09 to sense the temperature. In Fig. 4, the left-hand side shows the temperature situation without the heat pipe. The figures show the location of the senser. Location 04 is 180°C, while 01 shows 210°C. This shows much variation in temperature. The lower side of the sealer shows the same situation. The right-hand side shows the temperature situation with the heat pipe. As shown in the diagram, the largest temperature difference between 00 to 04 is within 10°C. The same applies to the lower block. By placing heat pipe in it, we can maintain the homogeneous temperature in a sealing surface.

Furthermore, by minimizing the temperature difference, we can improve the following situation. In the past, we tended to heat the area a little bit higher than required in order to attain the necessary temperature at the lowest point. But by narrowing the temperature difference, we could reduce the maximum temperature. We also covered the heater with an insulator. Here is one example. Before we used the insulator and the heat pipe, a CPP film of 35 μm thickness required 350 W in order to get a sealing strength of 300 g/15 mm. When we added the insulator, we could improve the situation by 50 W. When we use both the insulator and the heat pipe, we could improve the situation by 110 W. That means we could save more than 20% of the sealing energy on average.

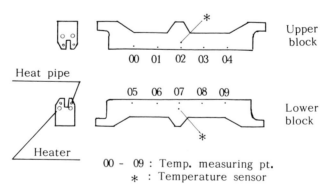

Figure 3. Heat-seal block.

Chapter 9. Recent Development of Packaging Machinery in Japan

In this particular experiment, we used a film that could be cooled down at room temperature by removing the heat block. However, some films require cooling while keeping pressure because of the particular property of the film. For this type of films, we use impulse seal. By minimizing the heat capacity of the heat block with heat pipe, or by using the heat pipe as a sealing panel, we might be able to give the heat to the sealing panel, and immediately take the heat out of it. Therefore, we might be able to seal the difficult type of films using the sealing panel in the future.

IV. INDUSTRIAL ROBOTS AND UNMANNED OPERATION

Besides computerization, recent topics in packaging machinery technology include industrial robots and unmanned operation. Robots have been widely discussed in Japan in the last couple of years. Formerly Versatran and Unimate, which were industrial robots, gave us a shock and increased the robot boom in Japan. Then, some impatient companies produced robots and most of them failed. Subsequently, the mass media paid no

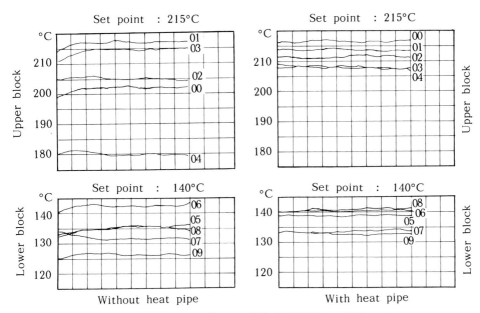

Figure 4. Temperature of heat-seal block (CPP 35 μm, 240 rpm).

attention to the robots. However, since that period there has been steady efforts for the development of robots. As a result, robots are presently being used especially in the auto industry and electric appliance industry. There has been manifestly successful utilization of industrial robots. It is said that more than 60% of the world's robots are operating in Japan.

The packaging industry is also discussing the possibility of the introduction of robots for unmanned operation. In the author's personal view, this is not an easy job to accomplish. Certainly, computers are spreading remarkably in the packaging industry. It is often said, "Should computers work as the brain, robots would work as the body. Consequently, the robots will diffuse throughout the packaging industry and finally lead to the unmanned operation." However, this conclusion is a hasty one.

It is quite an epoch-making idea to use robots as new machines working at beside humans, in contrast to traditional machines like a lathe or a press, and to control standardized hardware manufactured in the mass production system by software. In order to let this new machine function in place of a human being, we had to rearrange the peripheral system drastically. For example, consider an operation to pick up a product that is fed by belt conveyer. In order to let the robot perform the same function, we have to prepare a suitable setting in terms of positioning, timing, and orientation for the products. These peripheral devices turn out to be more expensive than the industrial robot itself. It seems more profitable to make an automatic feeding device than to make the peripheral devices for robots. Furthermore, from the standpoint of kinetics, there is no advantage in using a long arm to hold heavy items and move them quickly, because in the packaging operation the cycle time is generally required to be very short.

Nevertheless, the situation has changed in the last decade. Personnel expenditure has soared, while production cost has decreased mainly due to progress in control techniques. We also have successfully applied robots for some tasks like welding and painting.

However, it is hard for robots to do the "simple random work" that is still left to manual labor, and must be very expensive if it becomes possible. But supposing that robots will be mass produced within a few years and the price will decrease, will the packaging industry be able to utilize industrial robots? There seems to be much room for the study of robot utilization in some packaging work. The tasks, for example, include packing a gift pack, packing for shipment, bulk handling of products, and packaging materials supply, which need frequent change of the control value but have an operating cycle that is slow.

Some large companies have already started to study the introduction of unmanned operation. The largest technical difficulty in unmanned operation of packaging is the balance between reliability and inspection. In addition, since we handle a high volume of things in the packaging operation, it is necessary to study how to handle automatically both the supply of a large volume of packaging materials and the shipment of the finished products.

Unmanned operation will have a certain impact on the society. Therefore, we have to be cautious about the introduction of it.

PART V

New Trends in the Technology of Food Preservation

CHAPTER 10

New Trends in the Technology of Food Preservation—An Introduction

Michio Yokoyama
Food Science Laboratories,
Kureha Chemical Industry Co., Ltd.,
Tokyo, Japan

Recently, concern has grown about the safety and preservation of packaged foods. In the food industry around the world, a counterplan for food preservation has been devised, laying stress on *Clostridium botulinum*.

Food preservation is composed of food packaging technology, and sterilization and control technology of food microorganisms. When these two technologies are done perfectly, packaged foods can be preserved for a long period. This chapter describes food packaging technology and the behavior of microorganisms, and the sterilization and control of microorganisms in packaged foods.

I. FOOD PACKAGING TECHNOLOGY AND THE BEHAVIOR OF MICROORGANISMS

A. Vacuum Packaging

In general, molds growing on the surface of foods and aerobic bacteria cannot develop under anaerobic conditions. Vacuum packaging relies on this characteristic.

In food packaging, vacuum packaging does not provide a complete vacuum in the interior of packaged containers. The pressure there is usually 5–10 torr (760 torr = 1 atm). Therefore the definition of vacuum packaging is as follows: Vacuum packaging of food means to pack food in the container—mainly made of plastics—under conditions of the same or nearly the same internal pressure as the moisture pressure of the contents.

There are five technical methods for vacuum packaging:

1. Mechanical press fixing method.
2. Steam flash method.
3. Nozzle-style evacuation method.
4. Chamber-style evacuation method.
5. Skin pack.

However food is packaged with a packaging machine that draws a vacuum, if the proper methods are not used and if the appropriate transport and storage are not applied, putrefaction will occur. The microorganisms in packaged foods grow rapidly when the oxygen level reaches 0.5%.

Especially when cooked processed food is packaged in vacuum, complete evacuation is required. The lower the original amount of bacteria in food, the less there will be growth of them and the better the preservation will be.

B. Modified Atmosphere Packaging

Foods that undergo moisture separation or adhesion under vacuum conditions are often packaged with gas. Thin-sliced ham, cheese, fresh meat, coffee, dry milk, and oily confectionary goods are typical examples of modified atmosphere packaged foods. Sliced cheese, sliced ham, cooked processed food, and so on, which are proteinaceous processed foods, are packaged with mixed nitrogen and carbon dioxide. This modified atmosphere packaging prevents the oxidation of fat and pigments in meat and represses the growth of bacteria.

There are three modified atmosphere packaging technical methods:

1. Nozzle-style modified atmosphere packaging.
2. Chamber-style modified atmosphere packaging.
3. Gas-flash modified atmosphere packaging.

Tomioka et al. reported on the number of bacteria in modified atmosphere packaging beef as fresh meat. They reported that the increase of bacterial number was not observed when the meat was packaged with 100% carbon dioxide, but when it was packaged with air, the original number of 1.3×10 cells/g reached 1.0×10^5 cells/g after storage for 16 days at 1.1°C.

C. Packaging with Enclosed Free-Oxygen Scavenging Agent

In this method, oxygen in the packaged food is removed, for example by the reduction of iron oxide. The atmosphere of foods becomes closer to vacuum and the growth of bacteria and the oxidation of fat are prevented.

Especially in processed meat products, dairy products, and processed fish products, a free-oxygen scavenging agent is used to repress mold growth and prevent oxidation of pigments in meat. Packaging with enclosed free-oxygen scavenging agent is very easy to practice. The procedure is simply to put food into the packaging material and enclose the free-oxygen scavenging agent, and then seal the package. However, it is necessary to go through following method.

1. Choose the proper free-oxygen scavenging agent to remove the oxygen in the container completely.
2. Check the water content of food to be packaged. Choose the proper free-oxygen scavenging agent matching the water content of the food.
3. Use high-gas-barrier packaging materials.
4. After taking off the outer package of the free-oxygen scavenging agent, use it soon.
5. After putting food into the packaging container or material, then introduce the free-oxygen scavenging agent, and seal the lid of the container or the lip of the material perfectly.

D. Aseptic Packaging

The history of aseptic packaging foods began with aseptic canning. Using sterilized cans for aseptic filling of foods was tried in 1917, but it could not be accomplished then. After much trial and error for aseptic packaging, Dole Co. Ltd. established the aseptic system. From this opportunity, the aseptic canning of pudding, meat spread, and other products came about.

On the other hand, the application of the aseptic filling packaging system to paper cartons is comparatively recent. It originated in the research of long shelf life milk in Switzerland in 1951, and it was industrialized by Tetra-Pak Co. Ltd. in 1961. Then it attracted worldwide attention and made rapid expansion, with Europe as the center.

The Prime-Pak system for aseptic filling of pasteurized milk into completely sterilized plastic containers was established. Most packaged milk for coffee is filled with this system. Furthermore, the trial of aseptic packaging for solid foods was started in 1965, using sliced ham, and was followed by the introduction of bioclean rooms, which had been used for the assembling of rockets, as packaging room of foods.

Why was aseptic food born in Europe and America like this? There are various explanations. In Europe, as so many people were killed by illness in the Middle Ages, microorganisms growing in foods attracted increasing attention. The effort to exclude microorganisms, particularly pathogenic ones, growing on foods seems to have given birth to aseptic packaging of foods.

II. STERILIZATION AND CONTROL OF MICROORGANISMS IN PACKAGED FOODS

A. Sterilization of Microorganisms by Heat

With the growth of microorganisms, most watery processed foods are likely to putrefy. Fish paste products, processed meat products, daily meals, and so on are transported and marketed after killing the vegetative cells of microorganisms by the boiling of packaged foods. Canned foods, bottled foods, and thermoprocessed foods are sold at normal temperature after sterilizing microorganisms with high temperatures (110–120°C). The best way to improve the preservability of packaged foods is to eliminate microorganisms by heating after packaging.

The spores of *Clostridium botulinum* type A and B are killed at 100°C for 360 min and at 120°C for 4 min. In the case of spores of *Bacillus subtilis*, treatment is 120°C for 7.5–8 min; *Lactobacillus*, 71°C for 30 min; spores of molds, 65–70°C for 5–10 min; and spores of yeasts, 60°C for 15 min. For this reason, the vegetative cells of bacteria, spores of molds, and yeasts in common packaged foods are killed by boiling at 85–95°C for 15–60 min after packaging. Fish sausages weighing about 100 g and meat sausages are packaged with polyvinylidene chloride film and sterilized by retort at 120°C for 25 min. Foods packaged in retortable pouches like stewed beef and hamburger, are subjected to retort sterilization at 120°C for 30 min and at 118°C for 20 min, respectively. Short-time sterilization with ultra-high temperature (UHT) is used for aseptic packaging foods to kill the microorganisms in the contents.

B. Sterilization with Microwaves and Ultraviolet Rays

Microwaves, far-infrared rays, gamma rays, and ultraviolet rays are being used for the food sterilization. The electronic range, which is an application of microwave dielectric heating, is working as a home use cooking tool. Recently in the food industry, microwave (2450 MHz) dielectric heating has been applied to thermal processing of foods, defrosting, and sterilization of packaged foods and joined with new system of food packaging.

The world-famous Swedish maker of food sterilizing apparatus, Alfa-Laval Co., Ltd., has developed Multitherm, which is a continuous sterilizing system of solid foods with microwaves. This system is a series of packaging–thermal processing machines composed of the following processes: formation of containers from rolled film, filling of solid foods, evacuation and sealing, preheating to a temperature of 80°C, cooling down to 40°C, sterilization with microwave dielectric heating at 127°C in the water, and cooling down again.

An irradiation method with ultraviolet radiation sterilizing apparatus has been achieved as a way to sterilize foods. This apparatus is used to kill the microorganisms attached to food packaging materials, air-borne microorganisms in factories producing medicine and food, and microorganisms in water. Particularly for aseptic packaging foods, aseptic packaging systems combined with highly efficient ultraviolet sterilizing apparatuses are producing many kinds of aseptic products around the world. Brown Boveri Co. Ltd. (BBC) in Switzerland has developed highly efficient ultraviolet apparatus possessing an output of 1 kW for sterilizing packaging materials. The apparatuses are introduced in many aseptic packaging machines. The intensity of ultraviolet rays of this apparatus is 200 mW/cm^2. It has been reported that 99.99–99.999% of *Escherichia coli* is killed with the irradiation of ultraviolet rays at 50 mW/cm^2 for 1 s.

C. Control of Microorganisms at Low Temperature

After packaging, most fresh foods and processed foods are transported and marketed at low temperature. As proteinaceous foods is particularly apt to be polluted by microorganisms, it is sold under chilled or frozen conditions.

Microorganisms causing putrefaction or spoilage are hard to develop at low temperatures below 10°C. However, psychrophilic bacteria such as *Pseudomonas fluorescens* will double even at 5°C after 10.65 h.

Dairy products, dessert foods, and frozen foods are transported at low temperature, either chilled or frozen.

Dairy products, fish, and meat processed products and daily meals are transported at low temperature below 10°C. The foods transported at low temperature are classified as follows:

1. Cool foods: transport and market at 5–10°C.
2. Chilled foods: transport and market at −5 to 5°C.
3. Frozen chilled foods: produce under frozen condition, transport and market under chilled conditions.

D. Control of Microorganisms with Preservatives and pH

Preservatives and pH depress the growth of microorganisms in processed foods. Bacteria scarcely grow at pH below 4.0, above 10% salt concentration, above 30% sugar, and above 8% alcohol concentration. On the other hand, molds and yeasts can develop at pH 3.0 and at 40% of Brix degree but cannot grow at 8% alcohol. Microorganisms in processed foods are often controlled by the addition of organic acids. Some reports say that preservability of foods reaches a maximum at 50% alcohol, 48% water, and 2% citric acid or lactic acid.

It is said that the quality of salad made in Europe does not change after the storage at 10°C for 21 days. Most of these salads involve organic acids and a pH below 5.0.

CHAPTER 11

Retortable Packaging

Kanemichi Yamaguchi
Corporate Research & Development,
Toyo Seikan Group,
Yokohama, Japan

I. INTRODUCTION

Humans used to spend much time and labor to preserve food, and in modern ages when demand for greater quantities of better quality processed food has grown, we have seen development of a large food processing industry capable of meeting such demand both in quality and economy. It is no exaggeration to say that thermoprocessing procedures served as the basis for this food processing industry. The growth of large-scale industrialization of canned and bottled food is considered to be attributable, in part, to the fact that thermoprocessing was more suitable to meet food preservation demands, both in economy and quality, than were other means. While other food preservation means can only suppress or retard microbial growth, thermoprocessing can completely eliminate microbes from food, and such sterility can be readily and permanently kept once the containers are tightly sealed.

These features of thermoprocessing procedures for food preservation have been taken over by thermoprocess-resistant films emerging with the progress of the petrochemical industry, and such plastics combined with establishment of package manufacturing, filling and sealing, and thermoprocessing procedures have given rise to retortable packaging, which now plays a very important role in food packaging.

Food Packaging Copyright © 1990 by Academic Press, Inc. All rights of reproduction in any form reserved.

II. HISTORY OF RETORTABLE PACKAGING IN JAPAN

Retortable pouches were the first to come on the scene as retortable packaging. The history of flexibly packaged thermoprocessed food items (retort foods) is said to date back to 1940, but little is known about work conducted then. The concept of retortable pouches was originally implemented at the University of Illinois in the United States. There are, for instance, reports by Hu et al. (1955) and Nelson et al. (1956). As for retortable pouch packaging in Japan, it started around 1969 with pouches for curry, but attempts to use the basic principles of retortable pouch packaging, namely, to pack and seal food in flexible packaging material and to thermoprocess it for its preservation, began as early as 1950 with fish sausage, boiled beans, etc., using polyvinylidene chloride casing as packaging material. These foods were processed at temperatures below 100°C, and hence the preservation effect relied on food additives, but they may be said to have served as the forerunners of retortable pouches. In 1963, the cook-in-pouch (C.I.P) was developed as a boilable pouch, and in 1964, a foil-free retortable pouch was developed featuring an outer layer of biaxial oriented polyester and an inner layer of high-density polyethylene, laminated with isocyanate-type adhesive. These newly developed pouches were then commercially applied to packaging of Japanese hotchpotch and Chinese meat dumplings in 1967, marking the first and second retort foods in Japan. In 1967, the aluminum-foil-containing retortable pouches currently popular were also developed. With the commercialization of curry packed in foil-free retortable pouches in 1968 and in aluminum-foil-containing pouches in 1969 (Tsutsumi, 1972), full-scale production of retort foods was started. It is interesting to note that commercialization of retort foods happened to start around 1969 in most countries and that full-scale production began with the advent of aluminum-foil-containing pouches.

III. TYPES OF RETORT FOODS

Successful commercialization of foods in retortable pouches in Japan is widely known. The reasons for such success include the following:

1. While competitive products such as canned and frozen items have been fairly popular, consumers have not taken retort foods as alternatives to these but rather have taken them as prepared foods belonging to a different category.
2. The presence of foods pasteurized by hot water only and packed in

flexible pouches before the introduction of retort foods has made consumers less sensitive to retortable pouches.
3. Easy reheating and small size (one serving) packages have attracted the young generation in diversifying eating habits.
4. In Japan there are common tastes for foods and in particular for curried rice, which may be called a "national dish."
5. From the start of development, systems have been established to supply complete products.
6. The existence of a highly industrialized social foundation allows consumers to maintain high standards of living and high consumption of convenience foods.

Table I lists typical retort foods that are in production in Japan. They can be roughly divided into those characterized by convenience, thanks to easy reheatability, and those that are semicooked for use as cooking materials. Figure 1 shows the composition of retort food types produced in

Table I   Typical Retort-Pouched Products in Japan

Type of Product	Typical Products	
Sauces with meat and/or vegetables	Beef curry	Chicken curry
	Seafood curries (various)	Beef stew
	Beef stroganoff	Cream stew
	Hamburger in tomato sauce	Sukiyaki
	Meatballs in tomato sauce	Hash
	Mixed rice base	
Sauces	Prepared tomato sauce	Demi glace sauce
	Meat sauce (with and without mushrooms)	White sauce
		Neapolitan sauce
	Bouillabaisse sauce	Vongola sauce
Soups	Clear soup	Potage soup
	Corn soup	French onion soup
	Vegetable soup	Pumpkin soup
	Clear soup with corn and egg	Mushroom soup
	Miso soup (various)	Shark's fin
Meats	Hamburger	Meatballs
	Tuna meat in oil	Chicken meat in oil
	Seasoned fish fillet	Seasoned pork meat
Seasonings	Chinese seasonings (various)	Base of rice dishes (various)
	Base of egg dishes (various)	Teriyaki sauce
	Seasoning for buckwheat noodles	
Cereals	Boiled rice (various)	Pilaf
	Fried rice	Rice cake
	Porridge of rice and vegetables	Spaghetti
Desserts	Mousse essence	Sherbet base

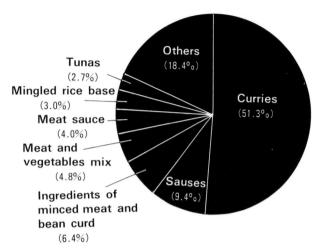

Figure 1. 1984 Retort food sales.

1984 (Japan Canners Association, 1984). As can be seen, curry products made up an overwhelming portion. The curry products are classified as a single category here, but actually they include various types. Seasonings for Chinese food and various soups are also important items. Tuna in oil is also growing for institutional use.

IV. PRODUCTION SYSTEMS OF RETORT FOOD

The retort food manufacturing system consists of food preparing equipment, filling and sealing machines, retorting equipment, sterilization racks, sterilization rack transfer equipment, cartoning machines, and outer casing machines. Figure 2 shows the major retort food production operation.

In Japan, preformed retortable pouches are often used, but some systems use pouches that are formed from roll stock film. Most semirigid containers are of the so-called precup type. The lids are supplied from roll stock film and sealed followed by trimming to shape, or preformed lids are used.

For filling pouches with food, the method adopted varies depending on the form of food, namely, solid food, mixture of solid and liquid food, or liquid food. For mixtures, solid and liquid foods are either charged separately or simultaneously as mixed. The former method is adopted when solids are bulky or a fixed amount of solids is to be charged. Liquid is

Chapter 11. Retortable Packaging

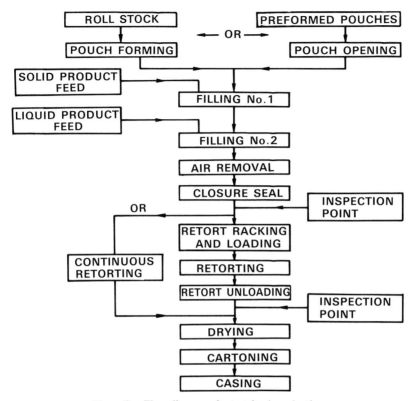

Figure 2. Flow diagram of retort food production.

supplied from a sanitary tank through sanitary piping to the hopper of the automatic filling and sealing machine for direct charging of fixed volume to pouches. In this process, the most important point is to ensure seal integrity. When filling and sealing, pouches are deaerated by physical flattening of the pouch and outward tensioning from the two side seals. In the case of pouches containing solids, vacuum sealing machines are also used to achieve vacuum packaging. For deaeration of semirigid containers, they are either filled completely to eliminate occluded air or sealed leaving some occluded air inside to prevent contamination of food onto the seal area. The sealed pouches are arranged in order on empty horizontal sterilization racks arriving from the sterilization rack feeding process. This is accomplished either by an automatic aligning machine or manually. The sterilization racks are automatically stacked on the tray stacker and conveyed on the retort car rails into the retort for sterilization. It is necessary to use a superimposed air-pressure-type retort in order to prevent damage to pouches during heat sterilization. In the case of semirigid containers with relatively large air content, a constant differential pressure-type retort is used to prevent container deformation. After adequate sterilization, the racks are unloaded onto the tray conveyor one after another by a rack unloader and conveyed to the automatic dewatering and drying machine. Dewatered and dried products then undergo checks by a weight checker or metal detector as necessary before they are inserted in individual cartons or are carried as they are to the casing equipment to complete the process.

A. Packaging Material and Packaging Design

1. Packaging Material

The basic requirement in the selection of packaging materials is heat resistance, because of the high-temperature processing they undergo. In addition, in the selection of package composition and package shape, consideration must be given to the nature of food to be packaged, required shelf life, and package cost. Packages of various composition have been proposed, and Table II lists representative material compositions of packages for retort foods currently in use (Yamaguchi, 1984). Package shapes can be roughly divided into three types: pouch, standing pouch, and semirigid container. Each of these types is available with various barrier material composition, such as aluminum foil, polyvinylidene chloride or saponified ethylene–vinyl acetate copolymer. In the case of large pouches for institutional use, four-layer composition is often adopted for higher

Table II Representative Examples of Packaging Materials for Retort Foods in Japan

Package Type	Material Composition
Aluminum foil pouch and aluminum foil standing pouch	12 μm Polyester/7–9 μm Al foil/70 μm polypropylene
	12 μm Polyester/15 μm nylon/9 μm Al foil/70 μm polypropylene (institutional use)
Plastic pouch and plastic standing pouch	15 μm Nylon/70 μm polypropylene
	12 μm Polyester/15 μm PVDC (or EVOH)/50 μm polypropylene (barrier type)
Aluminum foil tray	Body: epoxy resin/100–150 μm Al foil/50 μm polypropylene
	Lid: epoxy resin/50–100 μm Al foil/50 μm polypropylene
Plastic tray	350–450 μm Polypropylene/PVDC (or EVOH)/Polypropylene

drop strength. Standing pouches using aluminum foil usually adopt the four-ply laminate construction because of their inherently inferior drop strength as compared to four-seal flat design. Foil-free standing pouches often have polyester film as the outer layer in order to improve their self-standing nature. As for adhesive, in most cases, polyurethane type adhesive is used but modified polypropylene is also used. These adhesives have been selected for food hygiene consideration and have been approved by FDA.

Foil-free semirigid containers formed from polypropylene sheet have been in use, but they provide a rather poor barrier to oxygen gas. For this reason, multilayer containers using eval resin (saponified ethylene–vinyl acetate copolymer) or polyvinylidene chloride have come to be used. Semirigid containers using aluminum foil are often composed of aluminum foil laminated with polypropylene using epoxy resin as a protective layer, but due to high cost, this type is not in much popular use.

The lids of semirigid containers are, in most cases, of the same composition as the packages themselves. These lids are often required to be easy-open ends. There are two types of lid, peelable lid and easy-open end. In the case of a semirigid container with a polypropylene-based inside layer, the lid is usually of aluminum foil with polypropylene-based high-density polyethylene, low-density polyethylene, EVA, etc. used as sealant. In this case, the seal strength must be high enough to withstand retort processing and yet must be such as to offer peelability. For this purpose, resin type, melt index, blending method, sealant layer thickness, base material-to-sealant layer bonding strength, sealing condition, etc. must be controlled adequately. There are also easy-open lids with a scored portion provided inside the sealed section so that it may be broken by an opening tab (Hirota, 1984).

2. Package Design

With the exception of foil-free pouches, standing pouches, and large pouches for institutional use, other retort foods use individual cartons. This is to prevent damage or pin holes that could be caused to pouches during distribution and because of ease of stacking and attractive print appearance offered by such packaging. Individual cartons are designed such that the length and the width must be zero to 10 mm shorter than the respective size of the pouch to obtain drop shock resistance. In height, they are about 6–10 mm larger than the pouch thickness in order to obtain sufficient compressive strength (Toyo Seikan Kaisha, 1973a).

For solid contents, the pouch is sometimes bonded to the carton to secure the pouch in position. Round sealing of edge corners is also an effective way to prevent pin holes. When individually cartoned products are to be packed in a gross carton for shipment, horizontal packing will give higher strength than vertical packing. Large pouches for institutional use are also packed horizontally, but measures such as partitions must be taken to protect lower ones from the load of upper ones.

B. Filling and Sealing Equipment

1. Types of Filling and Sealing Equipment

Table III lists the filling and sealing machines in use in Japan for production of retort foods. In the case of retortable pouches, there are two types of machines in use, namely, those used when filling and sealing preformed pouches and those used when pouches are formed from laminated roll stock before they are filled and sealed. In Japan, the former type is used far more frequently. This type is classified into the rotary type, with eight stations operating intermittently, and the straight type. Figure 3 (Yokohama Automatic Machinery, Ltd., 1976) and Fig. 4 (Mitsubishi Heavy Industries, Ltd., 1980) show these two respective types. The production line standard speed of the former is 30 pouches per minute and that of the latter is 60 pouches per minute. In the latter type, higher capacity can be achieved by multiline configuration. A form–fill–seal machine offers package material cost reduction, but as it is originally designed for volume production of the same items, it suffers from time and labor loss for film replacement etc. and hence is not necessarily advantageous. For this reason, its application is limited to curry and other single products that are produced in a large volume or liquid seasonings. In the case of products with solid contents, either pouches are filled with solids together with some liquid and deaerated by an ordinary stretch method, or a regular vacuum sealing machine as shown in Fig. 5 is used (Nippon Polycello Kogyo, Ltd., 1985). The most serious drawback when the latter machine is used is low productivity. In an attempt to solve this, an automatic vacuum sealing system as shown in Fig. 6 has been developed that consists of an ordinary rotary type machine that fills pouches with food and another rotary type vacuum sealing machine that vacuum seals the filled pouches (Toyo Jidoki, Ltd., 1982).

Filling and sealing machines for semirigid containers are both mainly precup type, and form–fill–seal machines have not yet been adopted. As shown in Fig. 7, each semirigid container is placed in a metal mold and is filled with food. Then it is sealed by lid that is fed in the roll stock, which is then followed by trimming around the sealed periphery (Shinwa Kikai, Ltd., 1980). An alternative method is to use a preformed lid to seal the

Table III Packaging Machinery for Retort Foods in Japan

Container Style	Package Formation	Models	Capacity	Manufacturer
Pouch and standing pouch	Pre-formed Pouch	Rotary type		
		Y-77-A	30 ppm	Yokohama Automatic Machinery Ltd.
		TT-8C	30 ppm	Toyo Jdoki Ltd.
		TT-8CW	60–80 ppm (double)	
		Line type		
		PF-15	60 ppm	Mitsubishi Heavy Industries Ltd.
		PE-15D	120 ppm (double)	Toyo Seikan Kaisha Ltd.
		—	120 ppm	Toyo Jdoki Ltd.
		TL-A-4	200–240 ppm (4 lanes)	
	Form–fill–seal	Line and rotary type		
		TT-8CWZ	60–80 ppm	Toyo Jidoki Ltd.
		AFM-70	80 ppm	Izumi Food Machinery Ltd.
	Preformed pouch	Rotary and rotary-type mechanical vacuum, TVA-P2	30–40 ppm	Toyo Jidoki Ltd.
Semirigid container	Preformed tray and cup	NU-1Y2-148-2000	40–60 ppm (4 lanes)	Shinwa Machinery Ltd.

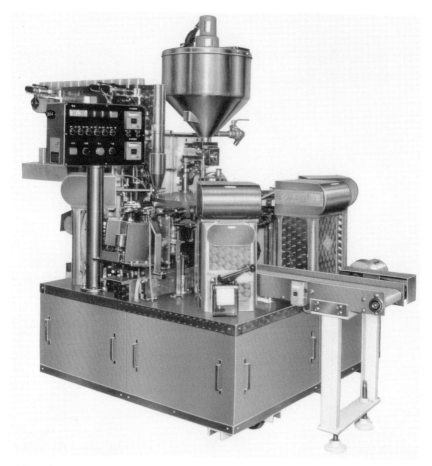

Figure 3. Rotary filler–sealer. (Courtesy of Yokohama Automatic Machinery, Ltd., Yokohama, Japan.)

container. As for sealing, all types adopt two heat-sealing actions and a cold pressing action to increase the reliability of the seal. The filling and sealing equipment is required to be constructed so as to prevent contamination of seal surfaces with food to provide seal integrity.

2. Defective Seals

Complete fusion of the opposing seal surfaces is essential for achieving a good seal. For this purpose, sealing must be done with temperature, time, and pressure suitable for the sealant in use. Even when sealing is done

Chapter 11. Retortable Packaging

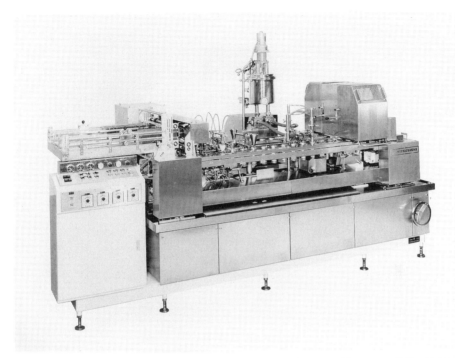

Figure 4. Mitsubishi retort pouch filling machine model PF-20D. (Courtesy of Mitsubishi Heavy Industries, Ltd., Tokyo, Japan.)

under such optimum conditions, the presence of product on the seal area will result in a defective seal. The most frequent causes for such defective seals include the presence of water vapor and contamination with food. Contamination with water vapor may occur when hot filling is adopted, and steam condenses into water drops on the seal surface and forms air bubbles in the seal when the pouch is sealed. These air bubbles grow larger and link with each other, depending on sealing conditions, and result in a seal failure in extreme cases. As a result, the products could be easily recontaminated with microorganisms. This problem has been solved by the introduction of a triple-seal method to filling and sealing machines. The causes of contamination of sealing surfaces with product include:

1. Contact when solids are filled in the pouch
2. Drip of liquid when solids are filled in the pouch
3. Leaks from filling pump
4. Drip from filling pump
5. Scattering by filling pump

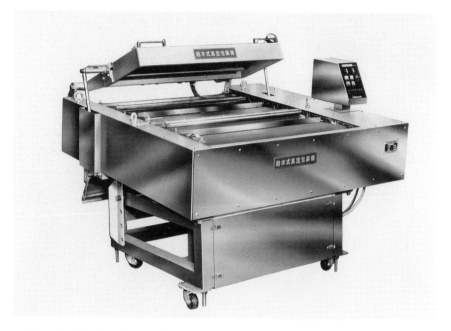

Figure 5. Kashiwagi automatic vacuum sealer (KV-110-4. (Courtesy of Nippon Polycello Kogyo, Ltd., Tokyo, Japan.)

6. Splash from filling pump
7. Rebound from bottom of pouch
8. Blow-up of inside product when sealing is done
9. Disturbance of inside liquid due to movement

These causes have to be eliminated through selection of optimum liquid temperature and viscosity, pouch size, adjustment of the machine to filling speed, measures to minimize dripping, fine adjustment of closing method, etc.

Methods for nondestructive checks for such defective seals have been proposed by Lampi et al. (1976) but have not yet been applied practically in Japan. In Japan, visual inspection of all products is adopted. In the destructive check, by immersion in methyl ethyl ketone or chloroform or by dissolving of aluminum foil by nitric acid, the aluminum foil and polyolefin layers are separated to allow macroscopic examination.

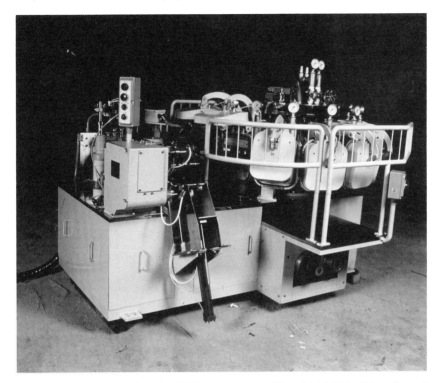

Figure 6. Toyo automatic packer TVP-A. (Courtesy of Toyo Jdoki, Ltd., Tokyo, Japan.)

3. Residual Air

In the case of pouches or semirigid containers, residual air is associated with seal integrity and there always remains some amount of air inside, and this influences the heat transfer, sterilization effect, and product quality. Techniques for air removal such as the snorkel method (Mayer and Robe, 1963), waterhead pressure method (Heid, 1970) and steam flash method (Schulz and Mansur, 1969) have been proposed, but they have their own merits and demerits when it comes to application on a commercial scale, and hence none of them have yet been successfully applied practically. In Japan, the stretch method is applied for pouches of standard size (170 mm × 130 mm) (Tsutsumi, 1972). Table IV shows the results, and with this degree of remaining air, no particular problem has been experienced (Yamaguchi et al., 1972). For large pouches, vacuum sealing is adopted.

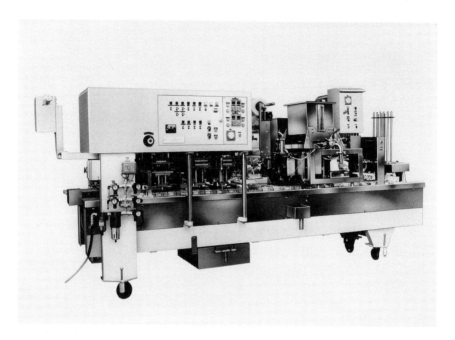

Figure 7. Shinwa filling–sealing machine model NWE-7Y12-4500. (Courtesy of Shinwa Kikai, Ltd., Saitama, Japan.)

Table IV   Volume of Residual Gases of Commercial Retort Foods

Product	Net Weight	Residual Gas (cm^3)	
		Destructive Method	Nondestructive Method
Curry	180 g	7.5	7.8
		2.9	3.5
		5.3	5.7
Cream stew	200 g	7.0	7.4
		7.2	5.8
		5.3	5.9
Hushed beef gravy	180 g	4.5	5.0
		7.0	7.5
		7.0	7.0

C. Retorting System

1. Types of Retorting Equipment

Features of retorting equipment for retort foods depend significantly on the characteristics of containers used. In particular, to achieve a hermetic seal by heat sealing requires careful pressure control as compared with canned or bottled items. Table V lists the types of retorting equipment currently in wide use for retort foods. They can be classified into steam–air type and hot water type by the heating medium in use or into static type, circulating type, and rotary type by product behavior in the retort. These types are similar in performance, and selection is made taking food properties and shape and packaging form into account.

2. Counterpressure Method and Constant Differential Pressure Method

When a retortable pouch or semirigid container is filled with food and sealed, followed by heat sterilization at 100°C or higher temperature, internal pressure will be developed in the pouch or semirigid container due to expansion of residual air and contents. While metal cans are strong enough to withstand such internal pressure, these pouches and semirigid containers undergo increase in their volume when an internal pressure occurs and sometimes expand to such an extent that their seals are broken. Reversion of inside and outside pressure due to cooling also could cause seal failure. Figure 8 shows the heating and cooling process. In order to prevent such failure, it is necessary to keep adequate counterpressure throughout the entire sterilization process. Retortable pouches, which are flexible and very stable against external pressure, can withstand a very rapid come-up rate or cooling, and this makes the retorting equipment quite unique. Figure 9 shows such equipment, and Fig. 10 shows the flow chart (Toyo Seikan Kaisha, Ltd., 1973b). Semirigid containers, however, are likely to have a sealed section broken by internal pressure built up; in addition, they require consideration as to possible deformation due to external pressure. When they have a large amount of residual air, in particular, its expansion or contraction during the sterilization process could cause permanent deformation of the containers. In order to prevent this, a system that controls the retort pressure in correspondence with container inside pressure, namely, the so-called constant differential pressure retorting system, is used. Because of pressure control in response to container deformation during the process, the come-up time and cooling time tend to be longer to gain easy control, and the heating and cooling patterns become closer than for a retorting system for canned or bottled items.

Table V Retort Equipment for Retort Foods in Japan

Method of Sterilization		Model	Number of Pouches[a]	Manufacturer
Steam	Static	H130-C300-120SA	1944	Toyo Seikan Kaisha Ltd.
		H130-C200-120/135SA	1296 (2 cars)	
		STR-301	1800	Fujimori Kogyo Ltd.
		H130-C300-120WA	2250	Toyo Seikan Kaisha Ltd.
Water	Static and circulating water	RCA-100	1500	Hisaka Seisakusho Ltd.
		RCA-120	2625	
		Auto Cooker-III	1200	Toppan Printing Ltd.
		Auto Cooker-IV	2625	
		UHR-W-301	2700	Fujimori Kogyo Ltd.

[a] Pouch size 130W × 170L × 15T (mm), 3 cars.

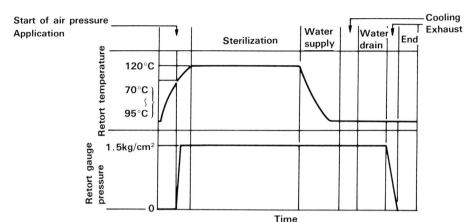

Figure 8. Pressurized heating–pressurized cooling procedures. (Sterilization at 120°C, 1.5 kg/cm².)

3. Heating Medium

A steam–air mixture or hot water is used as the heating medium, and it has been established that these two media are practically equivalent in terms of thermal characteristics that are associated with heat transfer efficiency (Pflug, 1964; Pflug and Borrero, 1967; Terajima, 1974). However, different heating media require different attentions in operation. When a steam–air mixture is used, for instance, removal of residual air from the retort is necessary in order to prevent air pockets from developing in the retort. For this purpose, it is essential to fully replace the inside atmosphere with heating medium before full introduction of the medium for heating. The closing time for the exhaust overflow valve and drain valve is determined with consideration paid to possible damage to containers. Closure at a too low temperature could lead to nonuniform temperature distribution. In the case of a steam–air mixture, the mixing ratio has to be approximately 85% of steam and 15% of air (120°C sterilization). When hot water is used, it has to be circulated in order to achieve uniform heating. In retort food manufacture in Japan, a steam–air mixture is traditionally used as the heating medium. However, there is also a trend toward adopting a hot water circulation system. This system is adopted when the thermoprocessing temperature is relatively low (105–110°C) or when it is expected that food that is likely to deform due to its own weight may be prevented from such deformation by the effect of buoyancy. In the case of food that is to undergo hot water pasteurization processing at around 100°C, and that loses its value if bubbles develop within, a hot water type system that can set thermoprocessing temperature and pressure separately is used.

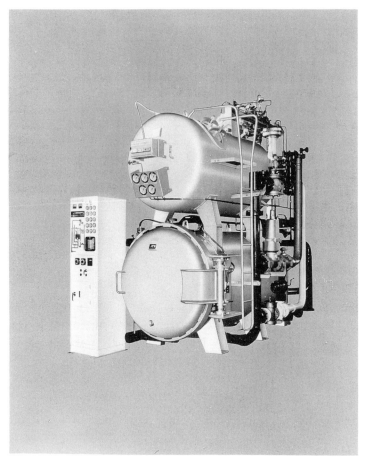

Figure 9. Fully automatic retort equipment H130-C200-120/135-SA. (Courtesy of Toyo Seikan Kaisha, Ltd., Tokyo, Japan.)

4. Sterilization Racks and Temperature Recorder with F_0 Value Measurement

When retort food is heated in the retort, sterilization racks are used to guarantee equal exposure of each container to the heating medium. In Japan, horizontal racks are used as shown in Fig. 11 (Toyo Seikan Kaisha, Ltd., 1973a). The racks, on which containers are arranged horizontally, have slots and small holes to ensure a smooth flow of heating medium. At four corners, the upper and lower racks are engaged and stacked up and

Chapter 11. Retortable Packaging

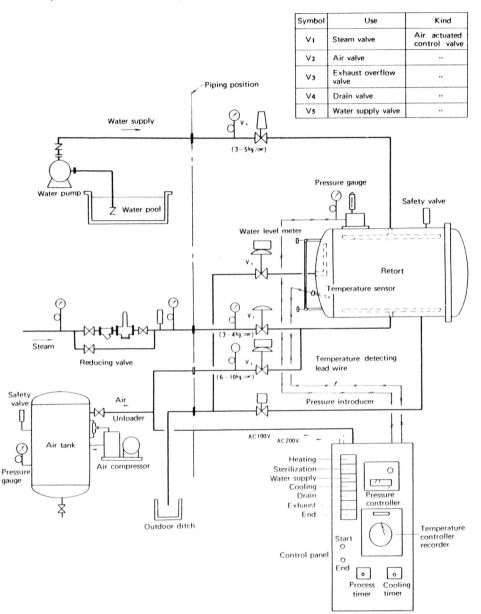

Figure 10. Flow sheet of overpressure heating, overpressure cooling retort for retortable pouch.

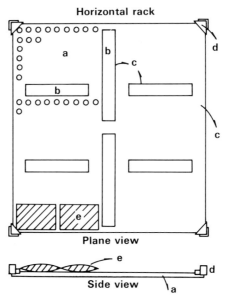

Figure 11. Sterilization rack: (a) punched aluminum plate, (b) steam pass, (c) side wall, (d) fitting for stacking, (e) food in retortable pouch.

are then loaded into the retort. This horizontal rack is far superior to the vertical rack in the ease of operation and automation.

Measurement of temperature distribution in the retort, temperature–time curve at the container center, and F_0 value is important for full control of sterilization. For these measurements, the temperature recorder with F_0 value measuring unit is widely used.

V. MICROORGANISM CONTROL IN RETORT FOOD

A. Manufacturing Standard in Japan

Retort food undergoes commercial sterilization in order to remove microorganisms that are likely to cause spoilage of contents, but selection of specific target microorganisms for sterilization purposes is impossible since raw material and contaminating microflora differ from food to food. Therefore, our manufacturing standards on retort food, as in the case of canned food, set the conditions necessary to inactivate spores of *Clostridium botulinum* type A, which can grow in the contents of a container and is an important toxogenic bacteria for food hygienie. In Japan, "Standards for pressurized heat sterilization of hermetically sealed packaged food" require that thermoprocessed food, heat processed at 100°C or higher and with pH value over 5.5 and water activity over 0.94, be sterilized such that the critical point of the container is heated to 120°C for 4 min or by an equivalent or superior thermoprocessing condition. This requirement corresponds to processing with $F_0 = 3.1$ minimum. However, this condition does not guarantee inactivation of other putrefactive microorganisms. No standards are in effect on food with other ranges of pH or A_w values. Thus, the sterilization conditions must be set with this $F_0 = 3.1$ as the minimum condition, and a taking compromise between food quality and zero spoilage rate and incorporating a certain safety factor.

B. HTST Sterilization Method

The progress of sterilization technique in the retort food industry is now taking place at an accelerated pitch with the introduction of the concept of high-temperature, short-time (HTST) sterilization, which means to process food at high temperature for a short time for efficient extermination of microorganisms, while minimizing influence of heating on food quality. The good heat transfer characteristics inherent to retort food and packages have made in-package HTST processing possible (Yamaguchi and Kishimoto, 1976). Figure 12 shows the heating temperature, time, and quality conceptually for canned food and retort food. In Japan, thermoprocessing of retort foods was initially applied with 115–120°C, 20–30 min sterilization on standard size pouches (content 180 g, 130 × 170 × 15 mm), and now the technique has been established to a allow heating temperature of maximum 150°C. There is also a move to combine the retorting method with other food preservation methods in order to keep better food quality. In this method, the sterilization conditions are set based on the acidity of food. In some cases, thermoprocessing can be completed at a temperature as low as 110–115°C and in a short time. Thus, by selection of heating temperature from 110 to 150°C according to the characteristics and heat conductivity of food, retort food can be manufactured with the highest quality and with high efficiency.

C. Factors Affecting Sterility of Retort Foods

In order to achieve sterility reliably, control over the entire manuf process is important. The important control points include:

1. Elimination of thermophilic bacteria from material and r initial number of bacteria

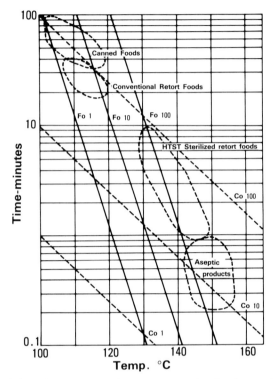

Figure 12. Relationship between heating temperature, heating time, and change in product quality in several thermoprocessing methods.

2. Specification of size and amount of solids and specification of liquid volume and head space
3. Sealing condition of containers
4. Arrangement of sterilization rack
5. Sterilization condition and construction and operation of retort equipment
6. Cooling-water condition
7. Product handling
8. Temperature recording

It will be necessary to check closely that these factors are correctly controlled.

It may appear that physical and chemical properties of food do not require control, since they are nearly constant once the specifications are set. For solid food with liquid, however, heat transfer to solids can be

Chapter 11. Retortable Packaging

faster or slower depending on their size. Therefore, it is preferable to specify the maximum size of solids, control their size, and specify the amount of solids and liquid packed in each container and control accordingly. In the case of viscous products, heat transfer is affected by the consistency. As the sterilization condition is set for food of given thickness, a thicker product could cause insufficient heat transfer to the inside, resulting in incomplete sterilization. The control of the amount of contents is essential for meeting product specifications, and in the case of retortable pouches it requires particular attention since deviation from specified amounts of contents causes change in the finished product thickness. According to a report on the influence of packaged food thickness on heat transfer, even 3 mm thickness variation could affect sterilization integrity (Pflug et al., 1963). In the case of liquid product the thickness of packaged product can be made nearly constant by controlling the content weight, but for solid products the same content weight does not mean the same thickness, making control difficult. In this case, control by the thickest solids in addition to minimizing solid thickness variations will be necessary. The possible influence of occluded air on heat transfer and sterilization integrity has also been studied. It has been known that 15 ml or more residual air in the case of a standard size pouch markedly interfers with heat transfer, significantly affecting the sterilization effect (Yamaguchi et al., 1977).

In view of the film processing techniques used, it is unlikely that plastic film has pinholes under normal condition and it is quite unlikely that there is any pin hole when many of these films are laminated. It has also been confirmed that even a single layer of film does not allow microorganisms to pass (Lampi, 1967). The problem lies rather in damage caused to the container surface by careless handling during manufacture of retort food. As described earlier, it is important to design the containers so that they can withstand without breaking or pinholing when they are exposed to drop impact and vibration during storage, transportation, and at stores.

Secondary contamination by penetration of water contaminated with microorganisms is one of the major causes of spoilage of canned food. In the case of retort food, the polyolefin of the sealed section is at high temperature and hence is soft in the initial stage of cooling, but since the packaging material itself is soft and flexible, it is expected that any impact acting on the container does not affect the sealed section but is instead absorbed by the container. Furthermore, the container's wide sealing width will give it by far higher sealing capacity as compared with the sealing compound of a canned product. At any rate, cooling water should be clean for food hygiene.

D. Confirmation of Commercial Sterility

In Japan, standards for retort food require that the product be free from leaks and swelling when subjected to an incubation procedure at 35°C for 14 days. It is also specified that when 5 ml of a one-hundredth dilution of the food subjected to the above incubation microorganism test is inoculated at 1 ml each in given TGC culture media, no microorganisms be found. The meaning and purpose of this test are that the product be free from microorganisms that may proliferate in the course of storage, distribution, and sales of retort food at ambient temperature.

VI. STANDARDS FOR PACKAGES

The standards of physical strength of packages include the heat-seal strength test, static load resistance test, and drop test. The required heat-seal strength is 2.3 kg/15 mm minimum. In the static load resistance and drop tests, the conditions and requirements have been set according to weight classes of contents. Figure 13 shows the static load resistance test method and loading conditions. For a container with contents weight bracket of 100g to 400g, for example, it is prescribed that no contents should leak when a static load of 40 kg is applied for one minute. Table VI

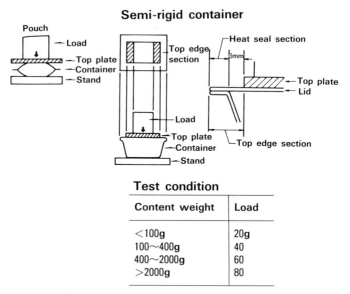

Figure 13. Static load resistance test and load conditions.

Table VI Regulatory Status for Drop Strength Test

Content Weight (g)	Drop Height (cm)
100	80
100–400	50
400–2000	30
2000	20

lists the drop test conditions. In this test, the sample is dropped onto a concrete surface from a height specified for its weight with its bottom or flat side facing down, and content leaks are checked after two drops.

VII. SHELF LIFE OF RETORT FOOD

The shelf life of retort food is said to be about 2 years for an aluminum foil pouch when stored at ambient temperature, about 2–3 months for a barrier-type nonfoil retortable pouch, and 1–2 months for an nonfoil retortable pouch. Figure 14 shows the results of an organoleptic test on quality change with time of hambergers packed in pouches of differing oxygen permeability and retorted (Ishitani et al., 1980). As can be seen, the

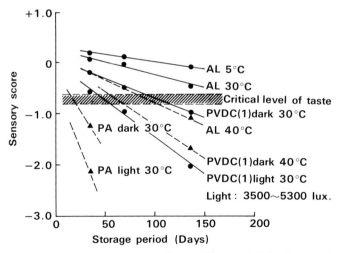

Figure 14. Changes in sensory score of retortable pouched hamburger stored under differing conditions.

sensory threshold level depends heavily on the oxygen permeability of the packaging material. Pouches using aluminum foil are seen to keep content quality for an extended period, giving preservation performance equivalent to that of a can.

VIII. CONCLUSION

The retort food industry with its accelerated growth rate and product diversification can be said to have entered a new stage of growth. Since the start of commercial production, the industry has accumulated a fairly large amount of output but no microbial problem has been experienced. Thanks to this stability, as food processing technology combined with diversification of packages taking place accelerated in growth recently, further development and growth of the industry are expected.

REFERENCES

Heid, J. L. (1970). Retort process minimizes residual headspace gases with "breathable" pouch. *Food Prod. Dev.* **4**(6), 82, 84, and 86.

Hirota, K. (1984). Easy open end suited for retorting. *Jpn. Packaging Inst. J.* **22**(4), 12–17.

Hu, K. H., Nelson, A. I., Legault, R. R., and Steingerg, M. P. (1955). Feasibility of using plastic film packages for heat-processed foods. *Food Technol.* **19**(9), 236–240.

Ishitani, T., Hirata, T., Matsushita, K., Hirose, K., Kodani, N., Ueda, K., Yanai, S., and Kimura, S. (1980). The effects of oxygen permeability of pouch, storage temperature and light on the quality change of a retortable pouched food during storage. *J. Food Sci. Technol.* **27**(3), 118–124.

Japan Canners Association. (1984). Canned food pack & trade statistics. *Canners J.* **64**(8), 2–185.

Lampi, R. A. (1967). Microbial recontamination in flexible films. *Act. Rep. Res. Dev. Assoc. Mil. Food Packaging Systems* **19**(1), 51–58.

Lampi, R. A., Schulz, G. L., Ciavarini, T., and Burke, P. T. (1976). Performance and integrity of retort pouch seals. *Food Technol.* **30**(2), 38–48.

Mayer, P. C., and Robe, K. (1963). Canning without cans. *Food Process.* **24**(11), 75.

Mitsubishi Heavy Industries, Ltd. (1980). "Mitsubishi Retort Pouch Filling Machine Model PF-20D," Brochure. Tokyo, Japan.

Nelson, A. I., and Steinberg, M. P. (1956). Retorting foods in plastic bags. *Food Eng.* **28**(1), 92–93.

Nelson, A. I., Hu, K. H., and Steinberg, M. P. (1956). Heat processible food films. *Mod. Packag.* **20**(10), 173–179.

Nippon Polycello Kogyo, Ltd. (1985). "Kashiwagi Automatic Vacuum Sealer KV-110-4," Brochure. Tokyo, Japan.

Pflug, I. J. (1964). "Evaluation of heating media for producing shelf stable food in flexible packages. Phase 1," Final Report, Contract DA19-AMC-145 (N). U.S. Army Natick Laboratories, Natick, Mass.

Pflug, I., and Borrero, C. (1967). "Heating media for processing foods in flexible packages. Phase II, "Tech. Rep. 67-47-GP. U.S Army Natick Laboratories, Natick, Mass.

Pflug, I. J., Bock, J. H., and Long, F. E. (1963). Sterilization of food in flexible packages. *Food Technol.* **17**(9), 87–92.

Schulz, G. L., and Mansur, R. T. (1969). "Sealing through contaminated pouch surfaces," Tech. Rep. 69-76-GP. U.S Army Natick Laboratories, Natick, Mass.

Shinwa Kikai, Ltd. (1980). "Shinwa/Packer Sealer," Brochure. Saitama, Japan.

Terajima, Y. (1974). Over-all heat transmission from the heating medium (steam and water) to the content of the retortable pouch. *Canners J.* **54**(1), 73–78.

Toyo Jidoki, Ltd. (1982). "Toyo Automatic Packer TVP-A," Brochure. Tokyo, Japan.

Toyo Seikan Kaisha, Ltd. (1973a). Brochure on RP-F. Tokyo, Japan.

Toyo Seikan Kaisha, Ltd. (1973b). "Fully Automatic Overpressure Retort for Retortable Pouches," Brochure. Tokyo, Japan.

Tsutsumi, Y. (1972). "Retort pouch"—Its development and application to foodstuffs in Japan. *J. Plastics* **6**, 24–30.

Yamaguchi, K., Komatsu, Y., and Kishimoto, A. (1972). Sterilization of foods in flexible packages. Part VI. Non-destructive method for determining residual air in pouches. *J. Food Sci. Technol.* **19**(7), 316–320.

Yamaguchi, K., and Kishimoto, A. (1976). "In-package high temperature short time sterilization of foods in retortable pouches," Presented at Internationale Konfernz uber den Schutz verderblicher Guter durch Verpackung, Munchen, West Germany.

Yamaguchi, K., Komatsu, Y., and Kishimoto, A. (1977). Sterilization of foods in flexible pouches. Part IX. Effect of internal residual air on food quality in retortable pouches. *J. Food Sci. Technol.* **24**(10), 501–506.

Yamaguchi, K. (1984). "Current status of containers for thermoprocessed foods in Japan," Presented at 4th International Conference on Packaging, East Lansing, Mich.

Yokohama Automatic Machinery, Ltd. (1976). "Automatic Filling-Sealing Machine Model Y-77-A," Brochure. Yokohama, Japan.

CHAPTER 12

Aseptic Packaged Foods

Michio Yokoyama
Food Science Laboratories,
Kureha Chemical Industry Co., Ltd.,
Tokyo, Japan

A wide variety of aseptic packaged foods are now manufactured by many food processors in all parts of the world. However, the term "aseptic packaged foods" has not yet been defined clearly. In this situation, food engineers and experts are at a loss to recognize what kinds of packaged foods belong to aseptic packaged foods when they read the literature.

In a broad sense, aseptic packaged foods (Yokoyama, 1984a) are defined as all the packaged foods in which the products are preserved in aseptic conditions after they have been packaged. In this sense, therefore, retortable foods may also be applicable to this category.

In the strict sense, however, aseptic packaged foods can be defined as the specially packaged foods in which the products are sterilized and are then packaged in sterile packaging materials under an aseptic environment without reheating for sterilization.

According to the degree of sterilization of the bacteria contained in foods, aseptic packaged foods are classified into two types. One is such foods as long-life milk (which can be stored at room temperature for a long period) and coffee milk, in which the products are completely sterilized at high temperature for a short time and are then aseptic fill/packaged in the containers sterilized with hydrogen peroxide or otherwise. The other is such foods as sliced ham, cheese, and rice cakes, in which the products are processed into almost aseptic products in the heat treatment and washing sterilization processes and are then packaged in sterile containers in a bioclean room. The former is generally called "complete aseptic packaged foods" or simply "aseptic packaged foods," and the latter "semi-aseptic packaged foods" or "commercialized aseptic packaged foods."

In addition, aseptic packaged foods can also be divided into two types by usages. One is for consumer uses and the other for institutional use. Furthermore, the latter is for either the dining-out industry such as restaurants and hotels and the bulk items for food processing factories.

Why have aseptic packaged foods have recently shown a rapid growth in every country throughout the world? (1) The content of salt and sugar in foods has gradually decreased with increasing demand by people for healthy foods, and consequently all foods have been inclined to degenerate in quality. (2) Consumers have experienced reactions to such food poisoning as botulism and subsequently they have come to desire the foods containing no microorganisms.

From the standpoint of food processors, aseptic packaged foods show no juice separation phenomena caused by the reheating operation, have improved taste, and reduce energy consumption. In addition, they can greatly save on both packaging material and shipping costs because they are mostly packaged in liquid-tight paperboard cartons replacing conventional glass containers.

I. RECENT TRENDS IN ASEPTIC PACKAGED FOODS IN OVERSEAS COUNTRIES AND JAPAN

A. The United States

Since 1919, when a trial was made by Dole Corporation for employing aseptic metal cans as food containers, the American food industry has aggressively grappled with the problem of aseptic packaging of foods. In particular, liquid-type foods (including coffee milk), puddings, and processed milk products make up the major portion of all aseptic packaged foods, production of which has greatly increased year after year.

In addition to this, some aseptic packaging techniques (Shibazaki and Yokoyama, 1983) were also early introduced to the fields of processed meat products as solid foods. In particular, the in-line aseptic packaging system developed by Oscar Meyer has caused widespread aseptic packaging of sausages all over the country.

Even in the United States, an advanced country in the aseptic food packaging area, the application of liquid-tight paperboard cartons to aseptic milk had not been approved by the Food and Drug Administration (FDA) for a long time. In 1981, however, the FDA approved several companies including Brikpak to sterilize microorganisms resident on paperboard cartons by making use of hydrogen peroxide. As a result, several kinds of aseptic packaged beverages have been introduced to the U.S. market. Representative examples include long-life milk from such dairies

Chapter 12. Aseptic Packaged Foods

as Real Fresh and Dairymen, grape and orange juices from such large food processors as Coca-Cola (Peters, 1982) and General Foods, and wine from Rideout Wine (Technical Reports, 1983). In this situation, however, the competition has become more severe among metal cans, glass bottles, and paperboard cartons in the beverage industry.

The aseptic fill/packaging technique has also been applied to package foods for institutional uses in the dining-out industry. For example, most tomato purees and condensed soups are aseptic packaged in bag-in-boxes. McDonald has very recently begun to aseptic fill/package raw materials for milk, ice cream, and shake mix on the aseptic packaging system from Asepak and freeze them to distribute them to retailers throughout the country. However, the company may distribute them at room temperature in the future.

In addition to consumer packaged foods, foods for institutional uses are also aseptic packaged widely in the United States. This example can be seen in tomato pastes aseptic packaged in large-size bag-in-boxes from Sholle and Fran Rica to ship them to food processing factories. This aims at considerable reduction in all the costs required for packaging and distribution through the use of aseptic fill/packaging methods.

B. Europe

The first experiment on aseptic milk packaging was carried out in Switzerland in 1951, and subsequently it was industrialized by Tetra Pak AB in Sweden in 1961. This is called "long-life milk," and has been widely produced in every country throughout Europe. Long-life milk made up 65% of all packaged milk shipments on the Continent and 40% on average all over the Western European countries in 1980.

An Italian company, La Parmalat (Yokoyama, 1982a), was the first food processor to try to aseptic package dairy products on the large scale. In addition to long-life milk, the company now produces aseptic packaged cream cheese, yogurt, and desserts. Furthermore, the aseptic packaging opertion for tomato pastes has also represented an active movement in Italy. For instance, tomato pastes are already aseptic packaged in sterilized metal drums to ship to food processing factories.

The West German food industry has also made efforts to introduce aseptic fill/packaging techniques. The result can be seen in long-life milk by the Tetra Brik and Combiblok systems and fruit yogurt and puddings by the Hamba and Gasti processes.

In Switzerland, various aseptic packaged milk and fruit drinks are now in the market. In particular, fruit drinks aseptic packaged in plastic cups are stored in the showcase at less than 10°C.

In France, the Multipack container (News/Trends, 1982) was developed by Mead Emballage. Different from Tetra Brik, the Multipak has three 1-litre containers aseptic packaged milk or fruit drink multipacked with a serrated cardboard. Thus, it is appraised highly as a package for home and institutional uses.

In Sweden where Tetra Pak was born, Arra succeeded in the application of aseptic packaging techniques to reduced-fat soft butter. This technique is now utilized in every country throughout the world. In Japan also, emphasis is given to the application of aseptic packaging techniques to such solid foods as meat products and daily dishes.

In Eastern Europe and the Soviet Union also, manufacturers have aggressively grapped with aseptic packaging of dairy products. For example, in the Soviet Union, an aseptic packaging plant for baby foods was constructed in Moscow in late 1982. The plant is said to produce aseptic packaged milk-based baby foods at the rate of 80 tons per day through the use of sterilizing equipments and aseptic fill/packaging machines operable for 24 h running from Alfa Laval and Tetra Pak, respectively.

As mentioned previously, aseptic packaged foods have been further diversified. Particularly in European countries, they range from long life milk, yogurt, and puddings to fruit drinks, baby foods, tomato pastes, and solid foods.

C. Japan

The history of aseptic food packaging (Yokoyama, 1982b) is rather brief in Japan. Nevertheless, various kinds of aseptic packaged foods have been rapidly developed in this country. Representative products now on the market include long-life milk produced on the Tetra Pak and Pure-Pak systems, coffee milk and puddings produced on the aseptic fill/packaging systems from CKD, Dai-Nippon Printing, and Bosch and Hamba, and fruit drinks, Japanese rice wine, and soy bean milk packaged aseptically in liquid-tight paperboard cartons.

In addition, ketchup, mayonnaise, and soft butter are also aseptic packaged in sterile containers. Two types of sterile containers now in use were originally developed in Japan. One is designed to be aseptic even at the room temperature because its mouth is airtight, sealed immediately after it is blow molded. The other is designed to be reconditioned aseptic in the packaging process because it is heat sterilized prior to the fill/packaging operation.

In aseptic packaging of solid foods, Japan is proud to have top position in line with the United States in the world. A wide variety of solid foods are aseptic packaged on the unique aseptic fill/packaging systems developed in Japan, including sliced ham, cheese, rice cake, and processed seafoods. In addition to consumer foods, institutional foods are also being aseptic packaged on a bulk basis. In fact, some aseptic packaged institutional foods are already marketed by food processors.

II. ASEPTIC FOOD PACKAGING SYSTEMS

In general, aseptic food packaging systems are composed of food sterilizers, aseptic fill/packaging machines, packaging material washing sterilizers, and bioclean rooms. The following description is of food sterilizers and new aseptic food packaging machines.

A. Food Sterilizers

Liquid foods such as milk are heat sterilized at high temperature for a short time to kill microbes before they are aseptic packaged. For example, in case of long-life milk, raw milk is processed at 135–150°C for 2–6 s on ultra-high-temperature sterilization equipment. This UHT sterilization process is divided into two types: direct and indirect heat sterilization processes. Table I shows the UHT sterilization processes (Yokoyama, 1984b) now available in every country of the world.

In the direct heating process, since the products are inclined to dilut due to the condensation caused by steam, water equivalent to condensed water must be removed from the products. This operation is usually carried out by flashing the products in a vacuum chamber. In Japan, this direct heating process is used in the manufacture of processed milk. On the other hand, tomato pastes and concentrated fruit drinks are sterilized on the Contherm from Alfa Laval, which is equipped with a scraper to prevent the foods from parching.

In recent years, continuous-operation and long-running types of sterilizers are being developed in every country throughout the world with progress in demand for aseptic fill/packaging of high-viscosity flowing foods. This can be seen in a UHT sterilizer for high-viscosity flowing foods ranging from 3000 to 200,000 cps developed in Japan. This equipment is well designed so that high-viscosity flowing foods can be instantaneously and evenly heated with mixed steam in the high temperature of 100–150°C and can then be immediately cooled through the evaporation of water in the vacuum system.

In the aseptic packaging operation of sliced hams, raw materials are sterilized on the continuous-operation ultrasonic washing sterilizer prior to the slicing operation to kill inherent microbes. In case of loin hams

Table I UHT Sterilization Processes Now Available Worldwide

Process	Brand	Manufacturer	Country
Direct heating			
Injection	Uperiser	APV	United Kingdom
	VTIS	Alfa Laval	Sweden
	Aro Bac	Cherry Burrel	United States
	UHT	Iwai Machine	Japan
Infusion	Pararistor	Paasch & Silkeborg	Denmark
	Thermo Vac	Brell & Martell	France
	Vac-Heater	Cremery Package	United States
Indirect heating			
Plate	Sigma	Cherry Burrel	United States
	Thermodule	Alfa Laval	Sweden
	Ultramatic	APV	United Kingdom
	Ahlborn	Ahlborn	West Germany
	Ster-in 3 UHT	Fran	Italy
	Steriglak	Sordi	Italy
	Crescent	Cremery Package	United States
	UHT	Hisaka Mfg.	Japan
	UHT	Iwai Machine	Japan
Tubular	Sterideal	Stor	Holland
	Thermutator	Cherry Burrel	United States
	Spiratherm	Cherry Burrel	United States
	CJ-Ste-Vac-Heater	Chester Jensen	United States
	CTA	Cremery Package	United States
Scraping	Contherm	Alfa Laval	Sweden
	Thermo Cylinder	Iwai Machine	Japan
	Thermutator	Cherry Burrel	United States
	Votator	Votator	United States
	Scraped Surface Heater	Fran Rica	United States
	Rototherm	Tito-Manzini & Figli	Italy

packaged with fiberous casing materials, common germs and bacilli will completely die out by washing the surface of hams with alkali-based surface activator and then sterilizing it with chlorite soda.

B. New Aseptic Fill/Packaging Machines

Worldwide aseptic fill/packaging systems can be represented by the Tetra Brik, Pure-Pak, and Combiblok paperboard cartons and the Hassia and Bosch (see Fig. 1) plastic containers. In addition to this, however, new different aseptic fill/packaging systems are also being used by a number of food processors in all parts of the world.

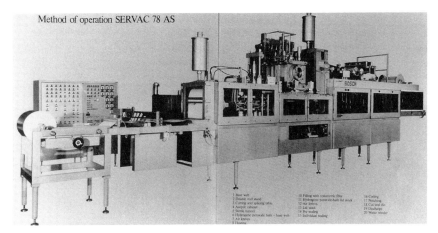

Figure 1. The Servac 78 AS aseptic fill/packaging machine from Bosch.

1. New Aseptic Fill/Packaging Machines for Consumer Packs

Table II gives representative new aseptic fill/packaging machines (Yokoyama, 1984a) for consumer packs. These new machines employ hot air, steam, alcohol, and ultraviolet rays as sterilizing media, replacing the conventional hydrogen peroxide. Of these, the most unique system is the Erca Conoffast (Aseptic Technology, 1982) developed jointly by a French company, Erca, and an American company, Continental Can. This system can be characterized by the following processes: (1) two-layer PE/PP sheets are firstly relased from multilayer coextruded PE/PP/PE/PVDC/PE sheets; (2) containers are next molded from remaining aseptic three-layer sheets; (3) the products are filled in sterile containers and sealed with aseptic lidding materials; and (4) consequently no hydrogen peroxide is required to sterilize packaging materials.

Vacuum and gas-flash packaging machines are also widely used in aseptic packaging of processed meats, dairy products, and rice cakes because they have bioclean mechanisms.

Worldwide aseptic packaging machines for processed meats can be represented by the two systems from Oscar Meyer and Omori Machine. In particular, Oscar Meyer's system is designed so that aseptic plastic films can be molded in the in-line process and the sterilized meat products can then be vacuum packaged in a continuous operation. On the other hand, the off-line aseptic packaging system developed by Omori Machine can be operated together with slicers, weighers, and inserters.

Table II New Aseptic Fill/Packaging Machines for Consumer Packs

System	Country	Packaging materials	Contents	Features of Machines and Sterilization Methods of Microbes
Dole	United States	Composite and metal cans	Pastes, puddings, and fruit drinks	Change from the conventional steam sterilization (221–226°C) to hot air sterilization (116°C); temperature of the heating source is 316°C
Metal Box	United Kingdom	Aluminum and plastic cups	Puddings, pastes, and fruit drinks	Molded cups are sterilized with 35% hydrogen peroxide sprays and then dried with hot air
Liqui Pak	United States	Paperboard cartons	Milk and fruit drinks	Formed paperboard cartons are first sterilized with 1 cm^3 of 0.1% hydrogen peroxide sprays, then sterilized with ultraviolet rays and then dried with hot air.
Erca (Conoffast)	France (United States)	Plastic multilayer sheets	Fruit drinks, puddings, and coffee milk	After two-layer sheets are released from the composite material, remaining aseptic several-layer sheets are molded into containers
Hassia	West Germany	Plastic multilayer sheets	Yogurt and desserts	Packaging materials are sterilized with steam at the high temperature of 130–150°C
Hamba	West Germany	Plastic multilayer sheets	Yogurt, puddings, and fruit drinks	Packaging materials are sterilized on the high-performance UV sterilizer
Thimonnier	France	Roll-type plastic films	Milk, fruit drinks, and pastes	Packaging materials are sterilized with both alcohol and ultraviolet rays
Serac	France	Glass or plastic bottles	Yogurt sauce and flowing foods	Bottles are sterilized with hydrogen peroxide and then dried at 138°C for 6 min; products are filled into sterile bottles in the aseptic chamber

2. New Aseptic Fill/Packaging Machines for Institutional Packs

Table III lists new aseptic fill/packaging machines (Yokoyama, 1984a) for institutional packs. The Asepak system can aseptic fill/package sterilized spaghetti sauces and ketchups with aseptic packaging materials. The Tito Manzini system is suitable for aseptic fill/packaging tomato pastes for the dining-out industry and other food processing factories. The Steriglen is an aseptic fill/packaging system in which the sterilizer for solid foods can be operated together with the aseptic fill/packaging machine at the same time. This allows the operators to continuously sterilize foodstuffs for institutional uses and aseptic fill/package them at speeds of 180 packs per hour in terms of 25 kg/pack simply by presetting a built-in microprocessor.

Both the Fran Rica and Sholle systems can aseptic fill/package sterilized tomato pastes, fruit drinks, and peanut pastes in 114- to 1140-litre gamma-sterilized bags. Representative packaging materials in use include 150-μm coextruded LDPE/PVDC/LDPE/PVDC/LDPE/EVA-LDPE films or aluminum-metallized PET bags.

The APV and Londreco (Herson and Shore, 1981) canning system can separately sterilize solid and liquid foods at a time, mix individual sterilized foods, and aseptic fill/package them together in sterile metal cans.

III. MANUFACTURING METHODS OF ASEPTIC PACKAGED FOODS

A. Complete Aseptic Packaged Foods

1. Dairy Products

Raw milk contains such bacteria as *Bacillus, Clostridium, Sporolactobacillus,* and *Desulfotomaculum* unless it is sterilized on the UHT equipment. Together *B. licheniformis* and *B. cereus* occupy almost 80% of all the *Bacillus* genus, which includes heat-resistant *B. stearothermophilus* (Kamei, 1982). For *B. Stearothermophilus,* 71.4% of all bacterial spores are reported to survive when raw milk is sterilized at 120°C for 10 min. However, if it is sterilized at 130°C for 10 min, they will completely die out.

In Western European countries such as West Germany and France, aseptic packaging opertions of liquid milk are widely carried out, where long-life milk is popularly sold in supermarkets (as can be seen in Fig. 2). Even if aseptic packaged milk continued to be stored in the room temperature (20–27°C) for a month, it would never degrade in quality.

In general, long-life milk is sterilized at ultra-high temperature (140–145°C) for a short time (2–6 s). In Japan, the indirect heating plate system

Table III New Aseptic Fill/Packaging Machines for Institutional Packs

System	Country	Packaging Materials	Contents	Features of Machines and Sterilization Methods of Microbes
Asepak	United States	Roll-type plastic tubular films	Meat sauces and tomato ketchups	Blown-molded tubular films are used without hydrogen peroxide and ultraviolet rays.
Tito Manzini	Italy	Plastic bags	Tomato pastes	Bags are sterilized with gamma rays; filling head is sterilized with germicides at 70°C for 30 s
Steriglen	Australia	Plastic bags	Block-type fruits and meat products	System is composed of sterilizers and coolers for solid foods, sterile bags, and aseptic fill/packaging machines
Fran Rica	United States	Plastic bags	Vegetable and fruit drinks, yogurt, and tomato pastes	Bags of 114 to 1140 liters sterilized with gamma rays are utilized; filling section is sterilized with steam for 2 min
Sholle	United States	Plastic bags	Tomato pastes and papaya purees	Bags of 19 to 1100 liter sterilized with gamma rays are utilized; filling section is sterilized with 80% ethyl alcohol or chlorite soda solution
APV and Londreco	United Kingdom and Italy	Metal cans	Meat- and vegetable-containing soups	Contents are sterilized on the DCAPV sterilizer from APV; metal cans are sterilized with steam at 221–226°C

Figure 2. Long-life milk aseptic packaged in liquid-tight paperboard cartons.

is mainly used in the sterilization of liquid milk. Both the Tetra Brik and Pure-Pak systems find an overwhelming application in the aseptic milk fill/packaging fields, where hydrogen peroxide is exclusively employed as a sterilizing medium for packaging materials.

Some coffee milk, puddings, and fruit juice-containing dairy products are also aseptic fill/packaged in plastic containers (as shown in Fig. 3). In this packaging process, plastic sheets are first sterilized with germicides, containers are then thermoformed from aseptic plastic sheets, the prod-

Figure 3. Aseptic fill/packaged fruit juice-containing dairy products.

ucts are aseptic filled in the containers, the filled containers are sealed with aseptic lidding materials, and the ears of the containers are finally trimmed off. Major suppliers of these packaging systems include Hassia (West Germany), Höffliger & Karg (West Germany), Bosch (West Germany), and Prime Pack (United States) in overseas countries and CKD and Dai-Nippon Printing in Japan.

2. Fruit Drinks

In fruit drinks there exist not only bacteria such as *Acetobacter xylium, A. malanogenus, Lactobacillus pantarum, Leuconostoc, Bacillus,* and *Microbacterium* but also molds such as *Penicillium expansum* and *Aspergillus nidulans.*

The aseptic fill/packaging method of fruit drinks is basically the same as that of liquid milk. However, since fruit drinks are less than 4.0 in pH, the objects of sterilization are yeasts, molds, and lactobacilli, among others. For this reason, the sterilizing operation is usually carried out at less than 100°C maximum. For example, Unshu-grown orange juices are cooled at about 20°C immediately after being sterilized at 93–100°C for 5–20 s, and they are then aseptic fill/packaged in liquid-tight paperboard cartons. Paperboard cartons are in most cases sterilized with hydrogen peroxide to kill resident microbes.

Major aseptic fill/packaging machines are represented by the AB-200 and AB-500 systems from Tetra Pak and the Aseptic Hypa-S system from Bosch. Particularly in the Hypa-S system, square-type composite cans are formed from foil-laminated paperboards and then sterilized with hydrogen peroxide, and in these, fruit juices with pH of less than 4.5 are aseptic fill/packaged (as can be seen in Fig. 4). It is said that the products packaged in the Hypa-S composite cans reach about 2 years in shelf life, even at room temperature.

3. Japanese Rice Wine (Sake)

Japanese rice wine is mostly packaged in glass bottles, metal cans, and PET bottles. In addition to this, however, it has also been aseptic fill/packaged in liquid-tight paperboard cartons (as shown in Fig. 5).

Japanese rice wine is an acid food with pH of about 4.5 and less microbes because it contains 10–20% alcoholic ingredients. Thus, in the aseptic fill/packaging process, a germ in question is *Lactobacillus plantarum,* which is one of the lactobacilli. This is because *L. plantarum,* which is represented by *L. heterohiochii, L. homohiochii, L. fermentum,* and *L. acidophilus,* continues to grow even in Japanese rice wine containing 20% alcohol. In this situation, Japanese rice wine is aseptic fill/packaged in

Chapter 12. Aseptic Packaged Foods

Figure 4. Aseptic fruit drinks packaged in Hypa-S composite cans.

liquid-tight paperboard cartons in the bioclean room immediately after being heat sterilized at 70°C on the heat exchanger and cooled in the cooler.

B. Commercialized Aseptic Packaged Foods

1. Soft Butter

There have frequently occured blood clots caused by overweight and cholesterol among people in overseas countries. In this situation, most people have eaten too much butter, but they are now changing to marga-

Figure 5. Japanese rice wine aseptic fill/packaged in liquid-tight paperboard cartons.

rine and soft butter with reduced fat. Especially in Sweden, soft butter called "L & L" is reported to show high consumption because it is reduced by half in fat and increased in water and protein compared to conventional butter. In Japan, two kinds of soft butter products are now in the market, including L & L produced domestically under a technical license from Sweden and Elm developed on the domestic technology. Either the L & L or the Elm is aseptic fill/packaged in sterile plastic containers in the bioclean room to prevent the growth of inherent microbes.

2. Processed Meat Produces

Of the processed meat products marketed in the United States, sliced sausages, sliced bacon, and frankfurters are not always reheated after they have been vacuum packaged in the aseptic conditions. In Japan also, most processed meat products are vacuum packaged in the bioclean room and then distributed at low temperature throughout the country because non-reheated products are better tasting.

A great number of *Bacillus* species are contained in ham products (Yokoyama, 1978), followed by *Staphylococcus, Micrococcus,* and *Lactobacillus*. To aseptic package these processed meat products, much consideration should be given to how to check microbes contained in foodstuffs and how to sterilize them in the smoking and slicing processes. Figure 6 illustrates sliced ham semi-aseptic packaged in the bioclean room.

C. Aseptic Packaged Foods for Institutional Uses

Emphasis is laid on aseptic fill/packaging of tomato pastes and papaya purees (Technical Reporting, 1982) for institutional uses in overseas

Figure 6. Semi-aseptic packaged sliced hams.

countries. For example, a papaya puree marketed by Amfac Tropical Products Corp. in the United States is processed in the following steps: (1) papayas are mixed together; (2) the mixed products are sterilized at 94°C for 2 s on the UHT sterilizer from Cherry Burell; (3) the sterilized products are rapidly cooled to 27°C; and (4) the cooled products are aseptic fill/packaged on the Auto Fill X-1 aseptic equipment from Sholle Corp.

Papayas must be heat sterilized at high temperature for a short time because they contain not only germs, molds, and yeasts but also ferments. Barrier-type PE/Al-metallized PET/PE laminated bags are mainly used as packaging materials.

IV. FUTURE TRENDS IN ASEPTIC PACKAGED FOODS

In general, all kinds of foods have decreased in preservation with increase in restraint in food additives and progress toward reduced salt and sugar. This further promotes the aseptic food packaging operation. Of all packaged foods, aseptic packaged foods are still lowest in market share. However, they are gradually increasing in percentage year by year. In the United States, an average annual growth rate of packaging materials stands at 3–4% for general-use foods but no less than 30% for aseptic packaged foods. In Japan, aseptic packaged foods are also expected to show a sharp growth in future.

In the aseptic packaging fields, liquid-tight paperboard cartons and plastic materials now make up the largest share of all packaging materials. In addition to this, metal cans, composite cans, and glass containers will also find a considerable application in this area. Furthermore, aseptic packaged foods will present much demand not only in the consumer market but also in the institutional market in the near future.

REFERENCES

Aseptic Technology. (1982). U.S. aseptic boom bred in Europe. *Food Eng.* **March,** 44–53.

Herson, A. C., and Shore, D. T. (1981). Aseptic processing of foods comprising sauce and solids. *Food Eng.* **May,** 53–62.

Kamei, T. (1982). Aseptic packaging for milk using paper carton. *Jpn. Food Sci.* **21**(5), 31–43.

News/Trends. (1982). Aseptic multipack wines French "OSCAR". *Packaging Eng.* **February,** 10–11.

Peters, J. W. (1982). Cost squeeze spurs ideas on food package options. *Packaging Eng.* **January,** 35–39.

Shibazaki, I., and Yokoyama, M. (1983). "Food Packaging Course." Nippo Co., Tokyo.
Technical Reports. (1982). Aseptically processed papaya. *Food Eng. Int.* **March,** 55–59.
Technical Reports. (1983). Wines aseptically packaged. *Food Eng. Int.* **December,** 10–11.
Yokoyama, M. (1978). Semi aseptic packaging technique and its packaging system for meat products. *New Food Ind.* **20**(7), 29–38.
Yokoyama, M. (1982a). Recent trend of aseptic packaged foods in the world countries. *Canzume Jiho* (The Canners Journal) **61**(11), 10–19.
Yokoyama, M. (1982b). Recent trend of packaging technique. 1982 Food Industry Technical Conference, Japan Food Machinery Manufacturers' Association.
Yokoyama, M. (1984a). Aseptic packaged foods and aseptic packaging system. *Chem. Biol.* **20,** 780–787.
Yokoyama, M. (1984b). The present condition of bactericidal equipment for aseptic packaging in the world. *Food Packaging,* **28**(5), 64–73.

CHAPTER 13

Free Oxygen Scavenging Packaging

Yoshihiko Harima
Mitsubishi Gas Chemical Company, Inc.,
Tokyo, Japan

I. INTRODUCTION

The oxygen absorber is a unique technique of preserving food, developed in Japan. Deterioration in food quality is caused chiefly by oxygen, which means that quality can be preserved fairly well if we can remove oxygen. Vacuum packaging and nitrogen gas packaging methods are already in use based on this idea. Oxygen cannot be removed completely with these methods, however, as they aim to eliminate oxygen physically. Methods to get rid of oxygen chemically were already being considered in the 1920s, and in the 1960s, the palladium-catalyzed oxygen scavenging system was developed in the United States. This method, however, is used only to culture anaerobes, but not for food preservation purposes, due to high cost and complex operations.

The oxygen absorber, which proved to be effective in food preservation, was first developed in Japan in 1976. In a decade, the oxygen absorber has spread throughout the food industry in Japan, to be used in various kinds of food packaging. In these few years, the use of the product has started in overseas countries such as Southeast Asian nations and the United States, and is currently attracting attention in Europe.

The technical development of the oxygen absorber has made considerable gains since it was first introduced, producing various types according to uses. Free oxygen scavenging packaging, including these new types of oxygen absorbers, is explained in detail in this chapter.

Food Packaging Copyright © 1990 by Academic Press, Inc. All rights of reproduction in any form reserved.

II. SUMMARY OF OXYGEN ABSORBER

A. Principles

The oxygen absorber is composed of easily oxidized substances packed in an air-permeable material (Fig. 1). By sealing this pack inside an airproof container with food, the oxygen inside the container is absorbed in about 24 h, producing an oxygen-free state (less than 0.01% oxygen concentration) inside the container. As a result, the food becomes completely free from the impact of oxygen during preservation.

B. Effects

The oxygen absorber completely eliminates molds and aerobic bacteria, as they cannot grow without oxygen (Fig. 2 and Table I).

The oxygen absorber is also effective in preventing damage by bugs, as they cannot survive without oxygen. After 2 weeks of oxygen-free conditions, the eggs are killed as well, so that they will not hatch even after the package has been opened.

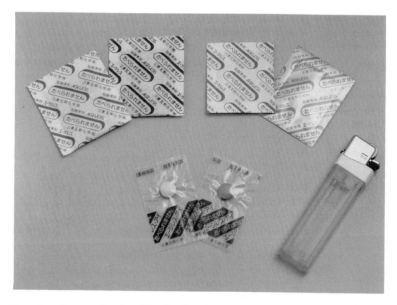

Figure 1. Oxygen absorber "Ageless" and oxygen sensing agent "Ageless-eye."

Chapter 13. Free Oxygen Scavenging Packaging

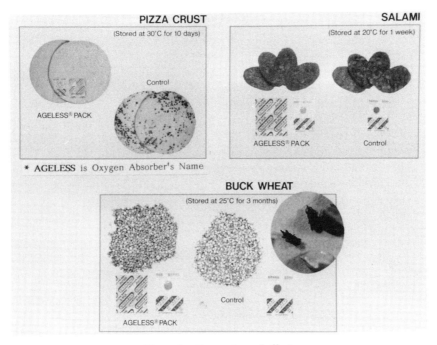

Figure 2. Comparison of effect.

Table I Growth of Mold on Bread and Oxygen Concentration (at 25°C)

Composition of Gas Filled (%)		Mold Growth with Time						
O$_2$	N$_2$	2nd day	4th day	6th day	8th day	10th day	12th day	14th day
0[a]	100	—[d]	—	—	—	—	—	—
0.2	99.8	—	—	—	—	—	—	+
0.4[b]	99.6	—	—	—	—	+	+	+++
0.6[b]	99.4	—	—	+	+++	+++	+++	+++
20.9[c]	79.1	—	+	+++	+++	+++	+++	+++

[a] Oxygen absorber level
[b] N$_2$ gas purge level
[c] The concentration of oxygen in ambient air.
[d] Key: —, no mold colonies found; +, some; +++, many.

Fats that oxidize easily, such as unsaturated fatty acids, do not oxidize in the absence of oxygen. Oxidation is the chief cause of destruction in nutritive values and resolution of taste and flavor components, which causes deterioration in taste and flavor (Fig. 3).

Change in color of foods is caused chiefly by oxidation. Light and heat only accelerate the process, so without oxygen, foods do not change color from oxidation when they are exposed to light.

Vegetables and fruits continue to ripen through their respiratory action even after they are harvested. This phenomenon can be prevented by lowering the oxygen concentration and temperature.

Apart from food, rusting of metal products can also be prevented, as the rusting process is also caused by oxidation reaction.

The combination of these effects helps maintain the quality and freshness of food, which contributes greatly to various advantages in marketing (Fig. 4).

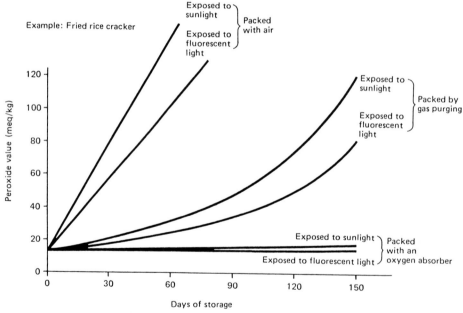

Figure 3. Effectiveness of an oxygen absorber in preventing deterioration of oil exposed to light.

Chapter 13. Free Oxygen Scavenging Packaging

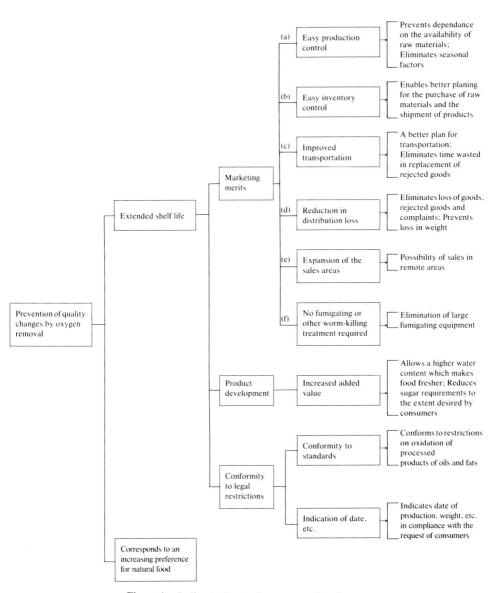

Figure 4. Indirect effects of an oxygen absorber.

C. CHARACTERISTICS OF OXYGEN ABSORBER PACKING

1. Causes of Quality Deterioration Eliminated

All conventional food preservation methods have been plagued by the influence of oxygen not being avoided completely. On the other hand, the oxygen absorber exterminates the bad influence of oxygen from the start by getting rid of the oxygen that causes deterioration in quality. This fact fundamentally differentiates this method of quality preservation from the others, which aim to avoid each phenomenon in a piecemeal way.

2. Perfect Oxygen Absorption

As oxygen is absorbed and removed chemically, the oxygen absorber can turn the inside of a container into a perfectly oxygen-free state, unlike physically conducted methods. Also, the inside of a container can be kept in the oxygen-free state for a long time, as the surplus capacity definitely absorbs the oxygen entering from the outside. This is possible regardless of the form of food (power form, grain form, sponge form, etc.).

3. Complete Safety

Unlike food preservatives such as antiseptics or antioxidants, oxygen absorbers are safe as they are not mixed with food to be eaten. Needless to say, the material itself if also safe. This method is safe also in the sense that it prevents the generation of peroxides and aflatoxins (from toxic fungi), which are thought to cause arterial sclerosis or cancer.

4. Unchanged Taste

Depending on the type or quantity, food preservatives tend to change the taste and flavor of food, and retort packing tends to change the flavor or color by reheating already processed food. The oxygen absorber, on the other hand, can maintain the original taste, flavor, and color, as the food does not undergo reprocessing.

5. Food with Natural Flavor

Salt and sugar not only adjust the taste of the food, but they also function in keeping down water activity as well as the breeding of microorganisms. Due to this, it has been difficult to reduce the salt and sugar content in foods as deemed necessary by public demand. However, salt and sugar can now be reduced due to the oxygen absorber's inhibitory action against aerobic microorganisms, which, as a result, enable the manufacture of more natural, fresh-tasting foods with more moisture and softness.

6. No Mechanical Equipment Required

Almost no mechanical equipment and hence no equipment costs are necessary, as the only process is sealing food after injecting in the food container. This method can be readily applied to products with large gaps between peaks and bottoms, or to a small-volume production. An automatic injector, which can be easily connected to existing packers, is available for large production.

D. Uses

The oxygen absorber can be applied to all sorts of food vulnerable to the adverse effects of oxygen. However, the product proves particularly effective in foods that are liable to mold growth, change in color, change in flavor due to oxidation, and production of toxic lipid peroxides.

III. TYPES OF OXYGEN ABSORBER

Oxygen absorbers can be classified into different categories as shown in Table II.

A. Classification According to Material

In theory, any material that combines easily with oxygen, or oxidizes easily in other words, can be used as the oxygen absorber. But because the oxygen absorber is used mainly for food preservation, the material is restricted by various conditions: it must be safe, be handled easily, not produce toxic substances or gas with offensive smell, be economically priced, compact in size, absorb a large amount of oxygen, and have an appropriate oxygen absorption speed. Iron powder, ascorbic acid, or catechol is used in most of the existing oxygen absorbers. Iron powder is most frequently used among these, as ascorbic acid is costly, and there are some safety problems with others, despite the fact that organic materials have an advantage in passing metal detectors easily. An oxygen absorber with hydrosulfite was once produced before the iron powder type was developed, but it was not commercialized because handling was difficult due to fast reaction speed and it produced ill-smelling gas such as hydrogen sulfide and sulfur dioxide.

Table II. Classification of Oxygen Absorber

A. Classification according to material
1. Inorganic—iron powder
2. Organic—ascorbic acid, catechol

B. Classification according to reaction style
1. Self-reaction time
2. Moisture-dependent type

C. Classification according to reaction speed
1. Immediate effect type
2. General type
3. Slow effect type

D. Classification according to use
1. For very moist food
2. For moderately moist food
3. For low-water food
4. For extra dry food

E. Classification according to function
1. Single function type—oxygen absorption only
2. Composite function type
    a. Oxygen absorption and carbon dioxide generation
    b. Oxygen absorption and carbon dioxide absorption
    c. Oxygen absorption and alcohol generation
    d. Oxygen absorption and others

B. Classification According to Reaction Style

Water is essential for oxygen absorption reaction. In the self-reaction type the water required for the reaction is added, while in the moisture-dependent type, the oxygen absorption reaction takes place by using the water transpired from food. The latter is easy to handle and highly workable, as it does not react by exposure to air. It also absorbs oxygen quickly after sealing, and oxygen can be absorbed in 1/2 day to 1 day in some products. It is not appropriate, however, for dried foods with low water activity, as oxygen absorption becomes very slow. It is designed to have water resistance for use in very watery food. Some products have the capacity to absorb oxygen even under water.

The self-reaction type should be handled carefully, as the oxygen absorption reaction starts as soon as it is exposed to air. The reaction speed can be controlled by adjusting the permeability of packing material, amount of material, and blending methods. The types vary from immediate effect type to slow effect type.

Chapter 13. Free Oxygen Scavenging Packaging

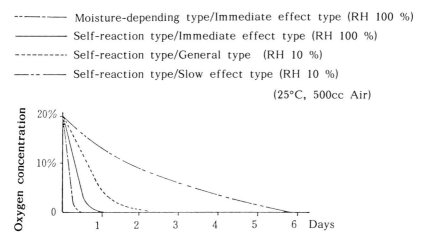

Figure 5. Deoxidization speed of iron type oxygen absorber.

C. Classification According to Reaction Speed

The average time it takes for currently market oxygen absorbers to absorb oxygen is 1/2 day to 1 day with the immediate effect type, 1–4 days with the general type, and 4–6 days with the slow effect type. The time depends on temperature, water activity of food, and whether it is used together with desiccants. In general, reaction of organic materials is slower than that of inorganic materials (Figs. 5 and 6). The impact of temperature is shown in Fig. 7). The oxygen absorption reaction does take place gradually under freezing temperatures. The capacity is restored after the temperature is returned to normal (Figs. 8 and 9).

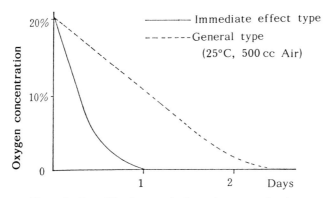

Figure 6. Deoxidization speed of organic oxygen absorber.

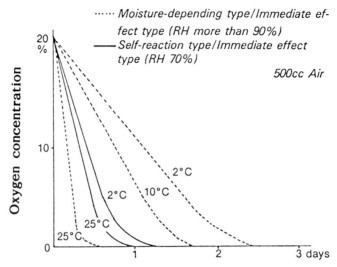

Figure 7. Change of oxygen-absorbing speed of iron type oxygen absorber with temperature.

In the past, oxygen absorbers have been used generally for foods that are distributed under normal temperature. As people became aware that preservation by freezing and refrigerating were not complete, we saw a notable move to cover this inadequacy by combining these methods with oxygen absorbers.

D. Classification According to Use

In general, foods with high water content are liable to mold growth and damage. The immediate effect type is therefore used in this kind of food, as

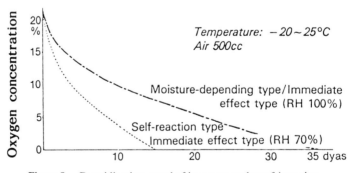

Figure 8. Deoxidization speed of iron type under refrigeration.

Chapter 13. Free Oxygen Scavenging Packaging

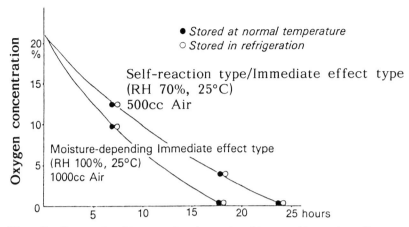

Figure 9. Oxygen-absorbing capacity when restored to normal temperture after oxygen absorber was stored under refrigeration for 20 days.

oxygen has to be absorbed quickly. Dried foods, on the other hand, have to be preserved under dry conditions since they are vulnerable to moisture but relatively free from damage by microorganisms. Under extremely dry conditions, however, the reaction may stop halfway after the water required for oxygen absorption is used up. To prevent this, types of foods with low water content and for dried foods have been developed. Although they have a slow reaction speed, they can be used under dry contitions, and are highly workable.

The medium moisture type is used whan a moderate speed of oxygen absorption is required but is not used under dry conditions. The relationship between materials, reaction style, reaction speed, and use is shown in Table III.

Table III   Interrelation of Types

Material	Reaction Style	Reaction Speed	Use
Inorganic	Self-reaction type	Immediate effect type	For highly moist food
		General type	For moderately moist food
			For low-water food
		Slow effect type	For very dry food
	Moisture-dependent type	Immediate effect type	For highly moist food
		General type	
Organic	Self-reaction type	Immediate type	For moderately moist food
		General type	For low-water food

E. Classification According to Function

Some types of oxygen absorbers have the sole function of oxygen absorption, while others incorporate other functions as well. Although we had only the single-function type at the start of development, those with composite functions have been developed in response to users' needs.

1. Carbon Dioxide Generating Type (Fig. 10)

Because oxygen makes up 20% of the air, the volume of a container decreases by one-fifth after oxygen absorption. This contraction may be regarded as a merit in many products: it serves as a criterion of oxygen absorption, the food is less likely to move inside the container and to be damaged, or it is easier to pack because of contraction. On the other hand, the change in volume is not welcomed in some products owing to poor appearance. To solve this problem, an oxygen absorber that releases roughly the same amount of carbon dioxide as that of absorbed oxygen was developed. Ascorbic acid and iron carbonate are used as materials.

There are two methods of carbon dioxide generation: The type where oxygen absorption and carbon dioxide generation are linked (carbon dioxide is not produced unless oxygen is absorbed), and the type where the reactions occur separately (production of carbon dioxide is not related to existence or absorption of oxygen). There is no problem with the former, but the latter has many disadvantages, as there exist cases where the produced carbon dioxide enlarges the outer packing even if preserved in an anerobic condition, or when surplus capacity remains after absorbing

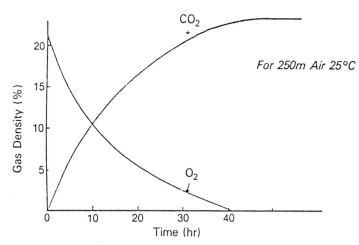

Figure 10. Change of gas composition of general carbon dioxide generating type.

oxygen. Thus caution is advised when using this type of product. The following points will also have to be taken into consideration.

a. Possibility of Accelerating the Growth of Anaerobic Bacteria The growth of anaerobic bacteria is generally accelerated with the existence of carbon dioxide. This is the reason why culture facilities of anaerobic bacteria in laboratories produce carbon dioxide while absorbing oxygen. Carbon dioxide is said to suppress the growth of bacteria, but this is truly only in the presence of oxygen. Under oxygen-free conditions, carbon dioxide may well accelerate the growth of anaerobic bacteria. Care must be taken when using this type of oxygen absorber in foods with high water content that are vulnerable to bacteria growth.

b. Possibility of Change in Taste Carbon dioxide is soluble in both water and oil, which indicates that when the product is used for foods with high water or oil content, the carbon dioxide could dissolve in these foods, causing changes in the food taste. Also, the amount of carbon dioxide produced could decrease through such solution, making it difficult to retain equal pressure.

c. Difficulty in Checking Decomposition Carbon dioxide is generally produced during the growth of microorganisms. Since the growth of microorganisms can therefore be detected at an early stage by checking the existence of concentration of carbon dioxide, this checking method has been widely used as a great convenience. The judgment becomes difficult, however, when this type of oxygen absorber is used, as carbon dioxide could be present from the start.

d. Use of Oxygen Detector is Impossible Oxygen detectors indicating the existence of oxygen through change in color have been developed. A detector, however, cannot be used jointly with this type of oxygen absorber since the color does not undergo a normal change with the existence of carbon dioxide. Along with the lack of contraction, this type of oxygen absorber deprives the detector of being able to check easily whether oxygen absorption has taken place.

2. Carbon Dioxide Absorption Type

This type of oxygen absorber, developed for the purpose of maintaining the aroma of coffee beans, absorbs not only oxygen but also carbon dioxide. When coffee beans are roasted, they discharge a large amount of carbon dioxide. If they were packed immediately after roasted, the carbon dioxide that comes out of the beans would expand and eventually break the package. To avoid this, coffee manufacturers have in the past made holes in the package to purge carbon dioxide or have exposed roasted beans to air until generation of carbon dioxide has stopped, which is called "aging." But aroma, the very essence of coffee, can be lost and oxidation can proceed with these methods. The old types of oxygen absorbers offered no fundamental solution to the problem, as they could be used only after carbon dioxide generation has stopped. So the new type of oxygen absorbers, which absorb carbon dioxide as well as oxygen, was developed.

A large amount of carbon dioxide is discharged over a long period of time, although beans differ according to the type and the degree of roasting. About 250 cm^3 of carbon dioxide is produced for 100 g of beans, as compared to 25 cm^3 of oxygen. This type of oxygen absorber is designed to absorb 10 times as much the amount of carbon dioxide as that of oxygen in a gradual manner. By using this design, the package is maintained in an airtight condition for a long time, since it continues to absorb carbon dioxide even after all the oxygen has been absorbed. Free from oxidation, the aroma of coffee cannot be spoiled and consumers can now enjoy the fresh aroma of roasted beans. Figure 11 and Table IV show the result of tests on coffee beans.

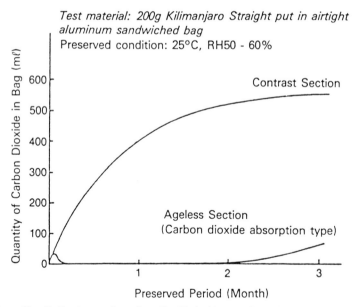

Figure 11. Daily changes in carbon dioxide inside container of roasted coffee beans.

Recently, packages with valves have been seen in the market and are said to preserve the coffee aroma. The package is designed so that the valve opens when the inside pressure rises, purging the generated carbon dioxide, but unfortunately the effect is not as remarkable as with the oxygen absorber. One reason is that the aroma is discharged outside together with the purged carbon dioxide. Second it cannot be completely free of oxidation even with vacuum packaging. The oxygen inside the roasted and now porous beans cannot be removed, while oxygen keeps on seeping in through the surface of the packing material, if not through the valve. As the oxygen absorber, on the other hand, removes oxygen by chemical methods, it also eliminates oxygen inside the food and oxygen that keeps entering from outside due to to its surplus capacity. This is the reason that the oxygen-free state can be maintained for a long time.

This type of oxygen absorber is used only for coffee at the moment, but its use is expected to cover other products that are damaged by oxygen and produce carbon dioxide (such as some types of antibiotics, vitamin tablets, and fermented products).

3. Alcohol Generation Type

This type of oxygen absorber, commercialized some time ago, is aimed at controlling the microorganisms that are relatively free from the effects of oxygen, such as yeast and bacteria, on which the ordinary oxygen absorber has little effect. It has not attracted much attention since it does not release much alcohol and is not as effective. In many cases, however, the quality preservative that only releases alcohol is used jointly with the oxygen absorber.

Table IV Organoleptic Test Result of Roasted Coffee Beans

		Preserved Period		
Classification	Item[a]	1 Month	2 Months	3 Months
Ageless section (carbon dioxide absorption type)	Aroma	3	3	3
	Taste	3	3	3
Contrast section	Aroma	2	1	1
	Taste	2–3	2	1

[a] Evaluation of aroma: 3; no oxidized smell; aroma of coffee is good and well preserved; 2; oxidized smell slightly, aroma of coffee becomes faint slightly; 1; oxidized smell, aroma of coffee is lost. Evaluation of taste: 3; remains nearly same taste as just after roasted; 2; somewhat oxidized taste; 1; taste has deteriorated.

IV. ADVANCES IN RELATED TECHNOLOGY

The essential conditions in the use of oxygen absorbers include packaging materials with a gas barrier effect and perfect sealing. Prominent technical advances can be seen in the development of related and secondary materials as well as these materials.

A. Airtight Containers

In the past, development of large and fixed-type completely airtight containers has been very difficult. In recent years, airtight drums made of stainless steel, airtight flexible containers, and super drums made of resin have been put to practical use to preserve or transport raw materials in large quantities. These containers are not used widely in industries (Fig. 12).

Figure 12. Ageles airtight drum (SUS-304, for 50 liters.)

B. Paper Eye (Oxygen-Sensing Paper)

An oxygen-sensing agent in tablet form was widely used at an early stage of development. It indicated the existence of oxygen through changes in color as a means to confirm the effect of oxygen absorber. The paper-type sensor, using the same substance printed on paper, was developed in 1985. As various designs and size are available, it fits naturally into the pack (Fig. 13).

C. Automatic Single-Pack Throw-In Machine

The automatic throw-in machine (the type that carries out charging while cutting the oxygen absorbers in belt-form) was developed in the past for large-scale use of oxygen absorbers. The automatic single-pack throw-in machine, in which oxygen absorbers are charged one at a time, was developed in 1985. The latter has an advantage over the former, as it is now free from the errors in cutting the oxygen absorbers in belt form (Fig. 14).

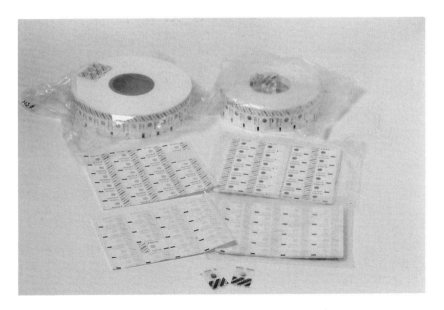

Figure 13. Oxygen-sensing paper (paper eye).

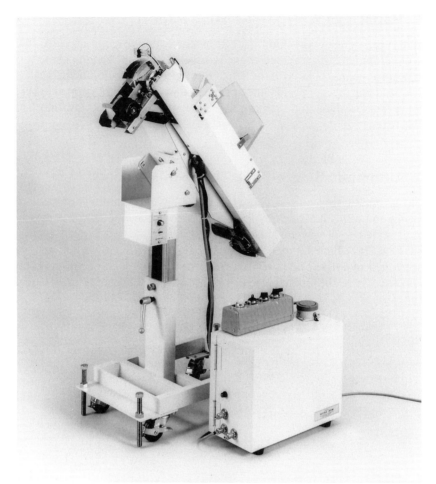

Standard Specifications of AG-B

Type	AG-B
Throw-in Capacity	Maximum 100 pieces/min.
Thrown-in Bag Form	Three-sided Four-sided seal Max. 60m/m x 60m/m
Dimension	L 715m/m x H 1260m/m x W 350m/m
Weight	40kg
Power Source	AC100V 50Hz/60Hz
Electric Power required	300VA
Air Pressure used	Not required
Air Quantity used	Not required

Figure 14. Automatic throw-in machine, AG-B for Ageless in pieces.

D. Fixing and Separating Oxygen Absorbers

As oxygen absorbers are inserted into the container together with the food, there is a possibility that they may mix into the food or even be cooked with the food. Although the material is completely safe, obviously it is not meant to be eaten. To prevent such mistakes, techniques to fix oxygen absorbers to packing materials or to separate them completely from food have been developed. In the former, the oxygen absorber is pasted to the packing material by a combination of automatic throw-in machine and hot-melt gun, and this technique is actually used in foods such as miso in cups, rice cake sweets, and coffee. Partitioned bags containing a thin film

Figure 15. Example of fixing of oxygen absorber.

Table V Growth of Fungi[a] That Can Grow under General Oxygen-Free Anaerobic Conditions, under Perfect Anaerobic Conditions

Group	Test Sample (Name of Fungi)	Media	Temperature (°C) and (Days)	Conditions of Oxygen Absorber Use					Existence of Growth				
				KON Film Packing		A1 Metallized Film Repacking		General Type Oxygen Absorber		CO$_2$ Generating Type Oxygen Absorber	
				Type	Number	Type	Number	KON	KON-AL, Metallized	KON	KON-AL, Metallized
Mold	*Mucor* sp. −2	YMA	28 (15)	S-100	2	S-200	2	+	−		
	Fusarium sp.	PDA	28 (14)	S-100	6	S-200	3	+	−	+	−
	Fusarium oxysporum IAM 5009	PDA	28 (14)	G-100	10	G-100	5	+	−		
	Arthrinium sp.	YMA	28 (15)	S-100	6	S-200	3	+	−		
Yeast	*Saccharomyces cerevisiae* IFO 0305	YMA	28 (14)	S-100	6	S-200	3	+	−		
	Candida utilis IFO 0396	YMA	28 (20)	G-200	1	G-100	2	+	−	+	−
	Hansenula anomala IFO 0130	YMA	28 (19)	S-100	2	S-200	2				

[a] Existence of growth: + growth recognized, − growth not recognized.

Table VI Varieties of Deoxidizers

	Variety Name (Color of Packaging Material)	Use Characteristics	Main Use	Deoxidation Time
Moisture-dependent type starts to absorb oxygen when in air with high humidity	Distinguished in waterproof property, FX type (sepia color)	Distinguished in waterproof property One side printing (absorbing oxygen from plain side) FX-L type with waterproof/oilproof property available (printing color is different) FY type available to absorb oxygen from both sides	Fresh cut rice cake, rice-flour dumplings, fresh crumbs, cupped bean paste, steamed bean-jam bun, fresh noodle, fresh wakame (seaweed), pizza crust (FX-L, and others	0.5–1 days
	Super mini size type; FJ type (sepia color)	Super mini size type same as FX (20 × 28 mm) FJ-20 size only		
	Absorption on both sides, F Type (brown color)	Absorbing oxygen from both sides		1–3 days
Self-reaction type, starts to absorb oxygen at contact with oxygen	Immediate effect type, S type (silver)	From high moisture food to medium moisture food in packaging form, for which FX cannot be used	Bean-jam bun, sweet jelly of beans, castella, baumkuchen, flakes of dried bonito, and others	0.5–1 day

(continued)

Table VI (*Continued*)

Variety Name (Color of Packaging Material)	Use Characteristics	Main Use	Deoxidation Time
For medium watery foods; ZN type (blue)	Good working capacity and can use even if food is watery to some extent.	Castella, bean-jam bun, sponge cake	1–2 days
Excellent in keeping aroma, ZK type (blue)	For alcohol-containing food or food that attaches importance to delicate aroma	Brandy cake, coffee, tea, etc.	1–3 days
For low moisture product, Z type (blue)	For low water-dried product (possible to use together with desiccant	Rice crackers, nuts, dried foods, tea, spices, medical supplies, metal products, and others	1–3 days
Carbon dioxide generating type; G Type (reddish brown)	In order to keep equal pressure inside container, generates carbon dioxide in same volume as oxygen absorbed As main ingredient is organic matter (food additives), it is not caught by metal detector		1–4 days
For coffee, E type (light green)	Simultaneous absorption of oxygen and carbon	Coffee	Varies according to the quantity of CO_2

	Table type, T type (light blue)	dioxide, distinguished in aroma preservation Absorbs quantity of carbon dioxide at the same time absorbs one-tenth as much oxygen The type to be affixed on the back side of bottle lid For low-water dried products (possible to use together with desiccant) Only T-10 size available	Medical supplies/ health foods	4–7 days
Freshness keeping agent	For fruits and vegetables, C type (green)	Absorbs harmful gas such as carbon dioxide, ethylene etc. generated from fruits and vegetables This is not reacting under usual open air and so it is contained in poly bag	Fruits and vegetables	
Oxygen sensing agent	Possible to see existence of oxygen at first sight, eye type	Sensing existence of oxygen at first sight by change of color (pink–blue) Special type for low moisture product available	Confirmation test in actual use Packaging control of products	

without a gas barrier effect are used in the latter. The film completely separates the food and the oxygen absorber (Fig. 15).

V. NEW KNOWLEDGE IN MICROBIOLOGY

In conclusion, we would like to introduce the latest knowledge we gained through development of an oxygen absorber that did away with the existing theories. The popular view in the past was that molds in general were aerobic, but that some special types of molds grew in anaerobic conditions, or that yeast, commonly an anaerobic organism, can grow in the absence of oxygen. We discovered, however, that such organisms cannot survive under conditions where no oxygen permeates from outside by the use of oxygen absorber, or under perfect anaerobic conditions, in other words. This means that the so-called oxygen-free state in the past has been inadequate to say the least. We were able to gain this knowledge through the development of the oxygen absorber, which, for the first time, allowed us to create a totally anaerobic condition. It was also made clear that oxygen absorbers have the effect of not only of preventing growth but also of sterilizing the aerobic organisms such as molds. It was commonly believed that fermentation could not be stopped by oxygen absorbers, although it was possible to prevent molds. On the contrary, fermentation can be controlled with the use of oxygen absorbers, depending on the conditions (Table V).

We can therefore see that the development of oxygen absorbers has made it possible to carry out what was impossible in the past, and has explained the undiscovered facts one by one. We are certain that this new method of food preservation, which was developed in Japan, will be used in many countries of the world in the not so distant future. It is a technology that embodies unlimited possibilities. In Table VI we have listed the oxygen absorbers that are currently available.

CHAPTER 14

Frozen Food and Oven-Proof Trays

Kentaro Ono
Corporate Development and Planning,
Snow Brand Milk Products Co., Ltd.,
Tokyo, Japan

I. OVEN-PROOF TRAY DEVELOPMENT

The continuing growth of microwave oven sales and increasing consumer use of microwave ovens for cooking processes as well as other conventional ovens has made development of dual-oven-proof trays an important proposition in packaging (Gitlin, 1984; Hunter, 1981). One of the largest markets for microwave oven-proof trays is restaurants equipped with microwave ovens for quick service (Woodward, 1979). Most plastic trays and paperboard trays with grease-proof coatings are available for microwave oven use only, since temperatures to which trays are exposed are lower than 100°C (boiling point of water) (Clarke, 1981).

With increasing demands for dual-oven-proof containers in the market, years of development efforts have been invested (Gitlin, 1984). Demand for dual-oven-proof containers increased in the frozen food industry, which is distributing products to retail stores and users who buy the frozen foods at the stores (Hunter, 1981). The users or consumers own microwave ovens and/or conventional, electric or gas ovens. Such factors, including social trends and changing life styles, making dual-oven-proof paperboard trays the current dominant products in the market, include (Clarke, 1981) (1) the growing popularity of convenience foods, (2) increasing institutional catering, (3) the development of the microwave oven, (4) the development of dual-oven-proof PET-coated paperboard, (5) the higher production cost of metal and plastic containers, (6) energy factors, and (7) environmental factors.

Food Packaging Copyright © 1990 by Academic Press, Inc. All rights of reproduction in any form reserved.

TPX plastic-coated board had been developed as a material for dual-oven-proof trays, but TPX-coated board has been introduced to the market only in Japan and other developments were abandoned. The heat resistance of TPX is almost sufficient as a coating material for dual-oven-proof trays; however, the processing is difficult and high in cost (Clarke, 1981). PET extrusion coating on SBS exceeds the heat resistance of other coating materials, which deteriorate at about 160°C, while cooking temperatures in conventional ovens are over 200°C (Clarke, 1981; Gitlin, 1984).

The advantage of paperboard and other new materials for microwave heating or cooking of food is a function of faster cooking through heating and cooking, which is provided from the microwaves passing through the paperboard. Also, the trays have sufficient heat resistance to be heated in conventional ovens—gas, electric, and convection ovens (Hunter, 1981).

II. VARIETIES OF DUAL-OVEN-PROOF TRAYS

A. Requirements for Dual-Oven-Proof Trays

The primary requirements of dual-oven-proof tray as a food packaging are as follows (Andres, 1981): (1) protect the food contained, (2) aesthetic appeal to the user, (3) ease of heating or cooking the food for serving, (4) operating efficiency in manufacturing, and (5) cost efficiency. Each requirement must be satisfied by the physicochemical functionalities or properties of materials used to compose a tray and the manufacturing. Changing from foil containers to other dual-oven-proof trays has also increased as a result of these factors. Foil containers cannot be heated in microwaving as quickly as other containers made from microwave-penetrable materials (Hunter, 1981), cannot be printed on like paperboard, and can be high in cost (Clarke, 1981). Containers with printing have stronger appeal for both frozen food manufacturers and consumers, since printed patterns or graphics on paperboard help it to resemble dishes to serve at the table (*Food Engineering,* 1984). New oven-proof tray materials are developed in conformity with these primary requirements and are growing in volume faster than conventional foil containers.

B. Varieties

Development of PET-coated paperboard as a material for dual-oven-proof tray opened a new age of ovenable trays. Today, PET leads and PET-coated paperboard is dominant in the market (Gitlin, 1984). Plastics have a long history of development in oven-proof trays, not only paperboard tray applications but also molded and vacuum-formed trays (Clarke, 1981; *Plastic Technology*, 1983). Aluminum foil containers also are dual oven-proof, according to the Aluminum Foil Container Manufacturers Association (AFCMA) (Rice, 1981). However, foil containers have some disadvantages in terms of heating efficiency with microwave ovens (*Quick Frozen Foods*, 1981). Typical trays available are listed on Table I. Basic materials used for dual-oven-proof trays are paperboard, plastic, and aluminum. PET is most popular as a material for plastic trays, because of the heat resistance and cost of resin and processing. A number of coatings have been examined and introduced for paperboard trays; however, PET, a combination of acrylic and silicon, and polybutylene terephthalate (PBT) in Europe are available today (Clarke, 1981; Peters, 1981). A variety of forming techniques are available as manufacturing methods. Pressed forming is dominant for the paperboard tray (Gitlin, 1984). Vacuum forming is available for the plastic tray (*Plastic Technology*, 1983) and the conventional pressed foil container is also available.

C. Dual-Oven-Proof Paperboard Trays

Dual-oven-proof paperboard trays are dominant in the market, especially the pressed trays, which are widely used for frozen food packaging (Peters, 1981; *Prepared Food*, 1982). Basically three techniques are available for tray manufacturing, and paperboard can be adapted to the needs of the users (Gitlin, 1984).

1. Requirements for Dual-Oven-Proof Paperboard

Dual-oven-proof paperboard has been developed primarily in conformity with the requirements for dual-oven-proof trays already described. De-

Table I Dual-Oven-Proof Trays[a]

Basic Material	Coating	Forming
Paperboard (pulp)	PET	Pressed forming
	Acrylic	Folded forming
	Silicone	Molded forming
	Polybutylene terephthalate (PBT)	
Plastic (PET)	—	Vacuum forming
		Molding (thermoset)
Aluminum	Plastics	Pressed forming
		Drawn forming

[a] From Rice (1981).

tailed requirements or advantages of dual-oven-proof paperboard for oven-proof trays are as follows (Clarke, 1981; Woodward, 1979).

1. Dual-oven-proof quality. Paper itself has a high heat resistance, which is reinforced by the PET coatings (Hunter, 1983). An oven temperature in cooking could be as high as 240°C, and heat resistance against 200–220°C for 30 min is sufficient for most applications (Clarke, 1981; Peters, 1981).
2. High temperature registance. Oven-proof trays used for frozen food packaging must have a high heat resistance, going from deep freezing at −40°C to reheating at 200–220°C for 30 min (Andres, 1980; Woodward, 1979).
3. Browning and odor development resistance. Browning of paperboard was a major problem to be overcome with PET coating on paperboard as a material of dual-oven-proof trays. Heat is accumulated at the outskirt of the tray flange, while the walls are protected from rising temperature by the heat-sink effect from the foods contained. Coatings should not be degraded to generate odor at oven temperatures (Clarke, 1981; Peters, 1981).
4. Good performance under deep freeze conditions. Frozen foods are handled and distributed under deep-frozen conditions at lower than −15°C, typically −40°C. Oven-proof trays must have sufficient properties as packaging for frozen foods at that temperature (Clarke, 1981).
5. Grease and water resistance. These are other important roles of coatings, and also are primary requirements for a packaging of frozen foods that will be used to heat and serve (Clarke, 1981; Hunter, 1981).
6. High in productivity. This refers to not only productivity of tray itself but also productivity at the packaging operation, which are very important in the cost performance (Clarke, 1981; Hunter, 1981; *Quick Frozen Foods*, 1981).
7. Printability. Ease of printing and esthetic qualities are distinctive properties of paperboard, especially with clay-coat SBS (Clarke, 1981; Hunter, 1981).
8. Heat sealability and gluability. These are primary requirements for food packaging to protect the contents. Coatings affect these requirements strongly, especially heat sealability, which is not original propertiy of paperboard (Andres, 1981; Clarke, 1981; Hunter, 1981, 1983; *Quick Frozen Foods*, 1981; *Food Engineering Int'l*, 1982).
9. Heat tolerance in physical properties. A tray with food, reheated for serving, must have sufficient physical properties for holding food and serving on the table without any troubles. Rigidity and grease and

water resistance are important properties (Clarke, 1981; Hunter, 1981).
10. Safety as food packaging materials. All the materials used should be in conformity with governmental requirements and be approved (Clarke, 1981; Hunter, 1983).
11. Microwave safe. Molecules often generate heat from electric oscillation of microwaves to degrade and may result in hazardous substances (Clarke, 1981).

2. Characteristic of Oven-Proof Paperboard Trays

Dual-oven-proof paperboard trays have advantages as frozen food packaging and in microwave cooking. The PET-coated tray has a sufficient heat resistance to resist for several minutes in an oven temperature in excess of 260°C, if food is filled in the tray (Woodward, 1979), and thus enables dual oven use for paperboard tray. Also, PET-coated paperboard trays can be heat sealed to protect contained foods from dehydration and other deteriorations, while pressed foil containers are difficult to provide with complete heat seals (Andres, 1980; Woodward, 1979; *Quick Frozen Foods*, 1981).

Microwavability of the paperboard will be another typical functionality for the paperboard tray. Microwave energy enters the tray though the wall, while the energy enters the tray only from the top opening for the aluminum tray, so that its heating time is longer than that of the paperboard tray (Hunter, 1981). Printability is an outstanding property of paperboard. Printed paperboards are used for pressed forming tray and folded forming tray. Clay-coat SBS used for folded trays has excellent printability to appeal to users. Molded formed tray also can be printed on the flange part and the bottom after forming. Printed paperboard trays offer multipurpose application from processing to serving on the table (*Quick Frozen Foods*, 1981).

Paperboard has comparatively high resistance against denting or impact from handling and distributing, which is often cost-effective on the production line of frozen foods. Paperboard is comparatively low in the cost and has potential for keeping lower in cost, while the cost of aluminum foils and plastics increases because of energy situations.

3. Pressed Formed Tray

Pressed formed paperboard trays share most of dual-oven-proof paperboard market in the United States, mainly used for frozen dinners (Gitlin, 1984). The development of PET-coated SBS offers to perform dual oven use for the paperboard tray. A typical manufacturing process of a pressed formed tray begins with a roll of PET-coated oven-proof paperboard as shown in Fig. 1. The paperboard is moistened to have good printability and formability (Hunter, 1983). Printing is performed after coating in the typical process, and sometimes before coating to provide two-side-printed paper (Peters, 1981), which has a better appearance and considerably higher cost. The printed paperboard is feed into a tray former where the paperboard is creased, cut with a die, and formed in a forming die with applied heat and pressure. Formed trays are stacked and packaged for shipment to the users (Hunter, 1983).

In-line forming is available from two supplying systems of material paperboard. The first is a roll feeding system as described above. The second is when die-cut tray blanks are shipped to the user and pressed to form trays in line (Peters, 1981).

4. Folded Forming Tray

The manufacturing process of folded forming tray is similar to that of folding carton manufacturing as shown in Fig. 2. Clay-coat SBS can be used for this system, while the pressed forming system has troubles from stickiness of the clay coat. Also, clay-coat SBS provides excellent prin-

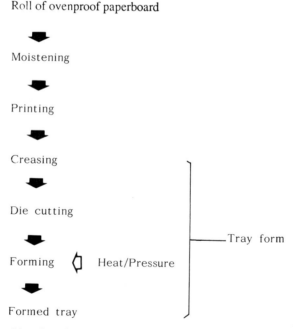

Figure 1. Manufacturing process of pressed forming tray. (From Hunter, 1983).

tability for offset and rotogravure printing, which is attractive to consumers. Printed paperboard is die cut into tray blanks. Tray blanks are stacked and shipped to the users, where the tray blank is erected with heat sealing, gluing, or locking into a formed tray prior to feeding into a filling and packaging system (Hunter, 1983). The cost of flat blank tray forming is considerably lower than of that of the preforming method (Andres, 1980).

5. Molded Formed Tray

The manufacturing process of the molded formed tray is totally different from those of pressed and folded trays and begins with mixing of a fiber substrate material, typically SBS, and a coloring agent, to provide a slurry material as shown in Fig. 3. The slurry material mixed with a bonding material is supplied into a forming mold. The molded tray is printed prior to PET film laminating. The PET film, typically 40 μm thick, is also printable (*Food Engineering*, 1984).

A good customer and consumer appearance can be available using printings and molded surfaces, which resemble real dishes. Also, multicompartment trays (*Food Engineering*, 1984) or trays with voids are avail-

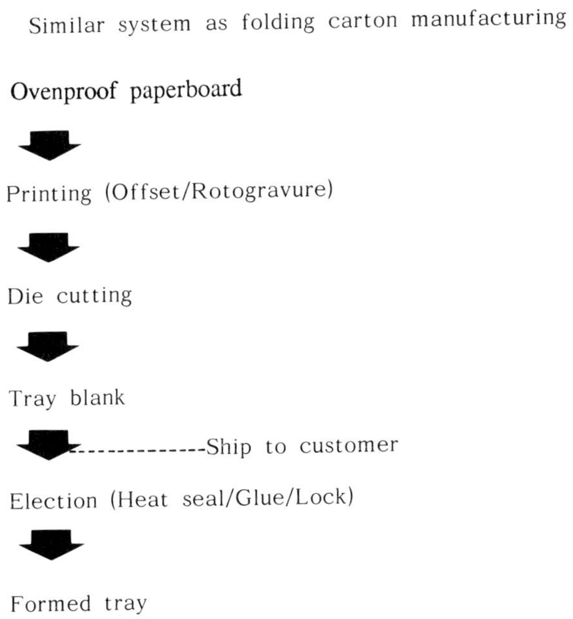

Similar system as folding carton manufacturing

Ovenproof paperboard
⬇
Printing (Offset/Rotogravure)
⬇
Die cutting
⬇
Tray blank
⬇------------Ship to customer
Election (Heat seal/Glue/Lock)
⬇
Formed tray

Figure 2. Manufacturing process of folded forming tray (system similar to folding carton manufacturing). (From Hunter, 1983.)

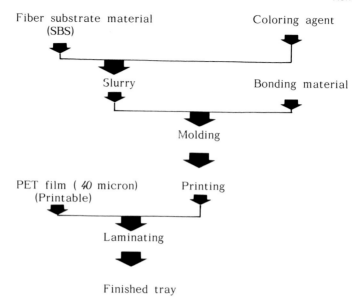

Figure 3. Manufacturing process of molded fiber tray. (From *Food Engineering*, 1984.)

able from the unique molding system (Peters, 1981). The tray surface without scoring or folding line is very important for quality food packaging and serving (Gitlin, 1984; *Food Engineering*, 1984). A tray consisting of two layers of pulp can provide different fiber substrate material and color (Peters, 1981).

The cost of the molded fiber tray is said slightly higher than that of the pressed formed tray and half that of the plastic tray (*Food Engineering*, 1984), although the cost of the multicompartment tray is lower than for the pressed tray. The molded tray requires a long production run with a given size to provide the best economy (Peters, 1981).

6. Coating for Dual-Oven-Proof Tray

Functions of coatings can expand the use of trays to dual-oven-proof use. Requirements for the coating plastic are as follows (Clarke, 1981):

1. Good runability properties. A coating layer is attached to the surfaces of the processing system and is transported on the surfaces.
2. Low in cost of resin. Resin cost is often given priority from a viewpoint of competition with other materials.

Chapter 14. Frozen Food and Oven-Proof Trays

3. High heat resistance at end-use state. Heat resistance over 200°C is required for dual-oven-proof coating.
4. Heat sealability. Easy handling, protection of contents, and convenience at end use can be provided with a heat-sealed lid for frozen food packaging.
5. Good barrier properties. Moisture and grease barriers are primary functions of frozen food packaging for cooking and serving.

Today, PET coating is most widely used for dual-oven-proof trays, and the coating weight is generally $40 g/m^2$, while PBT is the most favored material in Europe, because of governmental regulation. PBT is more expensive than PET and has almost the same properties (Clarke, 1981). Titanium oxide is used in white-pigmented coating to improve appearance by covering browning of the flange or hot spot of a tray from conventional oven cooking (Peters, 1981).

PET-coated SBS has a problem with removing baked products from tray. Acrylic or acrylic/silicone combination coating is used to provide easy-release properties for trays for baking application, which is a primary need (Hunter, 1983; Peters, 1981).

The runability of heated clay-coated SBS on processing machine is decreased by the stickiness of the coating layer. Clay-coated SBS is used for folded forming tray manufacturing and is not used for pressed forming tray manufacturing because of problems of sticking to the heated bottom of forming dies, while clay coating provides the surface of SBS with a good printability and a gloss surface of the package (Clarke, 1981).

7. Composition of Dual-Oven-Proof Paperboard

The most popular paperboard used for dual-oven-proof trays is SBS, with a caliper between 240 and 400 g/m^2. Typical compositions of oven-proof paperboard and their applications are as follows (Clarke, 1981):

1. Liquid packaging board with PET coating, which is used for a pressed forming tray.
2. Bleached SBS with PET coating, which is used for pressed and folded forming trays and a lidding material for oven-proof trays.
3. Clay-coated SBS with PET coating, which is used for folded trays, lock-form trays, and cartons.
4. Two-side PET-coated SBS, which is used for a heat-sealed tray.

Other compositions of SBS and silicone coating or PBT coating are also available specifically for dual-oven-proof applications (Peters, 1981).

8. Steps of Dual-Oven-Proof Tray Development

A number of steps are required to develop appropriate trays. The steps are as follows:

1. Marketing discussion. Prior to determining a direction of the development, the marketing potential of various types of container must be discussed (Peters, 1981).
2. Cooking test. The stability of container coatings and printings against food contained is determined on this step. Another important aspect is modification of food cooking instruction. Since the heating characteristics of paperboard trays in conventional ovens are different from those of foil containers, modification of cooking instruction is required to optimize the cooking (Peters, 1981).
3. Determination of container size. A number of standard sizes and shapes are available, and custom design will be provided upon economical conditions. The primary concern of selecting is the food contained, and another point is the adoptability to the packaging line and system, especially when changing from foil containers (Hunter, 1981; Peters, 1981).
4. Determination of paperboard caliper. The caliper of board used will be selected by considering (1) product weight and volume, (2) packaging line or system used, (3) distribution condition, (4) oven cooking condition, and (5) cost. Board selection is primarily involved in the desire of the end user for rigidity (Hunter, 1981, Peters, 1981). Changing of board thickness or caliper without changing the die causes dimensional change of top-in or top-out in pressed trays, which is derived from the difference in relaxation of paperboard. Uniformity of the trays is also determined by the manufacturing conditions, from the paperboard stock to the forming (Peters, 1981).
5. Determination of surface treatment. Surface treatment of the outside of the tray is sometimes required to be provided, such as grease resistance from fluorocarbon treatment or PET coating. PET coating of the outside increases the cost considerably (Peters, 1981).
6. Heat sealability. A number of oven-proof lid films are heat sealable to the flanges of the tray. The lid film must withstand the distribution handling and must be readily pealable at end-use (Hunter, 1981).
7. Governmental regulation. Oven-proof tray shall be in conformity with governmental regulation for safety as food packaging, such as the FDA and the U.S. Department of Agriculture (USDA) approval in the United States, the Food Hygiene Law in Japan, and similar regulations in each country (Hunter, 1981).

Chapter 14. Frozen Food and Oven-Proof Trays

8. Appearance. Paperboard and film for coating can be printed to improve the customer appearance and also to provide serving components that coordinate (Hunter, 1981; Peters, 1981).
9. Economics. The potential cost reduction in changing from a metal foil tray to an oven-proof paperboard tray is provided with lower tray cost and easy operation of packaging. In-plant manufacturing of oven-proof paperboard containers can reduce the cost considerably with extremely high volume production for pressed trays. Dual oven usage improves the marketing position by serving to consumers who have both microwave and conventional ovens (Hunter, 1981).

1. PET Tray

Dual-oven-proof PET trays have been developed for more than 10 years, and a few products have been introduced. PET has appropriate properties as a material for dual oven-proof trays, including cost of resin, heat resistance, mechanical properties, ease of processability, and compatibility with food products as well as availability of an amorphous and a crystalline state. PET can be extruded to sheeting and can be thermoformed at its amorphous state, and high heat resistance can be achieved in the crystalline state, which is a primary property of tray for conventional oven use (*Plastic Technology*, 1983).

Elegant china-like filled or thermoset trays are used for gourmet entrees and other high-end products; however, the cost reported is two or four times that of the thermoformed crystallized PET tray (Gitlin, 1984). The high cost of PET sheet and long forming cycle have restricted its popularity in the market. The crystallized PET tray was formerly much higher in cost than the PET-coated paperboard tray.

However, a new tray-forming technique has been developed recently to offer crystallized PET tray with conventional vacuum-forming or thermoforming system with high cycle speeds, 3.5 s per cycle. In this technique, proper crystallinity can be achieved by heating a forming PET sheet and formed trays on the mold. Crystallinity of 25–35% gives proper heat resistance for dual-oven-proof PET trays, while the maximum crystallinity is about 55%. The crystallinity is determined by heating a sheet at the thermoforming temperature of mold and for the actual time. The heat resistance of PET sheets is increased from lower than 81°C the glass transition temperature of PET, for the amorphous state, to 177–204°C for the crystalline state (*Plastic Technology*, 1983).

E. Dual Oven Usage of Foil Trays

According to the aluminum industry, the aluminum tray is microwavable; however, cooking instructions on food packages say to remove food from the foil container and put into an appropriate container for microwaving (Gitlin, 1984). Oven manufacturers warn against the use of aluminum foil containers in microwaving. And food processors are switching their aluminum oven ware to dual-oven-proof paperboard or plastic containers (Gitlin, 1984).

Potential disadvantages of foil containers in microwaving are (1) potential damage of the magnetron from microwaves reflected on the aluminum foil surface, (2) uneven or slow heating of food contained or limitation of depth of food because of lower heating efficiency by microwaves entering only from the top opening of container, and (3) arcing, which is a sparking phenomenon between the foil container and the oven walls when microwaving (Clarke, 1981; Rice, 1981).

The AFCMA developed cook-in-the-carton to start a reeducation campaign proving the safety and effectiveness of aluminum containers for consumers and food processors. Steps of cook-in-the-carton are (1) open the outer paperboard carton and remove foil container, (2) remove lid, (3) put foil container back into the outer carton, and (4) place the carton in microwave oven and heat. Advantages of the cooking-in-the-carton technique, according to AFCMA, are eliminating any arcing by protecting the foil container from direct contact with oven walls, and promoting the heating effect by steaming in the carton, which gives even heating of food contained, decreases of heat loss, and prevents dehydration when microwaving (Rice, 1981). The AFCMA study concluded that there was no effect on the magnetron tube from microwave cooking with an aluminum foil container.

III. LIDDING OF TRAYS

Various types of lids are available according to end use, including heat sealed, drop-in, crimp-on, and wrap-over, with a choice of materials and combinations (Woodward, 1979). Paperboard trays coated with plastic are heat sealable to provide advantages of functionality and attractive presentation to the consumers, preservation of the food from dehydration, easy handling and distributing, and preventing contamination (Andres, 1981).

Various types of heat-sealing systems are available, including conventional heat-sealing systems, using a rotary heat sealer and the lid film from rolled web (Andres, 1981). Also, a technique with cold PVA glue is available for sealing of PET-coated paperboard. The amount of glue used is about one-fourth as much as hot melt with spraying application to the board (*Food Engineering Int'l.*, 1982).

IV. USAGE OF TRAYS

Oven-proof characteristics of dual-oven-proof trays are varied depending upon material used, foods contained, oven type used to reheat, and other factors on heating efficiency. The Schiffmann report was submitted to the American Paper Institute (API) to determine characteristics of oven-proof trays in comparison test with each other, paperboard and nonpaperboard, in microwave and conventional ovens. The Schiffmann report determined some energy savings on heating of foods contained in paperboard trays (*Quick Frozen Foods*, 1980).

The amount of energy saved is different over differing conditions and performances of oven cooking, although the paperboard tray saves energy and time by 10–20% compared to nonpaper trays. Heating time savings were observed in conventional gas and electric ovens with paperboard trays. In microwave ovens they showed greater effects reducing by 2–5 min (20–50%) the preparation time for an entree (*Quick Frozen Foods*, 1980). In conventional ovens, the heat sink effect of food contained is influential to temperature rise in heating. Liquid foods are distributed evenly in the tray, and the temperature of the tray never rises above 104–107°C at an oven temperature excess of 200°C due to the high heat-sink effect. The contact of food with tray walls for solid foods is not enough to equalize the temperature rise on the tray wall. Uneven heating above 149°C causes the browning of paperboard (Peters, 1981).

End users know well the advantage of ovenable paperboard trays in microwave cooking. Heat resistance in conventional oven cooking has

Table II Cooking Time Difference[a] between Paper and Nonpaper Tray

	Oven		
Entree	Microwave	Electric	Gas
Small macaroni and cheese entree	2–5	0	1–2
Frozen creamed corn	2–4	1	1
Large macaroni and cheese entree	0–2	—	—
Large spagetti and meatballs	0–2	—	—

[a] Time difference (min) = (cooking time of nonpaper tray) − (cooking time of paper tray). From *Quick Frozen Foods* (1980).

Table III Achieved Temperature of an Entree in Various Trays[a, b]

Microwave Oven Brand	Type of Tray	Temp. Achieved (°C)
General Electric	Pressed paperboard	54.1
	Folded paperboard	60.7
	Nonpaper	53.3
Sharp	Pressed paperboard	60.4
	Folded paperboard	66.9
	Nonpaper	50.4
Amana	Pressed paperboard	58.1
	Folded paperboard	54.7
	Nonpaper	59.5

[a] From *Quick Frozen Foods* (1980).
[b] Entree: large spagetti and meatball entree. Microwaving: Heating for 5 min, stir, rest for 3 min, heating for 3 min more.

been a problem; however, higher heat efficiency in conventional ovens gives lower settings for temperature. The lower the cooking temperature is, the longer the paperboard trays can resist (Hunter, 1981).

The difference of temperature reached in dinner-style food varies from entree to entree in paperboard trays and nonpaper trays, and the cooking time savings in microwave ovens and in comparison with conventional ovens are shown in Table II. Also, makes of microwave oven and types of paperboard container are factors of energy saving or rising temperature in microwaving, as shown in Table III (*Quick Frozen Foods*, 1980).

Temperature distribution in food contained is not distinct from the average temperature of microwaving. Generally the top temperature is higher than the bottom temperature, although the temperature difference in nonpaper trays is larger than that in paperboard trays as shown in Table IV. Also, the difference is dependent upon makes of microwave oven (*Quick Frozen Foods*, 1980).

Hence uniformity of cooking instruction is sometimes difficult to pro-

Table IV Temperature Distribution of Spaghetti and Meatball Entree in Container Microwaved[a]

Type of Tray	Top Temp. (°C)	Bottom Temp. (°C)
Pressed paperboard	71.3	65.0
Folded paperboard	77.7	61.9
Nonpaper	75.0	46.9

[a] From *Quick Frozen Foods* (1980).

vide for packaged food and microwave ovens. The cooking time or directions might be changed from oven to oven, depending on makes, type, output power, mechanical design, and time used.

REFERENCES

Andres, C. (1980). *Food Processing* **41**(3), 44–45.
Andres, C. (1981). *Food Processing* **42**(13), 114–115.
Clarke, R. J. (1981). *Paperboard Packaging* **66**(1), 106–112.
Gitlin, R. (1984). *Paperboard Packaging* **69**(6), 24–28.
Hunter, D. L. (1981). *Quick Frozen Foods* **43**(12), 48–49, 69.
Hunter, D. L. (1983). *Package Engineering* **28**(4), 184.
Peters, J. W. (1981). *Packaging Engineering* **26,** 57–60.
Rice, J. (1981). *Food Processing* **42**(13), 116–117.
Woodward, S. (1979). *Frozen Foods* **32**(10), 8, 10.
Quick Frozen Foods (1980). **42**(12), 44–48.
Quick Frozen Foods (1981). **44**(3), 32–33.
Food Engineering Int'l. (1982). **7**(3), 90.
Prepared Food (1982). **151**(8), 76.
Plastic Technology (1983). **29**(12), 13–15.
Food Engineering (1984). **56**(8), 81–82.

CHAPTER 15

Gas-Exchange Packaging

Koji Satomi
Kureha Chemical Industry Co., Ltd.,

Tokyo, Japan

I. GAS-EXCHANGE PACKAGING AND ITS AIM

Packages in which an air atmosphere has been exchanged with some other gas composition (air in usual pressurized packagings) are generally called gas-exchange packaging or gas-pack in packed foods. These storage methods of food are also often called modified atmosphere (MA) and occasionally controlled atmosphere (CA), and it is frequently understood that the composition of gas at the packaging time is not intended to change in the former, although adjustment during storage may continue in the latter, but there is no strict discrimination in the terms.

The following are considered to be aims of gas exchange:

1. Preventions of processes such as oxidation, oxidation of lipids, discoloration, and fading, and protection of nutrients such as vitamins.
2. Prevention of the change and spoilage due to microorganisms.
3. Preservation of the red oxymyoglobin color (for red meat).
4. Physical protection of contents.

The prevention of oxidation is possible through vacuum packaging employing high-barrier packaging materials, through the removal of oxygen from the intrapackaging system; however, when the package is evacuated, its contents may be injured or deformed by high pressure, which are inconveniences that occasionally occur. In these cases, gas-exchange packaging is selected. The reason that potato chips or thinly sliced ham and sausage are processed by gas-exchange packaging is to protect physically the contents, together with prevention of oxidation or discoloration.

The color of fresh meat is maintained by exchange of gas combined with oxygen in order to keep the pigment oxymyoglobin a fresh red color. In this case, carbon dioxide including oxygen is combined, and this from is expected the bacteriostasis of carbon dioxide.

II. GASES FOR USAGE AND THEIR PROPERTIES

A gas is selected and used according to the above-mentioned aims, and as a rule, the gas should be noninnoxious and harmless. Moreover, it is desirable that the gas be tasteless and odorless as much as possible. At the present time, three kinds of gas, carbon dioxide, nitrogen, and oxygen, are used for the gas-exchange packaging, and these are employed alone or in combination.

It has been reported that among these three kinds of gas only carbon dioxide has an antibacterial action or bacteriostasis, and therefore, carbon dioxide is mainly used for the purpose of the inhibition of microorganisms. The activity of carbon dioxide against microorganism is determined by the kinds of microorganisms, concentration of carbon dioxide, atmospheric temperature, period of applying carbon dioxide, water activity of food, etc. It is generally said that carbon dioxide at 5–50% concentration is bacteriostatic for almost all bacteria, molds, and yeasts.

Nitrogen gas is mainly used to prevent oxidation through its inactive character. The influence of nitrogen gas on microorganisms should be paid attention to, because generally aerobic bacteria are slightly inhibited, while the growth of anaerobic bacteria is sometimes promoted.

III. FORM AND SYSTEM OF GAS-EXCHANGE PACKAGING

The packaging forms of gas-exchange packaging differ according to the food contents, and the following are mainly considered.

A. Pouch and Bag

The container made by bottom sealing one end of tube is called as a bag, and a bag-shaped form sealed on four sides or three sides (center and both ends) is frequently called as a pouch; in these cases, the formation and filling of bag, and the sealing are frequently discontinuous. The food is directly put into the bag, or there may be tray-in-bag (or pouch) packaging, putting a whole tray into a bag or pouch after arranging food on the tray.

B. Pillow-Type Package

This pillow-type package is occasionally classified as one kind of pouch, but here it is dealt with separately.

This package is a pillow-type form in which the center and both ends have been sealed. The filling and sealing of this package are conducted automatically and continuously due to the usage of roll stock. There are two kinds: vertical and horizontal pillow types. When gas exchange is performed, the spraying pressurized gas through a thin pipe into the film in cylindrical form is applied. It is said that this method has a slightly poorer gas-exchange ratio compared with the method of filling with gas after first evacuating the bag. The residual oxygen volume by this method is approximately 5%. When food is prearranged on a tray, products are occasionally packed in pillow packaging. The horizontal pillow-type package is applied in this case.

C. Tray Package and Deep Draw Package

Tray package is a method of filling with gas and covering after putting food into a plastic tray and evacuating, and there are two methods: a conventional type tray is used, or an upper cover is sealed in place due to gas exchange after forming a tray serially from plastic roll stock and filling with food (Fig. 1).

The deep-draw package is the same procedure as the serial method of tray package, except that its bottom material is soft film.

Gas exchange in these packaging formations is carried out by the vacuum gas-exchange method. This is a method inserting gas after first evacuating the chamber, and produces a high gas-exchange ratio.

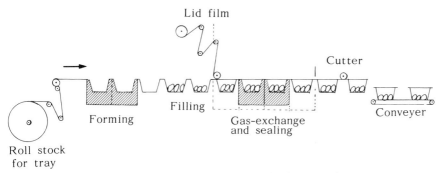

Figure 1. Form–fill–seal packaging machine for gas exchange.

IV. GAS-EXCHANGE PACKAGING AND PACKAGING MATERIALS

For gas-exchange packaging, for packaging materials which the gas permeability is as low as possible, namely, materials with high barrier, are required in order to maintain the gas composition of the inner region of exchanged packaging.

Packaging materials with high barrier include polyvinylidene chloride, ethylene–vinyl alchohol copolymer, acrylonitrile, etc., as simple plastic substances. Aluminum foil may also be used. Generally, these packaging materials are frequently used as layers with other films. Polyvinylidene chloride is also coated to polyamide, polyester, polypropylene, etc., as an emulsion. Aluminum is used for vacuum metallizing.

The most important characteristic of packaging material when gas-exchange packaging is conducted is the varied permeability of packaging material different to gases. Permeability of packaging materials to the three kinds of gas employed for gas exchange is highest for carbon dioxide, followed in order by oxygen gas and nitrogen gas, in the ratio of almost $3-5: 1:\frac{1}{3}-\frac{1}{5}$. When carbon dioxide was exchanged, a gradual vacuum package may be occasionally observed because gas escapes easily through packaging material and is easily soluble in the water of the food.

In addition, as a general character of packaging materials, gas permeability may changed due to temperature, and a relationship with an exponential function occurs between permeability and temperature. The humidity also influences the permeability of gas. The permeability of gas through a film of polyester and vinylidene chloride is hardly affected by humidity, while polyamide or ethylene–vinyl alchohol copolymer is highly influenced. Cellophane is similarly highly influenced.

V. CASES OF GAS-EXCHANGE PACKAGING IN FOOD

At the present time, the main foods packed by gas exchange in Japan are shown in Table I. In European countries, pizza, pastry, crepes, bread, confectionery, strawberries, etc., are packed by gas exchange.

Of these, some representative cases are explained as follows.

A. Ham and Sausage

Many sliced ham and sausage products are packed by gas exchange. Occasionally, sliced bacon, and Vienna sausage or frankfurters are also packed by gas exchange. Recently, the packaging of sliced ham or sausage

Chapter 15. Gas-Exchange Packaging

Table 1. Examples of Gas-Packaged Food in Japan

Product	Gases in Use	Purpose of Gas Exchange
Processed cheese	Nitrogen plus CO_2	Inhibition of mold growth
Flakes of dried bonito	Nitrogen	Preservation of color, flavor
Potato chips	Nitrogen	Inhibition of fat oxidation
Nuts	Nitrogen	Inhibition of fat oxidation
Sliced ham, sausage	Nitrogen plus CO_2 or nitrogen	Preservation of cured meat color
Green tea	Nitrogen	Preservation of flavor, color, vitamin C
Fish (young yellowtail)	Nitrogen plus CO_2	Inhibition of bacterial growth
Red meats	Oxygen plus CO_2	Preservation of cherry-red oxymyoglobin color

has been frequently conducted in bioclean rooms, and in addition, distribution at low temperature has also been established, so the possibility of spoilage due to adhered microorganisms has become low. Thus, the discoloration and fading become problematical, and oxygen influences it. Therefore, elevation of gas exchange ratio and removal of oxygen as much as possible become important. Nitrogen gas alone or nitrogen gas and carbon dioxide are concomitantly used. For the packaging form, sealing a package with an upper cover over arranged goods in a shallow tray, draw packaging, and pillow packaging are frequently used for sliced goods, thinly sliced goods, and Vienna sausages, frankfurters, respectively. For the packaging materials of the tray, PVC/EVOH/EVA and high barriers with PVC/PVDC/EVA or PVC/EVOH/EVA are used for bottom materials and cover materials.

B. Red Meat

Gas packaging of red meat has been mainly conducted in beef, using a gas combination of oxygen at 70–80% and carbon dioxide at 20–30%. The oxygen is used to keep the fresh red color of oxymyoglobin for meat pigment, and carbon dioxide is used to inhibit microorganism. Packages that are covered and sealed to a hard plastic tray are often used. PVC/PE (EVA) and PET/PVDC/PE or PVC/PVPC/EVA, etc., are used for bottom materials and cover materials, respectively. Usually, film with an antifog procedure is used for cover materials. Pillow packaging is also occasionally used. Meat is put into a tray with PVC, and this tray is accommodated in pouch. This form has been put to practical use as a package for businesses to retail stores. Materials such as PVDC-coated PET/PE, PA/PE, EVA/PVDC/EVA, etc., have been used for pouches.

For the tray, it is desirable that a bottom surface with unevenness or projections be made, so that the meat does not contact the surface of the vessel over a wide area. The oxygen is not sufficiently supplied to meat contacted the container, and the coloring of meat is not maintained. Similarly, contact between cover material and meat should be avoided, and space is necessary for the upper part. The temperature for storage of raw meat packed by gas is, as a rule, 0–2°C, and its quality can be favorably maintained at this temperature for 7 days. If there is a shelf life for 7 days, sales at retail stores become possible for 4 days, considering that 1 day is used for the delivery, and there are 2 days until consumption at home. However, full attention should be paid to temperature or hygiene of fresh meat at the time of gas-pack processing (from cutting up to packaging), as a matter of course in packaging. Meat should be cooled to 2°C before packaging because cooling after packing has poor transmission of heat and unfavorable efficacy. Thereafter, in order to keep this temperature, the cutting room should be kept at a low temperature, less than 5°C if possible. The initial number of bacteria of meat should be kept as low as possible; when the initial count of bacteria is high, the effect of carbon dioxide is not sufficient.

As shown in Fig. 2 (Bartkowski, et al., 1982), changes of psychrotrophs during storage at the time of gas packaging of beef were compared with those of the vacuum package and wrapping package (control).

C. Fish

Gas packaging of young yellowtail has been put to practical use in Japan, and some cases with salmon or trout have been reported overseas. The combination of gas in fish differs according to the kind of fish, and it should be chosen by experiment in each case. Moreover, few effects of gas packaging are noted in some fishes. In reddish fish, oxygen combined to protect the coloring of pigment in the meat, while in white fish the addition of oxygen is ineffective against oxidation of lipid. A combination of nitrogen gas/carbon dioxide at about 70/30 is used for yellowtail, and carbon dioxide, oxygen, and air are combined for salmon. The preservative temperature should be above freezing temperature but as low as possible. A shelf life of more than 2 weeks is possible. For the quality of fish packed by gas, retailers should consider a complete quality guarantee because judgment from the outer appearance without breaking the package is difficult for consumers. The importance of control of temperature and hygiene in the processes of processing and packaging is the same as for fresh meat. It

Chapter 15. Gas-Exchange Packaging

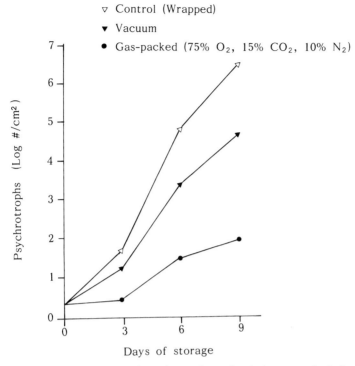

Figure 2. Change in number of psychrotrophs on longissimus muscle during storage under different atmospheres.

has been reported that the influence of elevation of temperature is larger than for fresh meat.

Gas pack is a technique with utility value in fishes in which the decrease of market value due to freezing is large. Figure 3 shows the results of investigating changes in bacterial number in gas-packed fishes (Parkin et al., 1982).

Generally, a form that directly seals the cover of film after putting fish into a tray, or putting a tray into a pouch is used, while a method of bulk storage in a gas including carbon dioxide, putting fish and ice into a large plastic bag, has been researched. Fey and Regenstein (1982) studied enclosing gas, putting red hake and salmon cleaned of head, bone, and viscera together with ice into a large bag, 160 × 170 cm. At the same time, the ice with sorbic acid was tried in order to elevate preservation. From the results, it was an exchanged gas consisting of 60‰ carbon dioxide, 20% oxygen, and 20% nitrogen gas, and ice with 1% sorbic acid potassium was

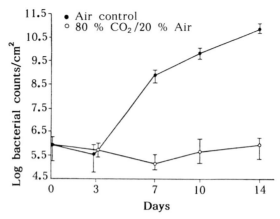

Figure 3. Aerobic plate counts of rockfish. Bars denote the range of counts.

most favorable, and favorable conditions physically were retained after approximately 1 month at 1°C.

D. Green tea

Hara (1986) reported in detail on research results of gas packaging of green tea. According to his theory, he reports that the main changes during the storage of green tea are changes of color, change of flavor, and decrease in vitamin C. The color varies due to the decomposition of chlorophyll or oxidation of catechin. It is said that the change of flavor may occur due to the oxidation of unsaturated fatty acid or carotenoid. These changes of green tea correlate well with the change of its vitamin C, and the changes of the quality of green tea are probably evaluated by the residual ratio of vitamin C. The degeneration is hardly noted when the residual ratio is more than 80%, while a large change can be noted when it is less than 60%.

Exchange packaging using inactive gas acts effectively to maintain the quality of green tea because these changes were naturally caused by oxygen. The gas employed is nitrogen. Discoloration and decomposition due to microorganisms in green tea are rarely observed because of its low water content.

Figure 4 (Hara, 1986) shows the results of measuring the residual ratio of vitamin C during storage of green tea by various packaging methods. By usual pneumatized packagings, its residual ratio is under 80% at over 1 month, while a residual ratio of about 90% was found in gas-exchange packagings after 6 months, and the efficacy of gas-exchange packaging is remarkable. With regard to materials for gas-exchange of green tea,

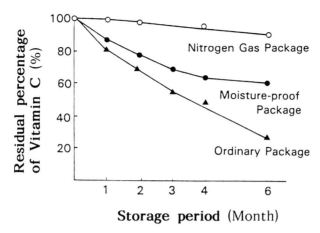

Figure 4. Relation between packaging method and residual vitamin C of green tea.

moisture-proof cellophane/PE/AL/PE/PVDC-coated OPP/PE, moisture-proof cellophane/OPP/Al/PE, etc., are used.

REFERENCES

Bartkowski, L., Dryden, F. D., and Marchello, J. M. (1982). Quality changes of beef steaks stored in controlled gas atmospheres containing high or low levels of oxygen. *J. Food Protect.* **45,** 41–45.

Fey, M. S., and Regenstein, J. M. (1982). Extending shelflife of fresh wet red hake and salmon using CO_2–O_2 modified atmosphere and potassium sorbate ice at 1°C. *J. Food Sci.* **47,** 1048–1054.

Hara, T. (1986). Research on gas substitution package of Japanese tea. *Packaging Japan.* No. 31, 73–76.

Parkin, K. L., Wells, M. J., and Brown, W. D. (1982). Modified atmosphere strage of rockfish fillets. *J. Food Sci.* **47,** *181–184.*

CHAPTER 16

Vacuum Packaging

Naohiko Yamaguchi
Food Research Institute, Aichi Prefectural Government.
Nagoya-shi, Japan

Through vacuum packaging, the chemical and biological deterioration of food in a package are greatly inhibited. Both deteriorations are concerned with oxygen and the oxidation of fat and oil; the discoloration of food and the rancidity of food with aerobic microorganisms are typical of these deteriorations.

I. VACUUM PACKAGING MACHINERY

This packaging machine is used for vacuum packaging with packaging materials having a extremely high gas barrier for the purpose of food preservation. Most of these machines seal the bag containing food in the vacuum chamber. The four types of machines are classified by mechanical faculty: nozzle, chamber, skin, and deep-draw type.

A. Nozzle-Type Vacuum Packaging Machine

A typical system of this type of machine is shown in Fig. 1. After air in the bag is evacuated through a nozzle, a mouth part of the bag is sealed by heater or impulse system. However, for blocks such as meat and meat products, the bag is mostly clipped with aluminum wire. The two processes of vacuumizing and sealing on most of this type of machines are automatically done, initiated by stepping on a foot swich. Accordingly, although the operation is easier than those of other types of vacuum packaging machines, the degree of vacuum of bags packaged by this type of machine is lower than for others.

Food Packaging Copyright © 1990 by Academic Press, Inc. All rights of reproduction in any form reserved.

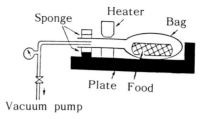

Figure 1. Nozzle-type vacuum packaging machine.

B. Chamber-Type Vacuum Packaging Machine

A typical system of this type of machine is shown in Fig. 2. The bag containing food placed in the chamber is vacuumized and then sealed by an impulse system in a vacuum-seal chamber. The degree of vacuum is usually about 0.5–8 torr for this type of machine, and the semi-automatic type machine may load several or dozens of bags in the vacuum-seal chamber at the same time.

C. Skin-Type Vacuum Packaging Machine

Skin vacuum packaging can be done by loading food and applying a lid over it. In this vacuum packaging machine, heated and softened upper film is applied skin-tight over food and a lower film (or tray or cardboard) in the vacuum chamber. The appearance of food vacuum-packaged by this type machine increases food value and display effect.

D. Deep-Draw Type Vacuum Packaging Machine

The methods of forming the bag and packaging in the vacuum chamber are shown in Fig. 3. The lower film (a) is warmed by a hot plate (b) and then

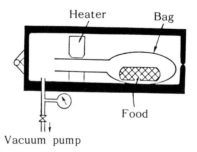

Figure 2. Chamber-type vacuum packaging machine.

Chapter 16. Vacuum Packaging

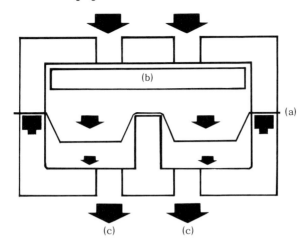

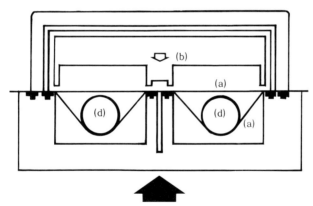

(a) : Film,   (b) : Hot plate,   (c) : Vacuum pump,
(d) : Food

Figure 3.   Deep-draw type vacuum packaging machine.

deep drawn by vacuumizing through a pump (c). After food (d) loaded in the bag is covered by the upper film (a), both films are sealed in the vacuum chamber. The newest machine of this type is controlled by computer, and deep drawing, loading, vacuumizing, sealing, trimming, etc. are set going automatically.

II. PACKAGING MATERIAL

Functions of packaging material are an important factor for vacuum packaging, and the requirements are as follows.

A. Gas-Barrier

The preservation of vacuum-packaged food is seriouly dependent on the gas barrier of packaging material, especially its oxygen barrier. By using packaging material with a high oxygen barrier, the high degree of vacuum in the bag containing food can be maintained for long time. Although the requisite oxygen barrier of packaging material for vacuum packaging is different with different kinds of foods, the degree of oxygen barrier generally must be below 15 $cm^3/m^2/\cdot 24$ h $\cdot$ atm. PT, PVC, PVA, EVAL, PET, etc. are mainly used as basic material. The laminated films used currently for vacuum packaging are K-coated films, OPP/EVAL/PE, ONy/PE, PET/PE, Ny/PE, OPP/PVA/PE, and so forth.

The relationship between the deteriorations of lard and soup vacuum-packaged by four films shown in Table I and their oxygen barriers are shown in Figs. 4 and 5 and Table II, respectively (Yamaguchi et al., 1986). The oxidative stabilities of lards vacuum-packaged by two films (EVAL and PET/OV) with high oxygen barrier were superior to others (Ny and KNy). The former did not show deterioration for 155 days during storage at 30°C. Similarly, the discoloration progressed more quickly in soups vacuum-packaged by films with a low oxygen barrier (Ny and KNy) than with those of the other two. However, the growth of aerobic microorganisms was detected only in soup vacuum-packaged by Ny, with the lowest oxygen barrier of the four films used in this test. As shown in these data, the oxygen barrier of packaging material strongly affects the shelf life of vacuum-packaged food.

Table I Oxygen Barrier ($cm^3/m^2 \cdot$ 24 h $\cdot$ atm) of Packaging Material Used

	0% RH	100% RH
EVAL	0.5	4.0
PET/OV	0.5	7.0
KNy	0.5	9.5
Ny	42	98

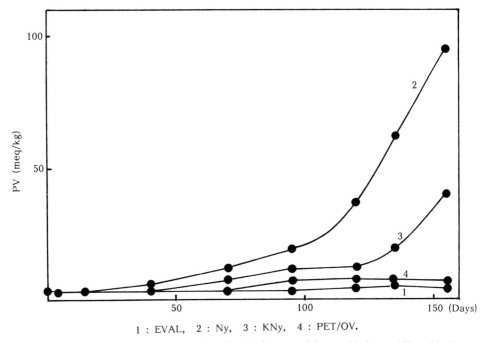

Figure 4. Effect of oxygen barrier of packaging material on oxidative stability of lard during storage at 30°C.

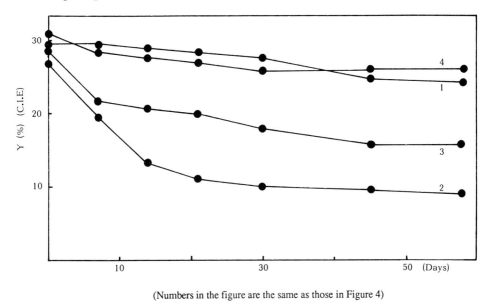

(Numbers in the figure are the same as those in Figure 4)

Figure 5. Effect of oxygen barrier of packaging material on discoloration of soup during storage at 35°C.

Table II Effect of Oxygen Barrier of Packaging Materials on Aerobic Microorganisms in Soup during Storage at 35°C (Bacterial Counts/ml)

	Days of storage					
	1	7	14	21	56	70
EVAL	≦30	≦30	≦30	≦30	≦30	≦30
PET/OV	≦30	≦30	≦30	≦30	≦30	≦30
KNy	≦30	≦30	≦30	≦30	≦30	≦30
Ny	≦30	1×10^3	9.5×10^5	5.3×10^5	8.2×10^6	5.1×10^6

B. Vapor Barrier

Many foods contain a particular moisture for the maintenance of their own qaulities. In order to protect against permeation of moisture from the bag containing food, packaging material with a high vapor barrier must be

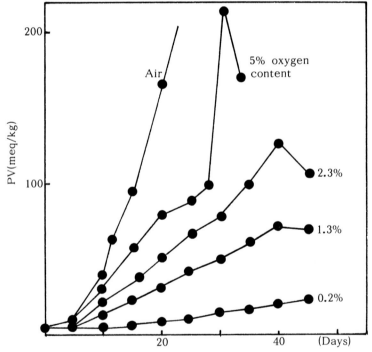

Figure 6. Oxidative stability of corn oil sprayed on wheat flour chips packaged by polyvinyl alcohol type laminated film at various oxygen content during storage at 60°C.

Chapter 16. Vacuum Packaging

used. Accordingly, the vapor barrier is generally necessary below $15g/m^2 \cdot 24h$ for vacuum packaging.

C. Relationship between Preservation of Food Oxygen Content in the Package

The oxidative stability of corn oil sprayed on wheat flour chips packaged with polyvinyl alcohol type laminated film (oxygen barrier = $8–12cm^3/m^2 \cdot 24 h \cdot atm$). at various oxygen contents at 60°C during storage in the dark is shown in Fig. 6 (Yuki and Wadaka, 1972). The relative rate of oxidation was recognized to decrease linearly with decreasing oxygen content, ranging from 5.0 to 0.2%. At the same time, oxygen content in each bag decreased along with the oxidation of oil (Fig. 7). Moreover, the relationship between peroxide value (PV), carbonyl value (COV), and oxygen

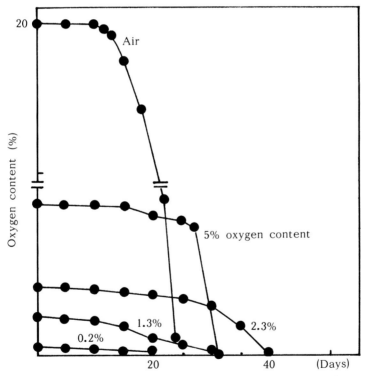

Figure 7. Changes of oxygen content in bag containing wheat flour chips during storage at 60°C.

consumption of wheat flour chips stored in the dark at 60°C was determined. As shown in Fig. 8, when the oxygen consumption ratio to the weight of oil is 0.1%, PV and COV can be reached 60 and 28 mEq/Kg, respectively. From these data, Yuki et al. suggested that the reduction of oxygen to an extremely low content may not be the essential condition, and lowering the ratio of oxygen to the oil content may be more important.

According to the following equations (Inoue, 1985), a highest value of PV can be calculated from the contents of both oxygen and oil in the bag.

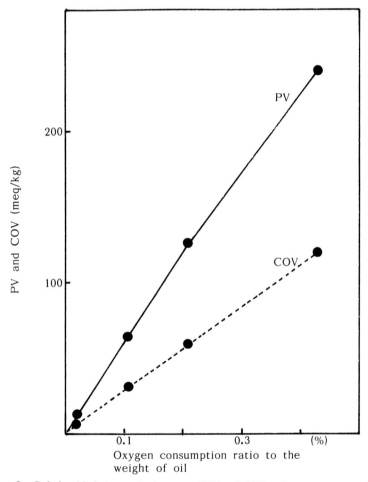

Figure 8. Relationship between the increase of PV and COV and oxygen consumption.

Oxygen content (mg) in 1 ml air = (32,000/22,400) × 0.2 = 0.286
Oxygen content consumed = 0.286 mg × A_{ml}

$$PV = \frac{0.286 \times A}{16} \times \frac{1,000}{G}$$

where A_{ml} is the content of air in the bag and Gg is the content of oil in the same bag.

The relationship between the growth of *Achromobacter*, *Pseudomonas*, and oxygen content is shown in Fig. 9 (Clark and Burki, 1972). Both aerobic bacteria grew rapidly above 0.5% oxygen. Moreover, Fig. 10 shows the influence of oxygen on the growth of molds at various content. The growth of these molds was suppressed below 0.5%. Therefore, it must be evacuated below 0.5% oxygen for inhibition of the growth of aerobic microorganism; furthermore, under these conditions the growth of microorganisms is remarkably inhibited by pasturization (Yanai et al., 1980).

It is well known that the amino–carbonyl reaction occurs without oxygen above about 50°C; however, oxygen accelerates this reaction at around room temperature. Hashiba (1976) has reported the oxygen-dependent browning (oxidative browning) of soy sauce: 1-ml aliquots of soy sauce were placed into two test tubes and one (a) was plugged with rubber (under aerobic condition) and another (b) sealed under vacuum

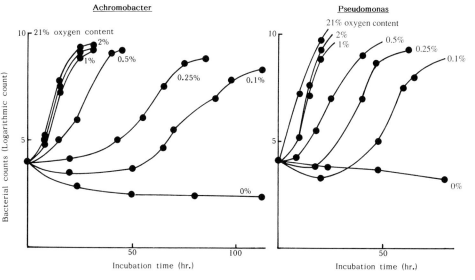

Figure 9. Relationship between the growth of aerobic bacteria and oxygen content.

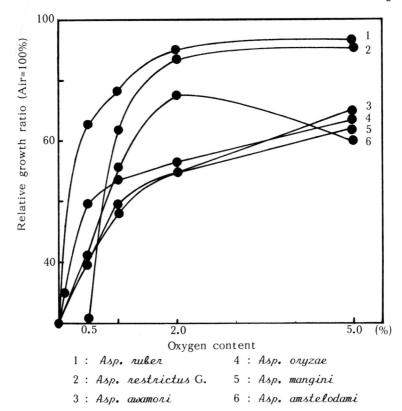

Figure 10. Effect of oxygen content on the growth of molds.

1 : Asp. ruber
2 : Asp. restrictus G.
3 : Asp. awamori
4 : Asp. oryzae
5 : Asp. mangini
6 : Asp. amstelodami

(0.3 mmHg). Both samples were held for 14 days at 37°C, and the color increase (E_{555}) was measured (Fig. 11). Soy sauce darkened rapidly in contact with atmospheric oxygen.

D. Application of Vacuum Packaging to Several Foods

Butter peanuts packaged under air and vacuum were stored for 150 days at 40°C at 91% relative humidity. The changes in PV and moisture content of both samples are shown in Figs. 12 and 13. The oxidation of vacuum-packaged peanuts progressed slowly, and its PV after 150 days was about 35 mEq/kg. However, the peanuts packaged with air reached 73 mEq/kg at that day. On the other hand, although both samples absorbed moisture, the

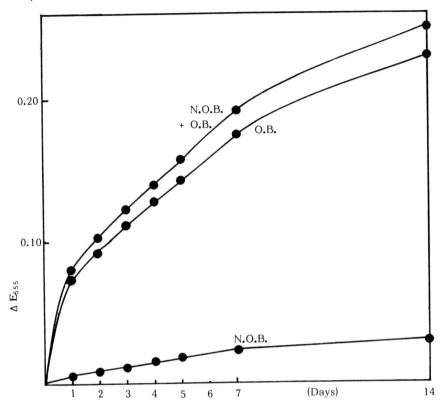

N. O. B. : Nonoxidative browning. (Under vacuum, 0.3 mmHG)
O. B. : Oxidative browning. (Under aerobic condition)

Figure 11. Oxidative and nonoxidative browning of soy sauce during storage at 37°C.

degree of moisture absorption of peanuts packaged under air with a laminated film consisting of PT/PE/CPP was more than for those vacuum-packaged with a film consisting of Ny/PE, and peanuts above 4.5% moisture lost their edibility (Uematsu, 1985).

The growth of aerobic bacteria in bacon packaged under air, carbon dioxide, and vacuum is shown in Fig. 14 (Baran et al., 1970). The growth of bacteria on bacon packaged under vacuum was more inhibited than on bacon packaged with air. However, its effectiveness was inferior to carbon dioxide packaging.

The discoloration of boiled corn beef vacuum-packaged (5 mm Hg) with

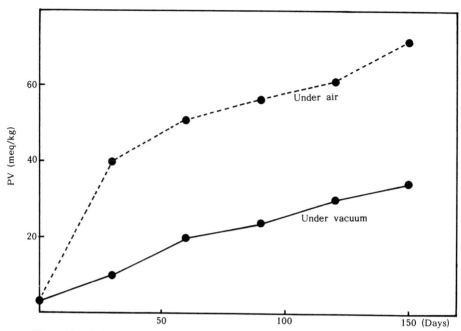

Figure 12. Oxidative stabilities of butter peanuts packaged under air and vacuum during storage at 40°C at 91% relative humidity.

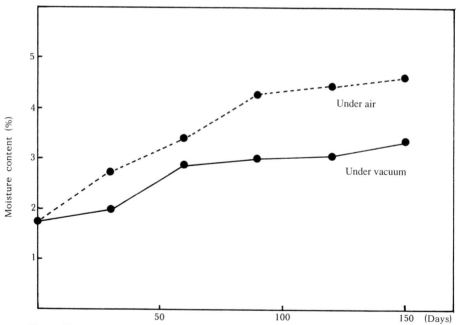

Figure 13. Changes in moiture content of butter peanuts packaged under air and vacuum during storage at 40°C at 91% relative humidity.

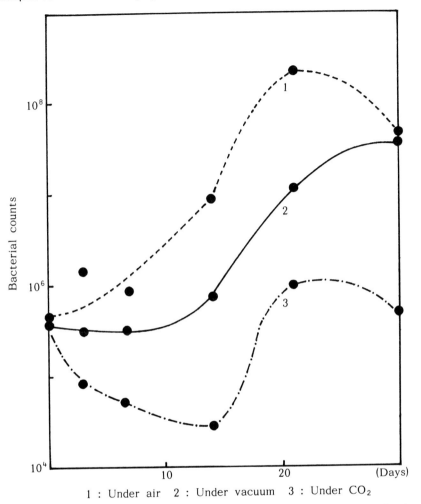

Figure 14. Growth of aerobic bacteria in bacons packaged under air, vacuum, and carbon dioxide during storage at 5°C.

four films is shown in Fig. 15 (Ikawa and Kubo, 1971). Both corn beefs packaged by OPP and PET with the low oxygen barrier faded greatly in comparison with two others (KONy and ONy) that had a high oxygen barrier in the dark and light.

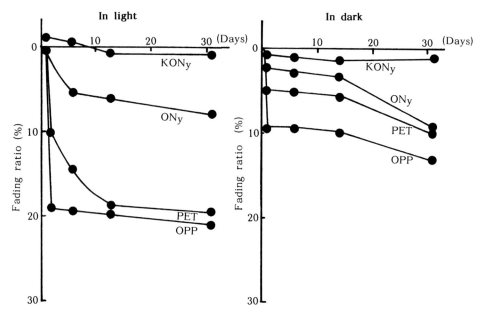

Figure 15. Fading of boiled corn beef vacuum packaged by four films in light and dark during storage at 5°C.

REFERENCES

Baran, W. L., Kraft, A. A., and Walker, H. W. (1970). Effects of carbon dioxide and vacuum packaging on color and bacterial count of meat. *J. Milk Food Technol.* **33**, 77–82.

Clark, D. S., and Burki, T. (1972). Oxygen requirements of strains of *Pseudomonas* and *Achromobacter*. *Can. J. Microbiol.* **18**, 321–326.

Hashiba, N. (1976). Participation of amadori rearrangement products and carbonyl compounds in oxygen-dependent browning of soy sauce. *J. Agric. Food Chem.* **24**, 70–73.

Ikawa, F., and Kubo, G. (1971). *Packaging of Foodstuff* **3**, 3–11.

Inoue, T. (1985). *Food Packaging* **7**, 74–79.

Uematsu, T. (1985). Quality preservation of Mamegashi. Report of Society for the Study of Food Quality Preservation.

Yamaguchi, N., Naito, S, Okada, Y. and Nagase, A. (1986). Effect of oxygen barrier of packaging material on food preservation. Annual Report of the Food Research Institute, Aichi Prefecture Government, No. 27, pp. 69–73.

Yanai, S., Ishitani, T., and Kojo, T. (1980). The effects of low-oxygen atmospheres on the growth of fungi. *J. Jpn. Soc. Food Sci. Technol.* **27**, 20–24.

Yuki, E., and Wadaka, H. (1972). Practical studies on the nitrogen gas exchange for preventing the oxidative rancidity of fatty food. *J. Jpn. Food Sci. and Technol.* **19**, 200–205.

PART VI

Packaging Fresh and Processed Foods

CHAPTER 17

Fruits

Akiyoshi Yamane
Food Technological Institute of Tottori Prefecture,
Japan

Fruits retain their physiological functions such as respiration and transpiration after harvest. There are several important components in fruit packaging: namely, packaging for freshness preservation and storage, shock-absorbing packaging during transport, and packaging to improve product value and exterior appearances. In this chapter, two important functions of fruit packaging will be discussed: packaging to keep the freshness of the fruit as well as its storage from a physiological point of view, and shock-absorbing packaging necessary during fruit transport.

I. PACKAGING FOR FRESHNESS PRESERVATION AND STORAGE

After the 1950s plastic film came into use to control respiration and transpiration of the fruits after harvest. In Japan, the use of plastic films came into use only after films (i.e., polyethylene, polyvinyl chloride, polystyrene) with the differing physical properties such as gas and water permeability were developed.

One of the methods for fruit preservation is controlled atmosphere (CA) preservation method. With this method, environmental gas can be controlled under low temperature, decreasing the O_2 content and increasing the CO_2 content of the container to control respiration to keep the good quality of the fruit for a long period of time. This method is mainly used for apples and pears.

When fruits are packed in polyethylene, the air composition inside the package changes due to the respiration of the fruit. This phenomenon produces effects similar to the CA preservation method. This is also called the CA effect and is the reason why polyethylene packaging is called easy CA preservation.

Usually, fruits have 85–90% of water content. When there is excessive transpiration, fruits shrink and the value of the fruit as a product decreases. Therefore, considering air permeability and other advantages, the use of adequate plastic film for fruit packaging becomes an absolute necessity.

A. Pears

For pears, plastic film packaging could cause decomposition to the core since some pears are very sensitive to CO_2. Chojuro (Japanese pear) is easy to spoil, but Nijusseiki (Japanese pear) is not.

Conditions for CA preservation for the Nijusseiki are as follows:

CO_2 content:	4%
O_2 content:	5%
Temperature:	+1°C
Humidity:	85–92%

Under these conditions, rotting was prevented and the best results were seen in the areas of taste, hardness, sourness, vitamin C, and pericarp color.

When O_2 concentration was raised higher than the above conditions, decomposition of the outside started early, and the color of the pericarp became yellowish. On the other hand, rapid changes were seen on the inside when CO_2 concentration was raised higher, and the taste became stale as well. However, the color was still in good condition. As far as plastic film packaging is concerned, polyvinyl chloride is more suitable than polyethylene at normal temperature. Polyethylene is more adequate than polyvinyl chloride at low temperatures, and the preservation time limit is about 4–5 months at 3°C.

Thick polyethylene wrapping could cause core decomposition and decoloration, but on the other hand, it has proved to prevent decomposition on the outside. It has been surmised that the ratio between CO_2 and O_2 has great effects on these changes, and the appropriate thickness of the film is found to be about 0.05–0.06 mm.

Furthermore, when fruits are packed in a box unit (15 kg per box) for practical purposes, 0.03–0.04 mm thick polyethylene showed the lowest decomposition rate (Fig. 1). The thinner films resulted in more decomposition on the outside, while the thicker ones showed changes on the inside. Shinsui (Japanese pear) can keep its freshness for about 5–7 days

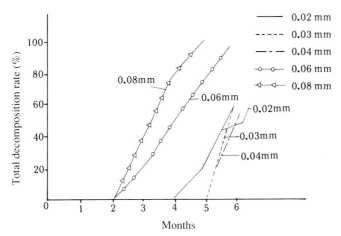

Figure 1. Polyethylene film wrapping preservation and total decomposition rate.

when harvested at the right time. However, if it is too ripe, with the green color of the pericarp almost gone, the freshness of the pear can be kept only for about 3–5 days.

It is possible to prolong freshness by sealing up-the-pear in a polyethylene (0.03 mm thick) bag together with 10 g activated charcoal (per kg fruit). Fruits preserved under this condition for 5–7 days can keep its freshness for 3–5 days after the package is opened.

B. Persimmons

Polyethylene wrapping showed the following results according to the thickness of the wrapping under chilling storage conditions (0°C): 0.03 mm thick polyethylene made the sarcocarp soft, changed its color to brown, and the fruit was too ripe. At 0.08 mm thick, polyethylene changed the color of the fruit to dark brown, and presented a characteristic sarcocarp quality that seemed to be physiological hindrance due to suffocation. However, 0.06 mm thick polyethylene showed the best results. Even after a 5-month storage, the quality of the persimmon was the same as a freshly harvested persimmon (Fig. 2). In this case, the respiration of the persimmon and the gas permeability of the polyethylene somehow kept a balance automatically (naturally) that helped form a suitable air composition for long-term storage.

More than 5% CO_2 content prevents the sarcocarp from softening; however, more than 20% CO_2 content causes carbon dioxide troubles.

More than 10% O_2 and less than 5% CO_2 lead to over-ripening and

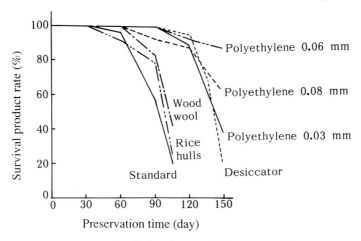

Figure 2. Survival product rate during preservation.

causes multiplication of microorganisms. The lower the O_2 concentration, the slower the decomposition of the fruit. As a result, the most suitable air composition seems to be 5–10% CO_2 and 5% O_2.

C. Apples

The CA preservation method is most widly used for apple storage. The air composition is around 3% for both CO_2 and O_2. However, gas composition varies according to the varieties. For plastic film preservation, 0.03–0.05 mm thick polyethylene can be used. The fruit should be cooled down before wrapping, and some countermeasures should be taken to prevent too much humidity. Freshness preservation effects such as ethylene adsorbent, deoxidation chemicals, active carbon, and surface tuning agents are not yet ready for practical use.

After CA preservation, the apple is sealed up with a deoxidation chemical to maintain an adequate CA condition during transport and storage.

D. Citrus Fruits

Generally, citrus fruits are not suitable for wrap-preservation because humidity causes the rind puffing. However, polyethylene and polyvinyl chloride are effective for citrus fruits. When individually wrapped, the Ama natsu and Hassaku (medium and late Japanese citrus fruits) can be stored until July or August at 5°C when wrapped with 0.018–0.02 mm thick polyethylene. When the fruits are packed all together in one bag, they rot due to the high humidity inside the bag, and long-term storage is not possible. Individual wrapping prevents Hassaku from getting oleocellosis as well.

E. Other Fruits

Grapes of American origin, such as the Campbell Early and the Delaware, are not suitable for storage since they are harvested during the summer season when temperatures are very high. For European grapes, it is recommended to fumigate the grapes with sulfurous anhydride gas for a certain period of time to prevent them from too much humidity caused by the plastic film wrapping.

Bananas are wrapped with an air-permeable film to prevent damage during long-distance trasports. It is reported that 0.02 mm thick polyethylene is best for keeping the freshness of the bananas.

II. SHOCK-ABSORBING PACKAGING

Rice hulls, straw, and sawdust were used inside wooden boxes as shock absorbers in the 1940s. However, after corrugated cardboard boxes became popular, since the 1950s, plastic foam with great shock-absorbing characteristics and little physical property changes due to temperature and humidity has come into use.

Two tests—a semistatistical compressive test and a falling impact test—were carried out to design a proper shock-absorbing package for the Nijusseiki. The following results on dynamical characteristics and damage were concluded from these tests.

1. Fruits have a tendency to soften if the harvesting time is delayed, and this may even cause damage on the outside.
2. From the semistatistical compressive test, the biological yield point, which seems to be the starting point of fruit decomposition, comes sooner as harvest time is delayed, causing the fruit to be easily damaged.
3. The maximum possible acceleration for a fruit not to damage is the G factor. This factor becomes smaller as harvest time is delayed. It also shows that fruits are not as strong against impacts.

When comparing the current shock-absorbing packaging for the Nijusseiki to other fruit packaging (Fig. 3), an exhaustion damage characteristic curve (S–N curve, Fig. 4) was obtained. Coefficients are summarized in Table I. The following facts were observed from the S–N curve. Although

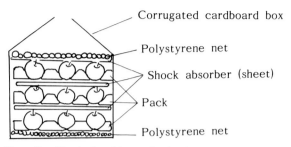

Figure 3. Shock-absorbing packaging for the Nijusseiki pear.

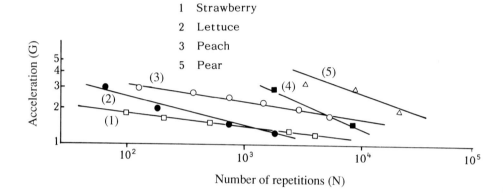

Filling method of Produces or packaging materials and the constants of α and β for S-N curve

Products	Filling method or packaging material	α	β
Strawberry	Filled in bulk into a PVC tray of 300 g capacity. Capacity per carton-box is 1.2 kg.	9.26	2.65×10^4
Lettuce	Filled in bulk into a ca ton-box of 10 kg capacity.	4.17	5.32×10^3
Grape	Filled in bulk into a carton-box of 10 kg capacity.	6.99	4.13×10^5
Peach	Filled in PVC pack. Capacity per carton-box is 5 kg.	2.44	2.46×10^4
Pear	Filled in PSP pack. Capacity per caton-box is 15 kg.	3.00	2.27×10^5

Figure 4. S–N curve of different products: (1) strawberry, (2) lettuce, (3) grape, (4) peach, (5) pear.

Chapter 17. Fruits

Table I Filling Method of Produce or Packaging Materials and the Constants α and β for S–N Curve

Products	Filling Method or Packaging Material	α	β
Strawberry	Filled in bulk into a PVC tray of 300 g capacity; capacity per carton–box is 1.2 kg	9.26	2.65×10^4
Lettuce	Filled in bulk into a carton–box of 10 kg capacity	4.17	5.32×10^3
Grape	Filled in bulk into a carton–box of 10 kg capacity	6.99	4.13×10^5
Peach	Filled in PVC pack; capacity per carton–box is 5 kg	2.44	2.46×10^4
Pear	Filled in PSP pack; capacity per carton–box is 15 kg	3.00	2.27×10^5

strawberries packed in polyvinyl chloride containers were able to resist only up to 100 vibrations under an acceleration of 1.8 G, Nijusseiki packed in a corrugated cardboard box with polystyrene pack can withstand up to 10,000 vibrations. It is clear that the current shock-absorbing package for the Nijusseiki is more than sufficient when compared with other fruit packagings.

Shock-absorbing packagings that were designed based on results from long-distance transporting tests after actually packing the fruits are expected to be examined from the fruit dynamical characteristic point of view in the future.

REFERENCES

Kawano, S., et al. (1984). *J. Soc. Agric. Machinery, Jpn.* **46**,(1), 627.
Kawano, S., et al. (1984). Report of National Food Research Institute, no. 45, 92.
Kitagawa, H. (1981). *Agric. Hortic.* **56**(1), 183.
Kudo, T. (1981). *Agric. Hortic.* **56**(1), 161.
Tarutani, T. (1960). *J. Soc. Hortic. Sci.* **29**(3), 212.
Tarutani, T. (1961). *J. Soc. Hortic. Sci.* **30**(2), 95.
Yamane, A. (1971). Special Report of Food Technological Institute of Tottori Prefecture, no. 1.
Yamane, A., et al. (1970). Report of Food Technological Institute of Tottori Prefecture, no. 20, 1.

CHAPTER 18

Vegetables

Masutaro Ohkubo
Seitoku University
Chiba, Japan

Only 25 years have passed since vegetables started to be packaged for freshness in Japan. During those 25 years, great progress has been made in the development of plastic film packages and packaging methods. In particular, packaging technologies to keep vegetables fresh at supermarkets have shown marked progress.

Around 1965, automatic selection and packaging machines were introduced for the first time into large vegetable-growing districts. However, the operation rate of the machines was low, and the price of vegetables was too low to pay for labor and other costs required for packaging. Accordingly, the machines were eventually put out of use, except those for packing green peppers.

On the other hand, precooling facilities for vegetables began to be introduced into vegetable growing districts around 1972. The operation rate of the facilities was high, and vegetables were priced according to their quality, which led to the introduction of the facilities in many vegetable-growing districts throughout Japan. Around 1978, when most vegetables were shipped to central wholesale markets after being precooled, the number of vegetables packaged in producing districts started to increase again. Possible reasons for this are as follows:

1. Packaging machines could be used for wider purposes.
2. Due to the greater variety of vegetables, they had to be packaged or precooled in their producing areas.
3. New film packages to meet such requirements were developed.

The development of preserving agents and their utilization methods is worthy of note. At present, there are several tens of preserving agents, including those under experiment. Different effects of the agents have been observed by each researcher. Some fruits and vegetables have produced a different effect, according to the kind, maturity, and concentration of oxygen and carbox dioxide within the package.

Recently, some new facts regarding vegetable packaging have been discovered.

I. OLD CONCEPTS REQUIRING REEXAMINATION

A. Vacuum and Partial-Vacuum Packaging

Since vegetables breathe, they need a proper amount of oxygen. Even under controlled atmosphere (CA) storage and air-tight packaging aimed a CA effect, a proper concentration of oxygen within the package is indispensable to the freshness of the vegetable. For this reason, vegetable vacuum packaging has rarely been studied by researchers. People with a poor knowledge of plant physiology are apt to think that vacuum packaging is good for long-term preservation because it controls the respiration of vegetables. However, the fact is that vacuum packaging accelerates the deterioration of quality of most vegetables.

Nevertheless, vacuum-packed yams are shipped successfully. According to experimental results, vacuum-packed yams keep fresh, with the cut section remaining white, no generation of smell, and good taste retention. However, it has not been made clear why vacuum-packed yams keep well and do not spoil in an extremely low concentration of oxygen.

Figure 1 shows variation in the concentrations of oxygen and carbon dioxide within vacuum packages of persimmons. The concentration of oxygen in the package is unexpectedly higher for the vacuum package than for ordinary packaging (air-tight packaging). Clearly, it is necessary to study further vacuum or partial-vacuum packaging of vegetables.

B. Utilization of High-Density Polyethylene Film

A 0.03 mm thick low-density polyethylene film has been the most widely used package film for fruits and vegetables, due to (1) gas permeability, (2) high strength, (3) low price, and (4) easy acquisition.

However, some vegetables are spoiled by the extremely high humidity and improper concentration of oxygen and carbon dioxide, because the gas and vapor permeability of the film is too low for them.

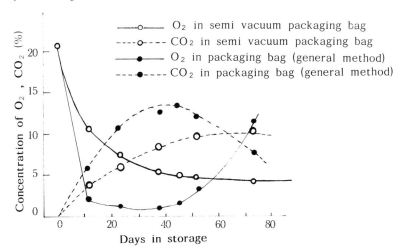

Figure 1. Changes of O_2 and CO_2 concentration in packaging bag of Japanese persimmon fruit (OPP_{20}/LPP_{40}). (From Furuta, 1984, with permission.)

On the other hand, high-density polyethylene film has not been used for packing vegetables, due to its low gas permeability.

Low-density polyethylene film, with a thickness of less than 0.02 mm, is suitable for wrapping vegetables due to its high permeability for gas and vapor; however it is difficult to handle because of its poor strength. Accordingly, the high-density polyethylene film and the low-density thin (under 0.02 mm in thickness) film have been thought to be unsuitable for packing vegetables, and selection of the wrong package film has occurred from the two films being mixed up.

Recently, good results have been reported from experiments with a 0.01 mm thick high-density polyethylene film for packing citrus fruit. This has prompted researchers to reexamine the high-density polyethylene thin film. Experiments conducted by Mr. Hasegawa and his coresearchers showed that the gas permeability of the 0.01 mm thick high-density film is equivalent to that of a 0.02 mm thick low-density polyethylene film, thus providing a moderate CA storage effect and keeping the proper concentration of oxygen. Therefore, the citrus fruit did not rot from the injury with unsuitable gas condition. The vapor permeability of the 0.01 mm high-density polyethylene film is slightly higher than that of the 0.02 mm thick low-density film. Consequently, if vegetables, which are easily damaged under a 0.03 mm thick low-density polyethylene film, due to low concentration of oxygen and extremely high humidity, are packed in the 0.01 mm thick high-density polyethylene film, CA storage and vaporization control effects can be produced.

II. NEW CONCEPTS

Packing materials for keeping vegetables fresh and preserving agents have originated from new concepts.

A. Humidity-Controlling Materials

A humidity-controlling material was developed by combining calcium chloride and a special resin (a copolymer between polyvinyl alcohol and maleic esters). Most conventional humidity-controlling materials prevent vegetables from being spoiled by extremely high humidity, due to the materials' strong hygroscopicity. The new humidity-controlling material can keep humidity inside the package suitably high if the amount of material to suit the volume of the package is put into the package. This is the most ideal method to preserve vegetables for a long period of time, because it can keep humidity inside the package constant.

B. Multifunction Film

Technology to produce a composite film made of polyethylene and activated silicon with an ethylene adsorption property has been developed. The film has both an after-ripening control effect (preserving effect) by removing ethylene and a CA storage effect. The application fields of the film are expected to expand in the future.

C. Utilization of Gas Control Agents

The main purpose of current preserving agents is to control the concentration of oxygen and carbon dioxide gases inside the package and to remove senescence (aging) promoting hormones, such as ethylene, from the atmosphere inside the package around fruits and vegetables. Many preserving agents were developed as an endermic liniment to keep vegetables from decay and prevent the propagation of molds, and many patent rights were obtained around 1965. These agents have not been widely used because of opposition to them from users with concerns for human health.

Meanwhile, substances that control the propagation of molds and bacteria were discovered from naturally generated gases. These substances are expected to appear on the market as new preserving agents in the near future. One of the substances is Hinokitiol, which is extracted from the Japanese hiba tree (*Thugopsis dolabrata*). Paper soaked in Hinokitiol extracted from scrap wood of the Japanese cypress is placed inside a package, so that Hinokitiol is gradually volatilized within the package, to prevent breeding of molds and bacteria in the fruits and vegetables. The next step after developing technology is to put it to practical use. Hinokitiol is a natural gas, which causes no safety problems for health. Besides Hinokitiol, there are other natural substances that generate gases with antibacterial and mold prevention functions. These substances are at the focus of attention as completely new preserving agents.

III. FUTURE TRENDS

Since precooling facilities have been introduced into most major vegetable growing districts, it is no surprise that most vegetables will be packaged at low temperature in the future. Sealed packaging by plastic films sometimes has a CA storage effect, but it sometimes accelerates deterioration due to the generation of gases. Most vegetables decayed by gases are packaged at high temperatures. Packaging vegetables at low temperatures helps keep them fresh and also stabilizes the CA storage effect. The distribution of half or quarter-cut vegetables and sliced vegetable salads for quick serving, which are very popular among working wives, is becoming possible through low-temperature packaging.

REFERENCES

Furuta, M. (1984). *Japan Technical Institute of Food Distribution* **55,** 11–16.
Okubo, M. (1980a). *Agriculture and Horticulture (Japan)* **55**(8), 288–291.
Okubo, M. (1980b). *Agriculture and Horticulture (Japan)* **55**(9), 292–296.
Okubo, M. (1982). "Keeping Method for Freshness of Vegetables." Yokendo (Tokyo, Japan).
Okubo, M. (1987). *Food Packaging (Japan)* **31**(9), 32–44.

CHAPTER 19

Fresh Meat

Yoshihiko Tomioka[1]
Packaging Division,
Technical Department,
Kureha Chemical Industry Co., Ltd.,
Tokyo, Japan

I. WORLD MEAT PRODUCTION

Meat is the most valuable source of protein supply for humans. The total world production of meat in 1987 was 158.8 million metric tons (m.mt) (FAO, 1987). This figure exceeded that for fish, at 91.5 m.mt (FAO, 1986), though it was smaller than for cereals and others:

Meat	158.8 m.mt (1987)
Fish	91.5 m.mt (1986)
Cereals	1786.6 m.mt (1987)
Roots and tubers	593.9 m.mt (1987)
Vegetables	421.0 m.mt (1987)
Fruits	323.6 m.mt (1987)

Among meats, the largest in quantity was pork, at 61.6 m.mt, followed by beef and veal at 48.3 m.mt, and poultry at 35.2 m.mt. The production amounts of mutton and lamb, goat meat, buffalo meat, and horse meat were relatively very small (Fig. 1).

---

[1] Present address Secondary Products Department, The Yamagishi Association, Mie-Ken, Japan.

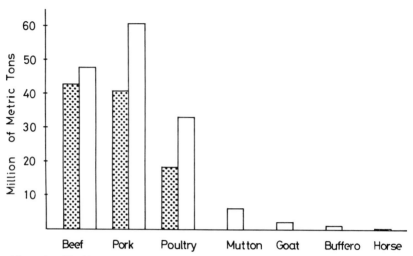

Figure 1. World meat production. The open bars show the production amount of 1987, while the dotted bars show that of 1975. (Data from FAO Yearbook.)

When we compare these 1987 figures with those of 12 years ago (average of 1974–1976 drawn with dotted bars in Fig. 1), we find the growth for poultry meat was the greatest (189%), and it was also very high for pork meat (151%), while beef meat was the lowest (113%).

However, the Japanese situation is somewhat different. The dynamic changes in Japanese meat consumption (production + import − export) are given in Fig. 2 (The Meat Journal, 1988).

In 1987, total meat consumption was 4.7 m.mt (pork 1.98 m.mt, poultry 1.63 m.mt, and beef 0.88 m.mt). A comparison between 1975 and 1987 shows growth of poultry at 216%, beef 212%, and pork 163%.

II. BRIEF HISTORICAL OVERVIEW

A. Traditional Packaging

Years ago, in Japan fresh meat was sold at butcher's shops, and sliced meat was wrapped in sheaths when delivered to the consumer. In some ways, such customs still remain even today; some processors wrap fresh frozen meat in bamboo sheaths as inner packaging (though they have now become very expensive packaging materials). Also, some meat stores

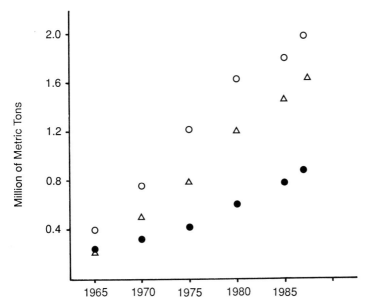

Figure 2. Japanese meat consumption. Figures indicate production + import − export. ○, Pork; △, poultry; ●, beef. (Data from *The Japan Meat Yearbook*.)

wrap fresh meat in meat paper printed with a pattern of bamboo sheaths and coated with wax.

The traditional packaging materials for fresh meat in Western countries were wrapping papers in the form of sheets or rolls. They were bleached or unbleached kraft paper, sulfite paper, vegetable parchment, greaseproof paper, glassine, and so on (Ramsbottom, 1970). They were sometimes waxed, oiled, or colored, to increase wet strength, to make them grease- and stain-proof, or to add antisticking properties when they were used for frozen meats (Romans and Ziegler, 1977). They were used for retail and wholesale purposes according to their physical properties.

B. Consumer Packaging

Packaging for meat can be classified into two categories: retail (consumer) packaging and wholesale packaging. Retail packaging worthy of the name can be said to have started after the advent of the supermarket. ("King Kullen," opened by Michael J. Cullen in New York in 1930, may be regarded as the origin of the supermarket.) One of the specific features of the supermarket is self service. In such stores, even fresh meat had to be prepackaged, to prevent dehydration, to maintain appearance, and to protect from discoloration.

For these purposes, cellophane as wrapping film and waxed paperboard trays as meat trays were used for packaging fresh meat. At the beginning, the work involved was performed manually, but it was later automated. In the early 1970s, such materials were superseded by a combination of stretchable polyvinyl chloride wrap film and clear polystyrene or foamed polystyrene trays.

These packaging systems remain basically unchanged today. Incidentally, it is said that in Japan prepackaged meat first appeared in 1957, at Seibu Department Store, Ikebukuro, Tokyo (Kato, 1980).

C. Wholesale Packaging

In the category of wholesale packaging, vacuum-shrink packaged beef is now familiar. Formerly, beef was distributed in carcass form (sides and quarters), but these days the carcasses are divided into parts (primals and subprimals), fabricated (deboned and trimmed), vacuum-packed in heavy plastic bags, placed in corrugated cardboad boxes, and shipped to the retailers. (This packaging system is referred to as "boxed beef" in the United States, and "chilled beef" in other countries.)

The original form of this packaging concept was developed by French scientists in 1932, to prolong the shelf life of frozen meat stored as military provisions in the Maginot Line fortifications. At that time, rubber hydrochloride film was used, but later, in 1946, heat shrinkable, high-gas-barrier polyvinylidene chloride (PVDC) film was developed by joint research of Dewey & Almy (the predessesor of W. R. Grace & Co.) and Dow Chemical Co. (Nishimura, 1981).

In 1948, turkey meat was vacuum-shrink packaged with PVDC bags. This was the first commercial application of the concept (Ramsbottom, 1970), and spread widely in the 1950s. However, it was not until 1966 that beef was centrally processed and distributed in the form of boxed beef (Perdue, 1979).

Also, in Australia, chilled meat was first vacuum packed by one supplier in the mid 1960s. In 1970, the first container load of chilled beef was exported to the Japanese market.

After that, this method developed rapidly in both the export and domestic markets (Husband, 1982). At the same time, this touched off an expansion of the chilled beef concept in the Japanese market.

Hereafter, we use the single term "boxed beef."

III. WHOLESALE PACKAGING

A. Boxed Beef

1. Advantages of Boxed Beef

The reasons that boxed beef has been preferred by distributors and retailers as wholesale packaging can be summarized as follows:

1. Weight loss and drying can be prevented: Bags for meat cuts have low moisture transmission rate.
2. It is hygienic: Meats are protected from contamination by dust or bacteria by packaging films. These films have very low oxygen permeability, thus preventing the growth of aerobic bacteria that are responsible for meat spoilage.
3. Maintenance of good meat color and good meat quality: If the meat is kept in the open air, the meat pigment myoglobin is oxydized to metmyoglobin, giving an undesirable brown color. But, if the meat is packaged with films with low oxygen permeability, myoglobin is maintained at a reduced level (purplish red color), and on exposure to the air, myoglobin is oxygenated to oxymyoglobin, which assumes a bright red color (Hood and Riordan, 1973).

In the same way, these packaging materials retain the other qualities of meat, such as flavor, texture, and juiciness, for a longer period than without packaging, thus providing prolonged edibility (Seideman and Durland, 1983).

4. Operations can be centralized: In a well-equipped plant, dividing and packaging can be concentrated in one place. This enables more efficient use of labor and by-products.
5. Rational distribution is possible: (a) Boxed beef reduces weight of shipment by some 20–25%, and thus reduces space and freight costs. (b) The sizable boxes are convenient for handling and stacking. (c) Standardized boxes bearing bar codes make possible movement through the plant and automatic palleting (Mans, 1980).
6. Ease of inventory control: Each box bears labels that clarify its production date, parts name, and weight. Packers can deliver products in order, according to the record (or computer memory), on shipping requirements. Store managers can utilize a necessary amount of meat cuts stored in their own refrigerator at any required moment.

2. Packaging Materials for Boxed Beef

Packaging materials used for beef are given in Table I (Hisazumi et al., 1988).

Table I Examples of Packaging Materials for Fresh Meat[a]

Film[b]	Thickness (μm)	Oxygen Transmission[c] (30°C, 80% RH)
PE	60	2000
PE/Ny/EVA	80	120
EVA/PVDC/EVA[d]	60	50
EVA/PVDC/Ionomer	60	50
Ny/EVOH/EVA	70	80

[a] From Hisazumi et al. (1988).
[b] PE, polyethylene; EVA, ethylene vinyl acetate copolymer; PVDC, polyvinylidene chloride; Ny, nylon; EVOH, ethylene vinyl alcohol copolymer.
[c] Units: ml/m^2 · 24 h · atm.
[d] Cross-linked.

Polyethylenes (PE), which are highly oxygen permeable, are used only in limited cases such as for temporary stocks or near-by shipment.

Nylon (Ny) based bags, generally coextruded laminates, were once used widely for their toughness. In Europe, small meat cuts have been packaged by thermoform vacuum seal machines that use nylon polyethylene (Ny/PE) roll stocks. However, the barrier property of Ny film tends to vary in with its moisture content (Rigg, 1979). It is estimated that Ny-based nonshrink bags were used in the U.S. market at a level of about 10% in 1984, while for Japan the figures were 20% in 1984 and 1.3% in 1987 (the remainder was PVDC-based shrink bags).

On the other hand, PVDC-based bags, which were originally single-ply film, were improved. Coextruded multistructure types appeared in the 1970s, and are now the most widely used packaging materials for boxed beef. These are heat-shrinkable bags, so-called "shrink bags" in some cases.

Figure 3 shows beef and pork meat cuts packaged with PVDC-based multiply shrink bags.

Recently, coextruded bags having layers of Ny and ethylene vinyl alcohol copolymer (EVOH) have been introduced in these markets.

3. Packaging System

For large-scale production, there are rotary clipping machines (Super Rotamatic, by Tipper Tie: see Fig. 4), or rotary chamber sealing machines (Cryovac 8300, by Cryovac division of W. R. Grace: see Fig. 5).

Chapter 19. Fresh Meat 315

Figure 3. Beef and pork meat cuts packaged with PVDC-based multiply shrink bags. (Photograph courtesy of Kureha Chemical Industry Co., Ltd.)

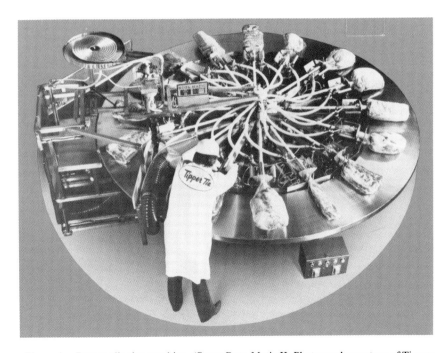

Figure 4. Rotary clipping machine. (Super Rota-Matic II: Photograph courtesy of Tipper Tie, Inc.)

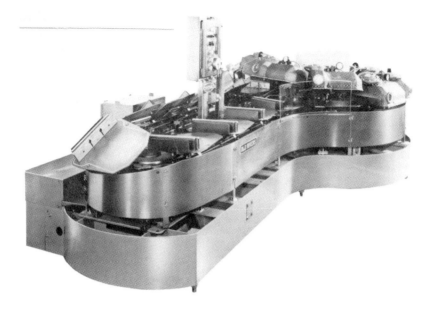

Figure 5. Rotary chamber sealing machine. (Cryovac 8300: Photograph courtesy of Cryovac Div. of W.R. Grace & Co.)

In Japan, meat packers generally use manual-type chamber sealing machines, such as Multivac, or conveyor-type chamber sealing machines, such as FVM by Furukawa Manufacturing Co.

For shrink bags, after vacuumizing, packaged meat cuts are taken to a hot-water shrinker, where the packaging materials are shrunk by hot water showering for a few seconds to conform to the meat surface. This smooths the bag wrinkles or folds and makes the film thicker, resulting in less purge and greater toughness. The hot water temperature varies according to the type of film, ranging from 75 to 90°C. After shrinking, the hot water is blown off with compressed air.

In Japan, after shrinking, chilled water (2°C, about 1 min) showering systems have been recently sometimes applied to cool down the meat surface quickly. This chilling machine reduces both the bacterial count and VBN value of the meats during storage (Nishino and Yasuda, 1986).

4. Packaging Process and Hygienic Care

After slaughtering, cattle are split into half carcasses, their hides are removed, and the carcasses are eviscerated. Washed with water, they are

Chapter 19. Fresh Meat

placed in a chill room for 24 h, then removed to a holding cooler, where they remain until sold.

These carcasses are then transferred to the dividing, boning, and fabricating line, and finally processed into boxed beef (McCoy and Sarhan, 1988).

The most important considerations throughout the line are sanitation and temperature control:

1. All plant facilities and equipment in direct contact with the meat should have been cleaned before starting operations, and should be kept clean during processing, to prevent microbial contamination.
2. All workers must be aware of the importance of sanitation, and their clothes, gloves, and boots should be maintained in a hygienic state.
3. From the chill room for carcasses to the holding cooler for boxed meat, strict temperature controls must be conducted.

B. Packaging of Pork

In 1982, the packaging method for fresh pork was 94% naked or parchment wrapping, plus packing in boxes. Among the remaining 6%, the bulk gas flushing method and vacuum packaging in shrink bags accounted for 5% and 1%, respectively, of the U.S. wholesale trade market (*Meat Industry,* 1983).

Vacuum packaging of pork subprimals in shrink bags is the same concept as boxed beef.

In 1986, mini-subprimals, which are half-cuts or third-cuts of ordinary subprimals, constituted nearly 10% of the fresh pork delivered to retail stores (Allen and Pierson, 1986).

In Japan, about 67% of fresh pork meat is processed into subprimals (the remainder is shipped in carcass form), and 86% of subprimals is distributed in chilled condition. About 33% of precessors package these chilled subprimals in high-density polyethylene (HDPE) sheet for inner wrapping and corrugated boxes for outer packaging. Another 61% of processors use a combination of corrugated boxes and returnable plastic containers for outer packaging (Japan Meat Grading Association, 1988).

Until recently, there was little use of the vacuum-packaged boxed pork concept, but the increasing amount of chilled boxed pork imported from Taiwan may stimulate a change in the packaging of pork in the Japanese market.

C. Packaging of Poultry

It may be said that the most developed field in the fresh meat industry has been that of poultry, from the viewpoint of production, processing, and packaging.

It is still developing, and the packaging applications have been changing rapidly.

From information we received, the poultry shipment methods several years ago (in 1983) in the United States were as follows:

Shipping Method	Percent of Shipment
Wet ice	50
Chilled/prepacked	30
Frozen	9
Gas flushed	7
Dry ice	5
Bulk tank	1

About 65% of "wet ice" and 35% of "chilled and prepacked" were in the form of whole birds. Recently, whole bird shipments have greatly decreased, except for roasting chicken, and the majority is shipped in the form of part meats. In case of "chilled/prepacked," poultry were vacuum packaged in polyolefin bags (LDPE, EVA; single- or two-layer), or stretch-wrapped on plastic trays. Recently, the movement toward branded, retail-ready packaging (tray overwrapped) has been increasing.

In Japan, broilers are divided up into individual parts and offals, and vacuum packed in PE bags in 2-kg units. Then six units are put into a plastic container with ice and shipped to the retailers in city areas. Direct shipping to the supermarket or convenience store is now undertaken, requiring, in addition to the standard 2-kg packaging, much smaller packs, or retail-ready tray-wrap packaging, both for fresh and further processed (ready-to-cook or cooked-deli) chicken products. The use of branded poultry merchandizing is also increasing in Japanese retail markets.

IV. RETAIL PACKAGING

As mentioned before, retail packaging for fresh meat is generally the overwrapped tray system. However, there are two major types of method that are now already practiced to some extent or are in a trial marketing situation. One is gas packaging and the other is again vacuum packaging, but in the form of deep-draw or skin packaging.

A. Gas Packaging

As compared with traditional gas packaging, whose concept is to extrude oxygen (O_2) for improving food preservation, the specific feature of gas packaging for fresh meat is to increase the O_2 above that of the atmosphere, to keep a bright red meat color, and at the same time to add carbon dioxide to achieve a bacteriostatic effect. This original concept is explained in two typical patents, that of Schweisfurth and Kalle AG in West Germany (British patent specification 1,186,978, 1970) and that of Unilever Ltd. (British patent specification 1,199,998, 1970).

Nowadays a gas mixture containing 50–80% O_2, 20–30% CO_2, and nitrogen (N_2) as the remainder is usually used. For packaging materials, 300–800 μm PVC/EVA, PVC/EVOH/EVA, or PS/EVOH/PE are used for tray materials, and 70–100 μm PVDC-coated PET/PE or PVC/PVDC/EVA and inner side antimist are generally used for lidding materials.

Packaging machines may be divided into two groups: in-line thermoform–fill–seal machines, which make trays on the machine from roll stocks, and off-line fill–seal machines, which are supplied with preformed trays. Among the former type, there are, for example, the Tiromat range (Krämer + Grabe GmbH & Co. KG), Multivac Series (Multivac Sepp Haggeenmüller KG), and FV series (Omori Machinery Co., Japan). Among the latter type, there is, for example, TG-1 (Nishihara MFG Co., Japan).

With this gas packaging method, the shelf life of fresh beef is doubled at 5°C, compared with the tray film overwrap method (Tomioka et al., 1983).

Some examples of gas-packaged fresh meat are given in Fig. 6.

B. Vacuum Packaging in Deep-Draw or Skin-Pack Form

The other major type of retail packaging is vacuum packaging, in the form of deep draw or skin pack.

Deep-draw packaging of fresh meat has been mentioned already under wholesale packaging, but recently it has been applied to smaller meat cuts, for retail-ready meat, especially in European countries.

Though skin pack is within the category of vacuum packaging, it uses soft upper films, and the pressure to the meat is small, resulting in less purge. The upper film is generally gas barrier type coextruded plastic film, heated to be more flexible, then evacuated to form a contour package of the meat on a shallow plastic (foam) tray.

Meats packaged by these methods, since the entry of oxygen is prevented, do not deteriorate to produce rancidity, which usually is inevitable

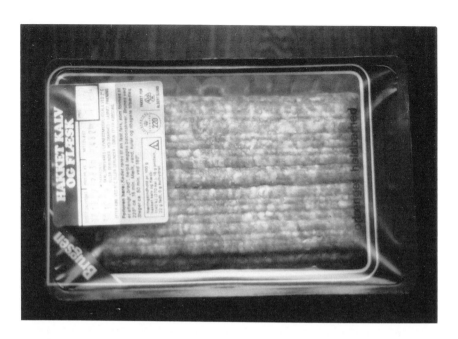

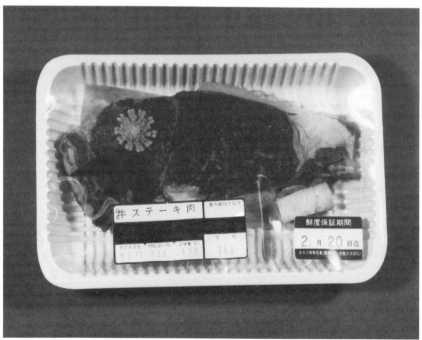

Figure 6. Gas-packaged fresh meat. Ground meat sold in Denmark (above), and beefsteak sold in Japan (below).

before spoilage with gas-packaged meat. However, the color of the meat, because of the lack of oxygen, becomes purplish red, which is very different from the hue of normal meat. If the consumer is educated to accept that the purplish color does not mean spoiled meat, but is rather a sign of freshness, the total shelf-life of skin pack is expected to be doubled or tripled compared with that of gas packaging.

In the United States, some supermarkets have been marketing branded fresh meat centrally skin packaged, combined with consumer-education promotions.

REFERENCES

Allen, J. W., and Pierson, T. R. (1986). In "Proceedings Future-Pack '86," pp. 5–20. Ryder Associates, Whippany, N.J.

FAO (1986). "Yearbook of Fishery Statistics 1986," Vol. 63, p. 20. Food and Agriculture Organization of the United Nations, Rome.

FAO (1987). "FAO Yearbook: Production 1987," Vol. 41, pp. 113–173, 253–266. Food and Agricuture Organization of the United Nations, Rome.

Japan Meat Grading Association (1988). "The Research Report Concerning about the Distribution of Meat Cuts in Japanese Market–1987," p. 70. Japan Meat Grading Association, Tokyo.

Hisazumi, N., Nishimoto, U., and Funabashi, S. (1988). Kureha Chemical Industry Technical Report.

Hood, D. E., and Riordan, E. B. (1973). *J. Food Technol.* **8,** 333–343.

Husband, P. M. (1982). *Food Technol. Aust.* **34**(6), 272–275.

Kato, S. (1980). *Freezing Air Cond. Technol.* **31**(368), 18–33.

Mans, J. (1980). *Processed Prepared Foods* **10,** 71–78.

McCoy, J. H., and Sarhan, M. E. (1988). "Livestock and Meat Marketing," 3rd ed., pp. 280–324. Van Nostrand Reinhold, New York.

Meat Industry (1983). **11,** 39–40.

Nishimura, K. (1981). *Japan Food Sci.* **20,** (6), 42–46.

Nishino, H., and Yasuda, M. (1986). Kureha Chemical Industry Technical Report.

Perdue, D. (1979). *Meat Processing* **18**(6), 36–44.

Ramsbottom, J. M. (1970). In "The Science of Meat and Meat Products" (J. F. Price and B. S. Schweigert, eds.), 2nd ed., pp. 516, 529. W. H. Freeman, San Francisco.

Rigg, W. J. (1979). *J. Food Technol.* **14,** 149–155.

Romans, J. R., and Ziegler, P. T. (1977). "The Meat We Eat," 11th ed., p. 602. Interstate, Danville, Ill.

Seideman, S. C., and Durland, P. R. (1983). *J. Food Qual.* **6,** 29–47.

The Meat Journal (Shokuniku Tsushin) (1988). "The Japan Meat Yearbook '89", pp. 506–507. Shokuniku Tsushin, Osaka.

Tomioka, Y., Yasuda, M., Tsuchiya, E., and Yokoyama, M. (1983). *J. Jpn. Soc. Food Sci. Technol.* **30,** 25–32.

CHAPTER 20

Meat By-Products

Yoshiyuki Tohma
Marudai Food Co., Ltd.,
Tokyo, Japan

During the more than 40 years since the end of World War II, meat packaging in Japan has seen enormous development. In fact, one can say without exaggeration that previous meat packaging techniques have been completely discarded and replaced with new ones, mostly based on the concept of aseptic packaging.

Meat packaging techniques have been profoundly influenced by packaging methods of medical products, and by now practically every meat processor in Japan uses "clean rooms" in which dust and bacteria are completely controlled. Some processors have gone so far as to convert their entire factories into a clean room, because for completely aseptic packaging practically each and every step has to be controlled.

So machines are now made of stainless steel instead of just steel, factory walls are covered with a smoother layer of paint, and floors are covered with hard tiles to make them all easier to hose down and keep spotlessly clean. While even a few years ago cooking was still done in hot water in big boiling vessels, now smoke-houses are used, less for smoking than for cooking.

A short description follows of how aseptic packaging systems were introduced into the manufacture of modern Japanese meat products.

1. Simple wrapping
2. Vacuum packaging
3. Double-sterilization packaging
4. Boil and steam cooking packaging
5. Retort sterilized packaging
6. Air-containing packaging

7. Oxygen-absorbing agent packaging
8. Aseptic packaging
9. Packaging of each product line
   A. Manufacturing process of roast ham and boneless ham
   B. Manufacturing process of wiener sausages
10. Summary

It should be kept in mind that when I explain modern Japanese packaging, it does not mean that each packaging method is always necessarily independent. Sometimes various methods are combined.

I. SIMPLE WRAPPING

Since World War II, many and various types of casing for the purpose of wrapping every kind of food have been developed in Japan. Initially, simple wrapping was the only means of wrapping. But simple wrapping was also the base on which modern packaging was developed (Fig. 1).

II. VACUUM PACKAGING

Simple wrapping means just wrapping in paper, cellophane, and other chemical films to keep the goods from dust, air, light, and contaminated

Figure 1. Wiener sausages are just wrapped in chemical films.

Chapter 20. Meat By-Products

human hands or to just protect them in the showcase environment. Formerly the shelf life of food was usually very short; in addition, home refrigerators were not as widely used as they are today. At that time many raw meats, raw fish, and other raw foods were apt to go bad during the hot and rainy seasons.

More than 25 years ago the technology of the fundamental style of vacuum packing was introduced from the United States or Europe and helped to prolong the shelf life. At that time, cooking temperatures of meat products were comparatively lower than today and thus unable to completely sterilize bacteria spores. But when the air in the packaging was taken out, most aerobic microorganisms could not germinate and shelf life of the products was prolonged. It has become fundamental to use vacuum packaging with many foods to keep them from spoiling (Fig. 2).

III. DOUBLE-STERILIZATION PACKAGING

Polyvinyl chloride films for wrapping of foods have made rapid progress since about 1960 according to production figures of the petroleum chemical industry. This packaging is similar to simple wrapping. When the inner air inside the packaging is taken out, it becomes the same method as vacuum

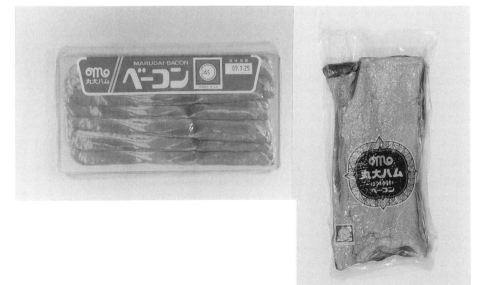

Figure 2. Vacuum packaging of bacons that did not have second sterilization.

packaging. But it has the defect that the shelf life is very short. Meat products in this condition have high bacteria counts, from 10^3 to 10^4/g at times, because these bacteria come from slicers, vessels, workers' hands, even the meat products themselves and the atmosphere. As the industry wanted to give the products a much longer shelf life and wanted to sell them in good condition, new sterilized systems were developed. On the other hand, if the method of preservation is inadequate—if, for instance, the chemical film has some defects—discoloration of the product from oxygen happens or shrinkage occurs, thus reducing the merchandise value. Many improvements of the packaging film have been made to ensure complete boiling of the products at 80°C for 20 min to preserve their condition. Double-sterilization packaging machines have made rapid progress, so that both heating by boiling water and cooling by chilled water are possible at the same time in the machine. As a result, good processed sliced meat products with 40–50 days shelf life even in the summer season have become possible (Fig. 3).

IV. BOIL-AND-STEAM COOKING PACKAGING

When ham and sausage are boiled in hot water, it is necessary that the packaging films have heat-resistant characteristics. Most Japanese meat products have been subjected to a cooking temperature between 60 and

Figure 3. Double-sterilization packaging of lump of roast porks and chopped ham in soft film packaging.

90°C at the center. The packaging film should be able to endure those temperatures and should not break down when the products are boiling. When the products are boiled in hot water, they need to be waterproof. On the other hand, steam cooking and smoking films are requested to pass air and smoke. Pressed ham and chopped ham are among the most popular and original products in Japan requested to be put in waterproof films. Ham and sausages of the European type, however, are made by a process of cooking and smoking, and the film must pass smoke and water during the process. The former are boiled in a vessel and the latter are finished in a smoke house. The film for boiling is made of chemicals like polyvinyl chloride. The film for smoking is made from fibrous casing through which smoke and moisture can pass. Then those products are sliced and wrapped with aseptic packaging (Figs. 4 and 5).

V. RETORT STERILIZED PACKAGING

In this group there are many types of small products from 50 to 80 g weight per stick, including the so-called finger type sausage. They are stuffed by

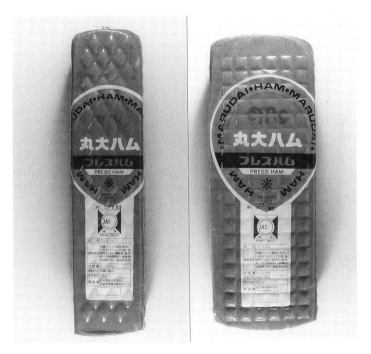

Figure 4. Boiled packaging, pressed ham and bologna sausage.

Figure 5. Boneless ham and roast ham, cooked and smoked in a smoke house.

high-speed filling machines at 200 sticks/min. Ground raw materials are filled into chemical film and are knotted with the first and second clips at the same time and then sterilized at 120°C. Such sterilized meat and fish products are sold without refrigeration at natural temperatures during all seasons. Shelf life is long, about 90–120 days. In Japan, since about 25 years ago, preservation food additives have been strictly controlled by the Ministry of Welfare, and about 10 years ago their use was forbidden at any time at all in meat products. After that, the shelf life of meat products became very short, under 1–2 weeks. Therefore, since then retort sterilized packaging systems became utilized for fish, ham, and sausage in Japan (Fig. 6).

VI. AIR-CONTAINING PACKAGING

This packaging system does not injure the sliced and pieced products inside. This is achieved by taking the air out of the pouch and thus creating a vacuum condition as the first step, and then inserting a mixed gas of N_2 and CO_2 at the ratio of 90% to 10% into the pouch. If the amount of CO_2 is higher than this ratio to N_2, CO_2 is absorbed into the protein of the meat

Chapter 20. Meat By-Products

Figure 6. Retort sterilized packaging of fish, ham, sausage, and hamburger.

product. Therefore it is important to preserve a product at the correct ratio of CO_2 and N_2. This packaging has the advantage of preventing discoloration of the product and preventing growth of aerobic microorganisms (Fig. 7).

VII. OXYGEN-ABSORBING AGENT PACKAGING

This packaging style looks like air-containing packaging, but here the inner air is not taken out but left as it is. When the meat product is wrapped into the packaging, an oxygen-absorbing agent is put in at the same time and the package is entirely closed. The oxygen is absorbed by this agent, so the inside of the film is kept in an anaerobic condition to prevent the growth of mold and other aerobic microorganisms (Fig. 8).

If the anaerobic condition is preserved, the color of the meat product does not change. In such a packaging system the meat product inside is not injured, the same as in air-containing packaging.

VIII. ASEPTIC PACKAGING

I think that aseptic packaging is the ultimate wrapping or tomorrow, at least for meat products (Fig. 9). For that kind of packaging system good aseptic condition of the packaging room is required. Most Japanese aseptic

Figure 7. Various types of air-containing packaging. The air was exchanged with a mixture of N_2 and CO_2. These aseptic products are made in a clean room.

rooms for meat products have a limit of dust count at 10,000/ft^3 of atmosphere. The air in the room is changed totally about 30 times per hour. The atmosphere in the room is kept under pressure, so outside air that is not sterilized cannot enter the room. When workers want to enter the room they have to remove their outer clothes, boots, and hats, and put on exclusive aseptic-room wear. They have to wash their hands thoroughly and can only enter the room through an air shower.

At the first step, the products are packed in film about 2 m long and 9 cm in diamter. The casing must be usable both for cooking and for smoking. Before entering the aseptic room, the long-size meat products are cleaned and pasteurized on their surface with a 1/500 sodium chloride solution until the bacteria on the surface are almost completely removed and the prod-

Figure 8. Oxygen-absorbing agent packaging of dry and semidry meat products. The agent is inside the product.

ucts are then pasteurized at 95°C for 3–5 min. Naturally, at such a short cooking time, the center of the product has still a bacteria count of about 10^2 to 10^3/g. Therefore the clean room is contaminated constantly by bacteria from the products, and thus control of the clean room is a very difficult. A bacteria inspector visits every day to check various places such as machines, conveyors, workers' hands, floor, and walls. He looks especially for colon bacillus, lactobacillus, and falling bacteria of the atmosphere. If the checker finds any bacteria in any place, the location is immediately cleaned and pasteurized until no more bacteria are found.

The shelf life of each aseptic package depends on the quality and the character of the meat product. The shelf life is marked on the surface of the wrapping casing, in accordance with the Hygiene Law of Japan.

If control of the clean room is not maintained well, the shelf life is immediately shortened. Control of the sanitary condition of the clean room depends also very much on hygienic processing before the products enter the clean room, but the clean room itself should be so designed that, for instance, the floors and walls are made smooth and easy to clean. The air conditioning of the factory should be adequate for every season. If the clean room is in good condition, the shelf life of, for example, wiener

Figure 9. Hams and wiener sausages in aseptic packaging.

sausages can be extended from 2 weeks to 6 weeks. So today in Japan most meat product factories have some kind of clean room.

IX. PACKAGING OF TYPICAL PRODUCT LINES

I would like to introduce some typical meat products in Japan and their packaging.

A. Roast Ham and Boneless Ham

Roast ham and boneless ham are comparatively high-quality products in Japan. After removing excess fat and any remaining small pieces of bone, the raw material is injected with liquid NaCl, $NaNO_2$, and mixed spices. Then they are put into a vacuum tumbler and cured for about 48 h in a refrigerator at 5°C. Casings of roast and boneless ham are made from artificial film which allows air and smoke to pass through. They are called fibrous casings. The meat is mixed in a vacuum mixer. Then they are filled into the casing by filling a machine semi-automatically. These fillings are about 2 m long and from 9 to 10 cm in diameter and thus easy to handle in the clean room and the right size to cut into consumer packs. Before entering the clean room the products are washed with sodium chloride and put into hot water (95°C) for 3–5 min. In the clean room, the surface casing is taken off and then sliced automatically into previously decided sizes. The sliced products go on the automatic chucker and are put into the packaging machine with robot hands, so they are not touched by human hands. Then they are vacuum-packed automatically into a completely aseptic package. Consumer packs of roast ham are made of two types of films. One is the smoke-passing type of film, most popularly the fibrous casing, which is used for long and large sticks for cooking and smoking. The other one is an artificial chemical casing that prevents any air, light, or moisture passing through. If such films are used in aseptic packaging, the shelf life of the sliced products is about 45 days, if the temperature is kept constantly under 10°C. Since the introduction of this system, spoilage of meat products has decreased dramatically.

B. Wiener Sausage

The most popular sausage in Japan is the wiener sausage. As far as casing and packaging are concerned, we have two types of wieners. One type is stuffed in natural casing, mostly sheep casing. The other type is stuffed in an artificial chemical casing, which is less palatable. Artificial casing is ideal for production of wiener sausage on a continuous basis. The yield is much better than with natural casing. But the feeling when biting is not as good as with natural casing. These wiener sausages are all packaged in clean rooms for aseptic packaging. They have a guaranteed shelf life at the supermarkets of between about 40 and 50 days, if they ar kept at 10°C. Most wiener sausages are of the European type, made from pork meat only and eaten fried or boiled. Besides that, we have in Japan a wiener sausage that contains fish meat. It has a light and plain taste and is original to Japan.

X. SUMMARY

Before World War II we had in Japan many kinds of diseases caused by microorganisms, which caused many deaths. Nowadays such instances are rare, since the hygienic circumstances have rapidly improved. I think that aseptic packaging wil also gradually spread to other food industries in Japan.

CHAPTER 21

Seafood Products

Takao Fujita
Central Research Laboratory,
Nippon Suisan Kaisha, Ltd.,
Tokyo, Japan

With progress in preservative techniques, seafood consumption has been increasing in the inland area and around the world, and seafood packaging has become an important problem. In Japan, seafood has been consummed traditionally in great amounts and many varieties, so the outline of seafood packaging will be explained mainly based on experience in Japan.

The first characteristic of seafood is its great number of species. Large fish such as tuna, shark, and halibut, medium-size fish such as salmon, cod, and flounder, small fish such as sardines, anchovies, and capelin, crustacea such as prawns, shrimp, and crabs, and mollusks such as squid, oysters, and mussels are mainly consumed. Many others, including echinoderms (sea urchin roe), coelenterates (jelly fish), sea mammals, and seaweeds, are used for food. When those marine products are processed by filleting, cuttintg, salting, smoking, pickling, drying, canning, and otherwise, the total of seafood items would reach an enormous number.

To cover such varieties of seafood, packaging is required to have several characteristics. For instance, small fish are packaged in bulk, regular-shaped packages are used for irregular-shaped seafood, large frozen packages of 10–30 kg, maybe used, and individual small packages for consumers all play important roles in providing better handling convenience or improved product characteristics.

The second characteristic of seafood is that it loses freshness much faster than other foods. Generally, seafood are juicy and susceptible to autolysis by enzymes and loss of freshness by bacteria. Even dried fish are susceptible to oxidation because of the high degree of unsaturated lipid components, causing discoloration and rancid flavor. To protect the contents from the deterioration of qualities such as color, taste, or odor is an important role required of packaging.

I. FRESH FISH

Fresh fish are usually distributed in iced or chilled condition from catch to consumption. Often, frozen fish are thawed and sold as fresh fish in retail store. Cultured fish, shellfish, and others are distributed as live fish.

Characteristic of those fresh fish is their highly juicy and mucous surface, which makes handling difficult and causes rapid loss of freshness. Therefore, a prerequisite to those packages is that they offer an easy disposal of drip, handling convenience, and hygienic protection of contents.

Formerly, wooden boxes with parchment liners were used for bulk packaging of fresh fish. Recently, plastic containers, plastic corrugated cartons, and metallic containers are used for fresh fish packaging. Fresh fish are often packed with ice in a foam polystyrene carton (Fig. 1).

Meanwhile, for consumer packaging in retail stores, the most popular

Figure 1. Sardines packed in foam polystyrene carton with ice.

package consists of a shallow tray and transparent film overwrap. The trays are fabricated from foamed polystyrene or clear polystyrene or wax-coated paper. For overwrap film, PVDC, polypropylene, etc. are used. Often, an absorbent sheet are placed on the inner bottom of the tray to soak up the drip. Fresh fish are also packed in polyethylene bags, small wooden boxes, and wax-coated paper.

Fresh fish are a difficult area for packaging, and the improvement on packaging material and the development of standardized packing form will be the future problems.

II. FROZEN FISH

A considerable amount of marine products is frozen for better shelf-life and is distributed in frozen condition with the development of freezer chains. Seafood are processed and frozen on-vessel immediately after the catch, or they are frozen on land at the factory after cleaning, beheading, gutting, shell removing, filleting, and cutting processes. They are frozen in freezing pans, in freezing pans with water, dry pack freezing, or IQF (individual quick freezing) by contact freezing, air-blast freezing, or immersion freezing.

The first characteristic of frozen seafood is their hardness and irregularity of shape, especially with protrusion of fin and spine and sharp edges. To cope with those conditions, good freezing resistance and antifreezing strength are required of the package. To adapt those requirements, for instance, shrimps are frozen into regular shapes in freezing pans with water and packed in wax-coated cartons.

The second characteristic is susceptibility to freezer burn. In cold and dry atmospheres, ice sublimes and the surface of frozen product is easily dehydrated, and rancidification occurs through oxidation. Therefore, packages are required to protect the contents from such adverse influence. Moisture resistance, gas-barrier capability, waterproofing, oil resistance, and airtightness are important requirements for the package.

Large fish like tuna are frozen in the round, and usually ice glaze treatment is given to the surface of tuna and it is distributed without further package. Medium-size fish are generally frozen by air blast individually and are packed in a polyethylene sheet or bag and, as outer packaging, in a water-resistant corrugated fiberboard box, plastic corrugated box, or foam polystyrene box (Fig. 2). For small-size seafood like shrimp and fish fillets of consumer size, packaging almost identical to those of frozen foodstuffs are applied, including polyethylene bag, vacuum packaging using laminated film, shrink package, wrapped tray package, paper box, etc. As an example of popular packaged frozen food, production process of "fish

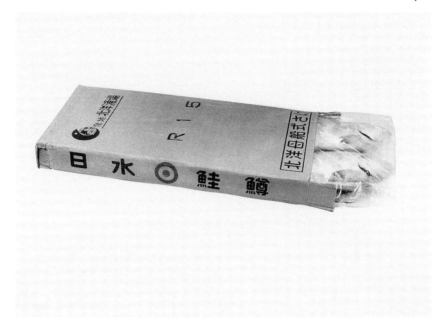

Figure 2. Salmon frozen individually are packed in polyethylene bag and corrugated fiberboard box.

stick" will be described. The fish fillets are packaged in a water-resistant fiberboard box and frozen by contact freezer on the vessel or at the land factory and stored. The frozen blocks are cut, breaded, fried, and frozen in IQF, then packaged in a consumer-size fiberboard box with an outer corrugated fiberboard box.

III. DRIED, SALTED, AND OTHER TYPES OF SEAFOOD PRODUCTS

There are various items in this category. They are dried, salted, seasoned, pickled, and smoked items, distributed in normal, cool, or freezing conditions. These processed seafood are packed, depending on type, in the following packages: wooden boxes, paper containers, plastic containers, polyethylene bags, vacuum packages, gas-replacement packages, packaging with deoxidative and desicating agents, and others (Fig. 3).

As an example of very strict requirements, the case of flakes of dried bonito will be described. These flakes are a favorite seasoning for cooking in Japan. Because the taste and flavor are very important but unstable, the

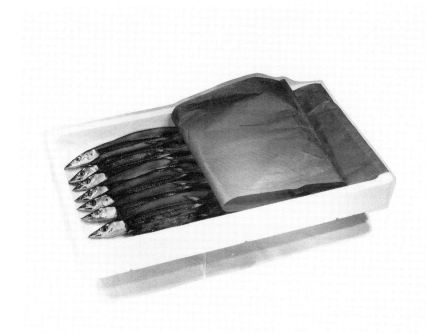

Figure 3. Processed saury (gutted, salted, and slightly dried), packed in plastic container with parchment paper.

item is usually packaged in a high-gas-barrier film such as EVAL (ethylene-vinyl acetate copolymer, saponified) with inert gas inside. The JAS (Japanese Agricultural Standard) calls for the packing material in dry condition at 20°C to not exceed an oxygen permeability of 1 ml/m$^2 \cdot$ 24 h.

Although it is not for food, a great amount of fish meal is made from fish. The fish meal is packaged in an airtight container such as a multiwall paper sacks with an inner liner of polethylene or three-layered high-density polyethylene cloth bags.

IV. CANNED FISH

Various species of seafood are packed in cans and glass bottle with brine, oil, seasoning, etc. Tinplate iron is usually employed for canning with various coatings on the inner surface, like oleoresinous type C enamel and phenolic type seafood-enamel coating. Aluminum and chrome-plated iron are also used. In canned crabmeat, chelating parchment paper is used inside for preventing struvite. Recently, deep-draw two-piece cans and pull-top cans, made of aluminum are often used. In addition to these rigid containers, the flexible retort pouch is also used, for instance, a three-ply laminate bag composed of an outer layer of polyester film, a middle layer of aluminum foil, and an inner layer of modified polypropylene.

Seafood vary in shape and are required in iced or frozen condition for distribution. These facts makes the packaging of seafood difficult. Solution of this problem with a new concept of container packaging that offers standardized handling conveniences and hygienic protection would be the future problem.

REFERENCES

Chai, T., Pace, J., and Cossaboom, T. (1984). Extension of shelf-life of oysters by pasteurization in flexible pouches. *J. Food Science* **49,** 331–333.

Gray, R. J. H., Hoover, D. G., and Muir, A. M. (1983). Attenuation of microbial growth on modified atmosphere-packaged fish. *J. Food Prot.* **46,** 610–613.

Josephson, D. B., Lindsay, R. C., and Stuiber, D. A. (1985). Effect of handling and packaging on the quality of frozen whitefish. *J. Food Science* **50,** 1–4.

Lannelongue, M., Hanna, M. O., Finne, G., Nickelson, R. II, and Vanderzant, C. (1982). Storage characteristics of finfish fillets packaged in modified gas atmospheres containing carbon dioxide. *J. Food Prot.* **45,** 440–444.

Mills, A. (1985). Fish: New guidelines for CAP. *Food Manuf.* **60,** 45–46.

OECD (1970). "Packages and Packaging Material for Fish." OECD, Paris.

Post, L. S., Lee, D. A., Solberg, M., Furgang, D., Specchio, J., and Graham, C. (1985). Development of botulinal toxin and sensory deterioration during storage of vacuum and modified atmosphere packaged fish fillets. *J. Food Science* **50,** 990–996.

Tejada, M., Borderias, A. J., and Moral, A. (1986). Stability of frozen trout II. Different trout preparations stored at $-18°C$. *J. Food Biochem.* **10,** 47–53.

Ward, D. R., Pierson, M. D., and Minnick, M. S. (1984). Determination of equivalent processes for the pasteurization of crabmeat in cans and flexible pouches. *J. Food Science* **49,** 1003–1004, 1017.

Williams, S. K., Martin, R., Brown, W. L., and Bacus, J. N. (1983). Moisture loss in tray-packed fresh fish during eight days storage at 2°C. *J. Food Science* **48,** 168–171.

CHAPTER 22

Fish Meat By-Products

Shigeyuki Sasayama
Tokai Regional Fisheries Research Laboratory,
Fisheries Agency,
Tokyo, Japan

I. FISH PASTE PRODUCTS (KAMABOKO)

The history of kamaboko is quite long. It was already in use in the sixteenth century as a special food for banquets.

The annual report on Marine Product Distribution Statistics states that recent annual fish catches in Japan average over 10 million tons, of which about 25% is consumed in manufacturing fish meat pastes, with annual fish meat paste production of about 1 million tons.

There are many kinds of kamaboko, according to heating methods and packaging, divided into steamed kamaboko, broiled kamaboko, boiled kamaboko, fried kamaboko, and specially packaged kamaboko.

Appearance, flavor, elasticity, and keeping properties determine the quality of fish paste products, requiring that they should be manufactured with production techniques that satisfy all these requirements. The main thing required of kamaboko is elasticity—it should be firm and yet be cut easily with the teeth (in Japanese, this elasticity is called *ashi*).

When fish meat is heated, the separated juice and the meat are converted into a brittle coagulant substance with low elasticity. Then 3% salt is added and the fish is ground. The meat is heated to convert into a sol, which is adhesive. Heating of this sol produces a gel with high elasticity. The reason why this occures is that when the fish meat is ground with salt, myofibril protein, which makes up 65–70% of the muscle protein, dissolves, causing the fish meat filaments to tangle with each other. Then heating makes the meat form a ternary network construction, which is a structure similar to that of the sponge. In the heat coagulation, part of the water separates. The separated water with the water enclosed in the ternary network construction helps to impart elasticity to the finished kamaboko (Table I).

A. Processing of Kamaboko

After fish are well washed and the scales, skins, and mucus have been removed, the head and internal organs are removed (Fig. 1).

Prepared fishes are placed on fish meat collecting machines to collect otoshimi. The collected fish meats are washed with water to clean and refine them. Otoshimi, the finely cut fish meats, are stirred after 5–10 times of the volume of the fish meat of fresh water is added. Then they are kept still for about 10–15 minutes. After meat is precipitated, the top portion of water is discarded. Water is added again and stirring is repeated. After stirring, the top portion of water is discarded again after meat has precipitated. This cycle is repeated three to four times. Following washing, fish meats are dehydrated. Basically, dehydration is performed by compressing fish meat kept in cloth bags. However, depending on the quantity of meat to be processed, such methods as hydraulic press, centrifugal dehydrator, or continuous screw press are used. After dehydration, fish meat is chopped by chopper. After chopping, the fish meat is ground. This process consists of three operations: *kara-zuri, ara-zuri,* and *hon-zure. Kara-zuri* is a grinding operation with only fish meats placed in the mortar (kneader). *Ara-zuri,* grinding of fish meat adding some salt, which causes hydration and dispersion of myofibril protein, produces sticky ground meat. During *hon-zuri,* starch, seasoning, bonding agent, and egg white are added. Ground meats, after grinding operation, are shaped into itatisuki-kamaboko, chikuwa, sumaki, casing-packed kamaboko, etc. Shaping may be performed manually in small-scale operations. However, recently, au-

Table I Production of Fish Paste Products (in Tons)

	1972	1977	1982	1987
Fish paste products	1,156,205	1,086,962	960,876	925,933
Kamaboko group	399,783	428,171	352,074	357,691
Kamaboko	305,984	266,216	212,171	170,952
Packaged kamaboko	30,032	77,651	56,364	57,990
Others	63,766	84,304	83,539	128,749
Broiled chikuwa	244,615	214,393	187,734	189,297
Fried kamaboko	326,623	303,224	289,361	271,488
Fish ham and fish sausage	162,398	125,088	95,152	89,146
Miscellaneous	6,381	16,086	36,555	18,311

tomatic shaping has become more popular using automatic shaping machine. After shaping, fish meat is heated. Heating gives the proper elasticity to ground fish meat and destroys bacteria. Together with the grinding operation, this is the most critical process for the quality of finished fish paste products. There are four fundamental methods for heating: steaming, broiling, boiling, and frying.

Steaming is done within a steaming basket with its temperature kept at around 80–95°C by steam. Major applications include production of itatsuki-kamaboko, sumaki, etc.

Broiling is heating by a broiling oven kept at 180–300°C using charcoal, gas, electric heaters, infrared lamps, etc. This method is mostly used for production of yakinuki-kamaboko, chikuwa, sasa-kamaboko, etc.

In the boiling method, products are boiled in water at 80–90°C. A major application is production of hampen, shinjyo, etc.

Frying is heating in edible oil heated at 160–180°C for preparation of various fried kamaboko. After heating, kamaboko is cooled in a clean room. Normally, it is cooled with a cool air recyling cooling system.

Almost all products are shipped properly packaged and then sufficiently cooled. As packing materials for fish paste products, materials that have excellent tensile strength, breaking strength, and low vapor and gas permeability, and that are highly waterproof, oil-resistant, and weather-resistant and with appropriate thermal contraction, are recommended. For the films used for wrap packaging and vacuum packaging of kamaboko, high heat-sealing capability is required. There are inner packing and outer packings for packaging. The inner packing that covers the product surface directly is particularly important to maintain product quality. Materials used as major inner packing materials include antimoisture cellophane, polyethylene, polyethylene/cellophane, polyvinylidene chloride, and nylon (Table II). Although outer packing is intended for protection of products and for advertisement, on many products it is mandatory to show various product details, such as date of manufacture, name of manufacture, or ingredients.

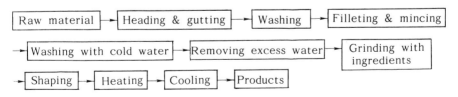

Figure 1. Process flow for kamaboko.

Table II Packaging Materials for Fish Paste Products

Product	Packaging materials

Simple wrapped products
  Broiled kamaboko
  Steamed kamaboko
	Plain cellophane, moisture-proof cellophane, polyethylene, polycello, pure white paper

Retainer-formed kamaboko (specially packaged kamaboko)	Inner packing: cellophane coated with saran, moisture-proof cellophane, polypropylene, polyethylene, polycarbonate Outer packing: polyethylene, polypropylene Display packing: moisture-proof cellophane, polyvinylidene choloride
Casing-packaged kamaboko (specially packaged kamaboko)	Polyvinylidene chloride
Fish ham, fish sausage	Polyvinylidene chloride
Vacuum-wrapped products (fried kamaboko, broiled chikuwa, sasa kamaboko, etc.)	Polycello, polyvinylidene chloride, nylon/polyethylene, polypropylene, polyester, polyethylene coated with polyvinylidene chloride/moisture-proof cellophane, nylon coated with polyvinylidene chloride/polyethylene, polypropylene coated with polyvinylidene chloride/polyester/polyethylene

II. FISH HAM AND FISH SAUSAGE

Fish ham was first marketed about 1936–1937, when tuna ham was introduced. However, at that time, due to a lack of appropriate packing materials, they could not be preserved for a long time. With development of flexible plastics film, full-fledged production started around 1954.

A. Processing of Fish Ham

The fish ham is prepared by adding binders to cured and seasoned fish meat blocks and by stuffing into casings and heating them after tightly sealing the casings. Major raw materials are tuna, bonito, cod, Alaska pollack, etc. Tuna meat, though less binding, with good flavor and color fixing to nice pink red color by curing, and the bonito meat, though thin in color, with good taste and stronger binding strength, serve as very important materials for the fish ham. However, recently, since these became much more expensive, Alaska pollack (frozen ground meat) is more popularly

Figure 2. Retainer-formed kamaboko.

used. Pork, beef, mutton, rabbit, and chicken may be added to the fish ham, as block meat or binder. Normally, as a part of block meat, pork fat is mixed with lard added as the binder. The Japanese Agricultural Standard (JAS), however, provides that the major materials shall be block meat and 50% or more of the block meat shall be block meat of fish and that the major binder material shall be ground fish meat.

Prepared meat is stuffed into casing by using a stuffer. For casing, flexible plastic film of polyvinylidene chloride is used. One end of the casing is bound with the packer, using an aluminum wire, and then after filling the stuffing meat, the other end of the casing is similarly tied. Stuffed meat is then placed in a retainer for shaping and heating (Figs. 2 and 3). It is determined by regulations that heating shall be performed by a method that allows heating of the product center portion at 80°C for 45 min (or by a method that has the equivalent or better effect) and that the finished products shall be stored at a temperature below 10°C. However, products (high, temperature sterilized products) to which a heating method that permits heating at 120°C for 4 min is applied can be distributed at room temperature. After sterilization heating, fish ham is cooled in cold water. Outer packing is intended for protection of products and advertisement; on many products it is mandatory to show various product details, such as date of manufacture, name of manufacturer, ingredients, etc. A JAS is established for fish ham.

Figure 3. Casing-packaged kamaboko.

B. Preparation of Fish Sausage

The fish sausage is a food product prepared by stuffing into casings, tightly sealing, and heating ground fish meat added with ground animal meats, chicken, fats, starch, spices, etc. (Fig. 4). Recently, use of frozen ground meats of Alaska pollack and Atka mackerel is increasing. Meat of fish or animal is cut into fine pieces by a chopper. The meat is then placed on a silent cutter or mixer and mixed for 10–15 min with 2–3% (by weight) of salt added. Then oil and fat, starch, vegetable protein, gelatin, seasonings, and spices or coloring matter is added, and the mixture is kneaded for 15–20 min. The JAS specifies that a minimun of 2% fat shall be added and that the maximum starch addition shall be 10%. Ground meats is stuffed into casings, one end of which is closed with an aluminum wire. After stuffing each casing, the other end is also tied with an aluminum wier to tightly seal the casing. Recently, a mass production system in which an automatic stuffer and packer, using flat film, carry out stuffing, packing, and separation is commonly used. As in manufacturing of fish ham, sterilizing heating is performed. However, no retainer is used, since heating causes the casing to shrink in a cylindrical shape. When heating is completed, casings are cooled in cold water. A JAS is established for fish sausage. On many products, it is mandatory to show various product details, such as date of manufacture, name of manufacturer, ingredients, etc.

Chapter 22. Fish Meat By-Products 347

Figure 4. Fish ham (bottom) and fish sausage (top).

Figure 5. Fully automatic packer.

III. SPECIALLY PACKAGED KAMABOKO

Casing-packaged kamaboko and retainer-formed kamaboko are classified as specially packaged kamaboko. For these kamaboko, it is mandatory that they should be heated so that the center portion can be heated at 80°C for over 20 min and that they should be stored at a temperature below 10°C, as applicable to other kamaboko. However, those heated in a way that brings the center portion to 120°C for 4 min can be distributed at room temperature.

Casing-packaged kamaboko is made by stuffing ground fish meat into plastic casings (such as using polyvinylidene chloride as the casing material) and heating tightly sealed casings by tying both ends with metal wires.

Retainer-formed kamaboko is made by shaping ground fish meat by heating with steam. The ground fish meat is placed on a board by the forming machine and then the plastic film packs the board as if it were wrapping the entire board. Then, the wrapped board is inserted into a kamaboko shaped retainer, and fixed and then heated. During heating, films are welded to complete tight sealing. Figure 5 shows a fully automatic packer of steamed kamaboko.

CHAPTER 23

Dairy Products

Kentaro Ono
Corporate Development and Planning,
Snow Brand Milk Products Co., Ltd.,
Tokyo, Japan

I. YOGURT AND FRESH CHEESE

A. Development of Yogurt Packaging

With the growing yogurt products market in Japan, designs or types of package for yogurt products have been developed (Niwa, 1982). Since fresh cheeses such as cottage cheese and quark are sometimes used and consumed like fermented milk products such as yogurt, and principles of designing packages for them are same as yogurt, these fresh cheeses are included in this section as part of fresh cultured milk products from the viewpoint of packaging.

Yogurt packaging and merchandising have a long history in Japan, back to the 1950s, beginning with packaging in glass bottles (Wada, 1982). Pasteurized yogurt products with fruits were introduced with a paper cup. Then plastics were introduced as materials for packaging, and have taken over from glass and paper packaging because of such characteristics as durability, printability, formability, productivity, and cost (Makino, 1981). A recent boom of so-called "plain type" yogurt started in the mid 1970s with paper milk cartons (Hoshino et al., 1982). New packages have been developed with growth of the market, and the dominant one is the 500-ml cup-type package, using plastics and composites of paper and plastics (Makino, 1981). The 500-ml cups are in a striking contrast to gable-top milk cartons, which are low in cost while poor in functionality as yogurt packaging. The Food Hygiene Law involves governmental requirements for yogurt packaging in terms of materials and physical properties (Aoyagi,

1981). Thermoformed plastic packages have been used for fresh cheese packages (Ono, 1981).

B. Requirements for Yogurt Package

Yogurt products are classified into three types, which include in-package fermented yogurt, prepackage fermented yogurt, and pasteurized yogurt. Some requirements are common to all, but some are specific to each type. Common requirements are as follows:

1. For end use: Functionalities related to (a) easy opening, (b) scoopability, (c) storing in the home refrigerator, (d) easy handling, (e) appeal, (f) reclosability, (g) hygiene, and (h) printability are major concerns on packaging designing for the end use (Aoyagi, 1981; Hoshino and Tothuka, 1982).
2. For handling and distribution: Major concerns in designing the packaging are (a) size and shape of package, (b) physical strength, (c) space saving, (d) stacking at the retailer, and (e) hygiene (Aoyagi, 1981; Sasaki, 1983).
3. For productivity: Major concerns on designing of the packaging are (a) cost of container, (b) productivity of packaging operation, and (c) packaging system used (Aoyagi, 1981; Hoshino and Tothuka, 1982).

Specific requirements for each type are as follows:

1. In-package fermented yogurt: Fermentation control through packaging design must be studied in terms of temperature control and oxygen supply, which are primary concerns to provide quality yogurt fermentation (Hoshino and Tothuka, 1982).
2. Prepackage fermented yogurt: Maintaining the quality at the packaging operation is required through preventing contamination and other deteriorations such as water separation and excess acidification (Wada, 1982).
3. Pasteurized yogurt: Aseptic filling systems are available as the packaging system for dessert-type yogurt products. High-barrier materials are available to prevent deterioration from oxygen (*Food Processing,* 1982).

C. Materials Used for Yogurt Packaging

Glass bottles and milk cartons have been used as a yogurt package at the beginning of the introduction. Cup-type packages are the dominant yogurt package in the market. The yogurt packages are composed of two components, lid and bottom or body (Makino, 1981).

Plastics and paperboards with plastic coatings are the materials used for the bottom or body. The plastics include polystyrenes such as HIPS and GPPS, which are thermoformed with injection molding or vacuum-forming technique (Makino, 1981). Polyethylene-coated paperboards are widely used for yogurt packaging because of the excellent printability, the lower cost, and easy disposal (Hoshino and Tothuka, 1982; Niwa, 1982). Composite materials and assembling are also available to provide the advantages of each material (Wada, 1982).

Polyethylene-coated paperboard is composed with a polyethylene skeleton to provide physical strength, excellent printability, easy disposal, and light weight (Totsuka, 1982). Thermoformed polystyrene sheets are heat sealed with a composite paper label in the forming system to provide physical strength, lower material cost, and prevention of postcontamination (Wada, 1982; *Food Processing,* 1982).

Plastic-coated aluminum foils or paperboards and those composites are available as lidding materials that offer easy openability and prevention of postcontamination by heat sealing. Plastics coated are polyethylene, the copolymers, and waxes (Suzuki and Hoshino, 1982).

II. NATURAL CHEESE

A. Development of Natural Cheese Packaging

Natural cheeses have been cut and sold in cheese shops. However, with increased retailing through self-service shops, development of packaging to extend shelf life has been demanded to provide natural cheeses with easy handling and distribution, keeping the initial quality and productivity for nation-wide package distribution (Ono and Kinoshita, 1981). Semihard and hard-type cheeses such as cheddar and Gouda are sold in precut form, simply wrapped with cellophane or thin plastic film, at retail shops. These simple wrappings are observed commonly in European countries, while cheeses in hermetically sealed flexible packages are observed in Japanese and American markets. However, imported ones are also sold in the simple packaging in the United States (Ono, 1981).

The flexible packages are evacuated or gas flushed to prevent mold growth or deterioration and provide longer shelf life for nationwide distribution through the current supermarket system. Lactic bacteria are active in natural cheeses to disintegrate fat and protein to mature the cheese. Hence, carbon dioxide gas is generated from the activity or respiration of the bacteria, whose activities are varied with the variety or strain and environmental conditions, such as oxygen supply, existence of carbon dioxide gas, and temperature. A number of packaging materials are available to control the packaging parameters, as we can see in the retail shops (Chandan, 1982; Ono and Gilbert, 1980; Ono, 1981).

Another development of packaging for natural cheese will be a heat-inactivating technique for the mold-ripened cheeses such as camembert (Nakae, 1980). A process for maturing is very soon to be sold through a nationwide distribution system. Enzymatic activities of the mold are inactivated by heat in hermetically sealed packages.

B. Requirements for Natural Cheese Package

Major requirements for natural cheese packages to offer nationwide distribution will be (a) prevention of mold growth, (b) prevention of swelling or excess evacuating of packages, and (c) durability during handling and distribution (Chandan, 1982; Ono and Gilbert, 1980; Ono, 1981; Ono and Kinoshita, 1981).

Mold growth can be prevented by eliminating oxygen in the package with evacuation or gas flushing of inert gas or carbon dioxide, and by intercepting oxygen supply from outside the package with an oxygen-barrier material.

Swellings or excess evacuatings are caused by unbalanced packaging conditions or misdesigning. The amount of carbon dioxide generated from a cheese in a package should be balanced with an amount of carbon dioxide discharged through the packaging wall. The larger the amount generated is, the faster the swelling is. The larger the discharge is, the faster the evacuating is. Respiration also is controlled by the barrier properties of the packaging material. Excess carbon dioxide penetrability may also means excess oxygen supply or mold growth on the cheese in the package.

The durability of the flexible package depends upon the properties of packaging material used and the magnitude of swelling or evacuating of package. The swelling package has less problems in durability in terms of pinholes, which allow the mold growth and contamination. Wrinkles or creases from excess evacuation may cause pinholes on the package. However, a bulging package may be avoided by consumers, who view it as deteriorated.

C. Materials Used for Natural Cheese Package

Natural cheese packaging involves the nationwide distribution of packages, for extending the shelf life of semihard or hard-type cheeses and mold-ripened cheeses and as retailer wrapping for point of sale handling. Flexible films, most of which are laminated, are used for the gas-flushed or evacuated package. The gas permeability of the film used should be varied with the variety of cheese, the gas composition inside the package, and the environmental temperature (Chandan, 1982; Ono, 1981; Ono and Kinoshita, 1981).

Dependency on these parameters must be studied carefully prior to determining the packaging materials or packaging system to be used. Popular flexible films used for natural cheese packaging are PVDC-coated nylon or PET/PE or EVA copolymer, sealant layer, and nylon or PET or OPP/sealant. For thermoform–fill–seal packaging, cast nylon, sometimes coated with PVDC, is popular as the bottom film (Chandan, 1982; Ono and Kinoshita, 1981).

Respiration or carbon dioxide generation from a cheese in the package is controlled by the gas permeability of film used. The higher the gas barrier property is, the lower the respiration rate is. Gases used for gas-flush packaging are nitrogen, carbon dioxide, and a mixture of these. The amount of carbon dioxide discharged through the packaging wall also can be controlled by the partial pressure of the gas in the package (Ono and Gilbert, 1980; Ono, 1981; Ono and Kinoshita, 1981).

Metal cans are commonly used as a packaging for the heat-inactivated mold-ripened cheeses. Aluminum drawn cans with inside plastic coating are widely used (Chandan, 1982), while high-barrier plastic containers with coextruded sheet such as PP/EVOH/PP are making inroads on the market with new retort technique commonly applied to the retort pouches. Both can be provided easy-open features with easy-open ends or peel-openable films.

III. PROCESSED CHEESE

A. Development of Processed Cheese Packaging

J. L. Kraft developed the technique of process cheese making in 1912 at the dawn of processed cheese packaging. Since then, development of processed cheese packaging has kept in step with the development of packaging material, packaging machinery, and processing systems. A number of packagings have been developed, and some are available today in retailer shops. The major ones are classified into four types: sliced cheese, portion cheese, chunk and tube, and candy cheese (Makino, 1981; Nakae, 1980; Ono, 1981; Ono and Kinoshita, 1981).

Individually wrapped cheese slices are sold most widely today. Several types of packaging systems are available today, including a hot filling system and cast cheese packaging system. Some cheeses contain dehydroacetic acid or sorbic acid/potassium sorbate as preservatives, while some are in a gas-flushed package with a high-barrier film (Makino, 1981; Ono, 1981; Ono and Kinoshita, 1981).

Portion, chunk, and tube-type cheeses have a long history in processed cheese packaging. The portion-type cheese was developed for a single serving and the contents are 20–30 g, while the chunk was developed for slice serving. However, the market shares of these types are shrinking and giving way to slice-type cheese (Makino, 1981; Nakae, 1980; Ono, 1981; Ono and Kinoshita, 1981).

Candy-type cheeses were developed and introduced at the end of the 1970s in the Japanese market. Cast cheeses are wrapped in a plastic film with a twist wrapping machine generally used as a candy wrapper. This type of cheese is served for a snack or appetizer (Makino, 1981; Ono, 1981; Ono and Kinoshita, 1981).

B. Requirements for Processed Cheese Packaging

Essential and common requirements for the tightly wrapped package, such as portion, chunk, and tube, are prevention of postcontamination and microbial or chemical deterioration from oxygen, while prevention of microbial growth on the cheese is required for the loosely wrapped package such as sliced cheeses and candy cheeses (Chiba, 1984; Ono, 1981; Ono and Kinoshita, 1981).

Easy-opening features or easy-serving functionalities must be provided for end use. An easy-opening tape is provided for the portion type of aluminum foil package and the tube-type package. A packaging material for the chunk type must provide sealability and easy openability by peeling open the film or easy sliceability over the wrapping film (Makino, 1981; Nakae, 1980; Ono and Kinoshita, 1981).

C. Materials Used for Processed Cheese Packaging

PET, PVDC-coated cellophane, or PP/PE laminate film are used for individual wrapping film of sliced cheese (Makino, 1981; Ono, 1981; Ono and Kinoshita, 1981). The outer wrapping film should have a high gas barrier property, if preservatives are not added to the cheese to prevent microbial deterioration. Inside of the package also must be gas flushed with carbon dioxide, nitrogen, or a mixture of these to eliminate oxygen in the package. PVDC-coated nylon, PET, or OPP laminated with a sealant layer is used for the outer wrapping film. These are also applicable for an outer wrapping film with the gas flushing technique for the candy-type cheese packaging.

Aluminum foils coated with lacquers or other heat-sealing materials are available for the portion-type cheese packaging. PVDC-coated cellophane or aluminum-laminated materials are used for the chunk-type cheese packaging.

PVDC copolymer is used for the tube packaging. Aluminum-metallized materials making inroads in packaging with their excellent appeal to the consumers and high gas barrier property for extending shelf life, while they have problems in cost and stability.

REFERENCES

Aoyagi, H. (1981). *JPI Journal* **19,** 443–447.
Chandan, R. C. (1982). *Dairy Record* **83**(2), 141–142, 144.
Chiba, Y. (1984). *Japan Food Sci.* **23**(3), 72–78.
Food Processing **43** (1), 38, 41 (1982).
Hoshino, Y., and Tothuka, S. (1982). *JPI Journal* **20,** 514–518.
Makino, T. (1981). *Nyugikyo Shiryou* **31**(3), 2–14.
Nakae, T. (1980). *Chikusan no Kenkyu* **34,** 1382–1386.
Niwa, S. (1982). *New Food Industry* **24**(5), 11–15.
Ono, K. (1981). *JPI Journal* **19,** 136–139.
Ono, K., and Gilbert, S. G. (1980). *JPI Journal* **18,** 162–165; also presented at the 1979 AIChE Anual Meeting in Boston, Mass.
Ono, K., and Kinoshita, H. (1981). *Japan Food Sci.* **20**(9), 70–80.
Sasaki, H. (1983). *JPI Journal* **21,** 103–105.
Suzuki, H., and Hoshino, Y. (1982). *Japan Food Sci.* **21**(9), 74–80.
Totsuka, I. (1982). *Japan Food Sci.* **21**(1), 22–31.
Wada, A. (1982). *JPI Journal* **20,** 630–634.

CHAPTER 24

Cakes and Snack Foods

Shinichi Minakuchi
Hachiro Nakamura
Toppan Printing Co., Ltd.,
Tokyo, Japan

I. INTRODUCTION

In the present food market, demands for quantity generally are not expected to increase. It is therefore necessary to produce a variety of products and to produce small-lot products for the market. In the field of cakes, this tendency is conspicuous, and a merchandise design to meet with consumers' needs is quite important.

On the other hand, from the standpoint of food packaging, it is necessary to perfectly pack all the products regardless of mass production, variety, or small-lot production. It is a serious problem for the packaging industry to pack diversified foods without spoiling their features.

There are a number of functions required in the packaging, including:

1. Protection and preservation of foods
2. Operation of packaging
3. Value of Commodity
4. Convenience
5. Sanitation
6. Economy

In order to satisfy these requirements, it is necessary to study various factors such as the shape of products, characteristics of products, quality deterioration of products, consumers' tastes, effects of design, etc. This chapter explains the status quo of quality preservative packaging and new packaging systems in the field of cakes.

II. CLASSIFICATION AND REQUIRED QUALITY FOR PACKAGING OF CAKES

Cakes are generally classified into two categories according to:

1. Water content and preservation of cakes:
   a. Moist cakes with water content of over 30% immediately after being produced
   b. Semimoist cakes, of 10–30% water content
   c. Dry cakes, of less than 10% water content
2. Water activity (AW) indicating the quantity of free water content:
   a. Very moist cakes, with AW of more than 0.9
   b. Moist cakes, of approximately 0.65–0.90 for AW
   c. Semi-moist cakes of less than approximately 0.65 for AW

The latter means of classification is quite proper from the standpoint of preservative packaging of the food in terms of quality deterioration due to biological and physical factors, including water evaporation.

In terms of antioxidation packaging of oil and fat, cakes can be classified according to the oil and fat content.

Table I shows a classification of cakes and required quality of the packaging material, Table II water content and oil and fat content, Table III water content and AW, and Table IV minimum AW for growing microbes.

III. QUALITY DETERIORATION OF CAKES

There are two major categories of quality deterioration. One is due to water, and the other is due to factors other than the water, such as oxygen, radiation (light), temperature, physical damage, damage by insects, and so forth.

A. Quality Deterioration Related to Water Content

The water content, which is originally contained in cakes, is quite significant since it shows the characteristics of the cake. AW, which indicates the quantity of free water that can be used for chemical reactions and the growth of microbes, is an important element for preserving the cakes.

Potential effects on quality due to variation of the water content, including the moisture absorption or release, are as follows:

1. Quality deterioration from moisture absorption

Chapter 24. Cakes and Snack Foods

Table I Requirements of Cakes and Packaging Materials[a]

	Requirements for Packaging
Moist and semimoist cakes	
Sponge cakes	1. Damage prevention
Castella	2. Humidity control
Pies	
Cold cakes	1. Cold-proof
	2. Water-proof
Bread cakes	1. Oil-proof
Cream puffs	2. Gas barrier
Dessert cakes	1. Cold-proof
	2. Waterproof
Sweetened adzuki beans	1. Gas barrier
	2. Humidity-proof
Sweet jelly of beans	1. Humidity control
Bean-jam-filled wafers	2. Wrapping aptitude
Bean jam buns	
Dried cakes/cookies	
Biscuits	1. Humidity-proof
	2. Damage prevention
Chocolates	1. Radiation interception
	2. Flavor preservation
Candies	1. Humidity-proof
Chewing gums	2. Damage prevention
Crackers	
Rice crackers	
Rice-cake cubes	
Other snack cakes	
Cakes with oil and fat process	
Potato chips	1. Oxygen interception
Doughnuts	2. Humidity-proof
Fried rice cakes	
Fried rice crackers	
Fried dough cookies	
Peanut butter	

[a] From Watanabe (1980).

    a. Degradation of starch
    b. Denaturation of protein
    c. Physical change of structure
    d. Damage by the growth of microbes
    e. Liquefaction phenomenon
2. Quality deterioration from moisture release
    a. Caking
    b. Cracking

Table II Water and Fat Contents of Major Cakes/Cookies

Moist Cakes	Content (%)		Semimoist Cakes	Content (%)		Dry Cakes/Cookies	Content (%)	
	Water	Fat		Water	Fat		Water	Fat
Japanese muffin containing bean jam (*imagawayaki*)	45.5	1.2	Sweetened adzuki beans (*amanatto*)	23.5	0.6	Toffees (*amedama*)	2.5	0
Sweet rice jelly (*uiro*)	54.5	0.2	Castella	26.9	5.1	Rice-flour cake (*rakugan*)	3.0	0.4
Green-powdered bean-jam bun (*uguisu-mochi*)	40.0	0.4	Millet dumpling (*kibi-dango*)	24.4	0.3	Millet-and-rice cake (*okoshi*)	5.0	1.0
Rice-cake wrapped in oak leaf (*kashiwa-mochi*)	48.5	0.5	Starch paste (*gyushi*)	36.0	1.4	Oshikiri cookies	10.0	0.5
Rice-cake with sweet-boiled beans mixed in (*kanoko*)	34.0	0.4	*Kirisansho*	38.0	0.4	Fried dough cake (*karinto*)	3.5	24.8
Sweetened jelly of yam and rice flour (*karukan*)	42.5	0.3	*Kingyokuto*	28.0	0	Confeitos (*kompeito*)	1.6	7.8
			Taruto	30.0	3.1	*Gokabo*	10.0	5.9
			Chatsu	22.5	4.1	Rice cracker		
Kintsuba	34.0	0.6	Castella bun with bean jam filling (Castella *manju*)	27.9	1.8	*Iso-senbei*	4.2	0.8
Spitted dumpling (*kushi-dango*)	50.0	0.6				*Kawara-senbei*	6.0	3.5
Bean paste rice-cake wrapped in cherry leaf (*sakura-mochi*)	40.5	0.5	Chesnut-jam bun (*kuri-manju*)	24.5	1.3	*Maki-senbei*	3.5	1.3
						Nanbu-senbei	3.3	11.1
Rice-cake stuffed with bean jam (*daifuku-mochi*)	41.5	0.6	Bean-jam filled wafer (*monaka*)	29.0	0.4	Rice-cake cubes	4.7	4.2
						Hina-arare	4.7	4.2
Rice-dumpling wrapped in bamboo leaf (*chimaki*)	62.0	0.3	*Yubeshi*	22.0	3.6	*Arare*	4.4	1.4
						Rice cracker		

Item			Item			Item		
Bean-jam pancake (*dorayaki*)	31.5	2.6	Bean jelly (*mizyokan*)	26.0	0.2	Age-senbei	4.0	20.4
Nerikiri	34.0	0.4				Amakara-senbei	4.0	1.3
Bun with bean-jam filling (*kuzu-manju*)	45.0	0.2	Moon cake (*geppei*)	20.9	8.7	Salt-senbei	5.0	1.4
			Cup cake	20.0	21.0	*Matsukaze*	5.2	2.0
Chinese bun with bean-jam filling (*to-manju*)	36.5	2.3	Donut	20.0	22.7	Yam (*Yatsugashira*)	1.8	0.7
			Marshmallow	18.5	4.0	Wafer	2.3	23.3
Boiled bun with bean-jam filling (*mushi-manju*)	35.0	0.5				Cracker	2.5	23.6
						Puff pie	2.5	35.7
Soft adzuki-bean jelly (*mizu-yokan*)	45.2	0.2				Hard biscuit	2.5	12.9
						Soft biscuit	2.0	21.2
Boiled adzuki-bean jelly (*mushiyokan*)	39.5	0.3				*Boro*	4.5	2.3
						Russian cake	4.0	23.6
Chinese bun with bean-jam filling (*an-man*)	35.5	6.3				Caramel	8.0	11.8
						Drops	2.0	0
Chinese bun with meat filling (*niku-man*)	40.0	7.3				Chocolate	1.2	32.5
Apple pie	45.0	19.7						
Cream puff	55.0	14.6						
Short cake (strawberry)	31.0	13.2						
Waffle	46.0	8.9						
Bean-jam bun	35.5	2.1						
Cream bun	36.0	1.8						
Jam bun	32.0	2.1						

[a] From Yoshii et al. (1976).

Table III Water Content and AW of Cakes/Cookies

Item	Water Content (%)	AW
Chocolate	1	0.32
Biscuit	4	0.33
Grape sugar (dextrose)	9	0.48
Cracker	5	0.53
Jelly	18	0.65
Cake	25	0.74
Orange marmalade	32	0.75
Sweet jelly of beans	25	0.87
Bread	35	0.93
Jam		0.94–0.82

[a] From Igawa (1978).

    c. Shrinking
    d. Disappearance of tone of color and luster
    e. Crystalling and precipitation of the water soluble components

As a countermeasure to prevent quality deterioration due to water evaporation, moistureproof packaging, explained in Section IV, has been adopted.

B. Quality Deterioration Caused by Factors Other Than Water

Quality deterioration can also be caused by factors other than water. Oxidation of oil and fat and discoloration are accelerated due to the existence of oxygen and the effect of ultraviolet rays. Furthermore, growth of molds and harmful insects is accelerated due to the existence of oxygen, depending upon AW and preservation temperature.

As a countermeasure to prevent quality deterioration, the following methods have been adopted: gas barrier packaging, vacuum packaging, gas substitution packaging for the purpose of intercepting the oxygen, and as a new method of preservation, packaging with oxygen absorbent and packaging with gas substitute agent to chemically get rid of oxygen from the packed products or substitute another gas for the oxygen. These methods and examples of packaging will be detailed in Sections IV–VI.

Chapter 24. Cakes and Snack Foods

Table IV Minimum AW for Growing Microbes

Microbe	Min. AW
Bacteria	
Pseudomonas	0.97
Achromobacter	0.96
Escherichia coli	0.96–0.935
Bacillus subtilis	0.95
B. mycoides	0.99
Clostridium botilinum	0.95
Aerobacter aerogenes	0.945
Salmonella newport	0.945
Streptococcus faecalis	0.915–0.930
Micrococcus roseus	0.905
Staphylococcus aureus	0.86
Halophilic bacteria	0.75
Range of common bacteria generally	0.94–0.99
Yeasts	
Torulopsis utilis	0.94
Beer yeasts	0.94
Schizosaccharomyces	0.93
Candida utilis	0.94
Bread yeasts	0.905
Mycoderma	0.90
Saccharomyces cerevisiae	0.895
Rhodotorula	0.89
Endomyces	0.885
Willia anomala	0.88
Saccharomyces rouxii	0.60–0.61
Range of common yeasts	0.94–0.88
Molds and imperfect microbes	
Mucor	0.93–0.92
Rhizopus	0.94–0.92
Penicillium	0.83–0.80
Aspergellus flavus	0.80
A. candidus	0.75
Botrytis	0.93
Oospora lactis	0.895
A. chevalieri	
A. repens	
A. ruber	0.65
A. amstelodami	
Xeromyces bisporus	
Range of common molds	0.80

[a] From Ishitani (1978).

[b] Minimum AW of molds shows the value of limit of germination of a spore.

[c] Called xerophilic mold.

In order to intercept radiation, especially ultraviolet rays, which are reported to accelerate oxidation and discoloration, it is necessary to pack products with packaging materials that have the ability to intercept a ray. Aluminum foil laminated packaging materials and aluminum metallized film laminated materials are generally effective and frequently used. However, these packaging materials are costly and the contents cannot be seen from outside. In many recent cases, packaging materials printed with ultraviolet ray absorbing ink are frequently used. Some types of ink are almost transparent and highly advantageous because the contents can be seen from outside.

IV. MOISTURE-PROOF PACKAGING AND GAS BARRIER PACKAGING

A. Moisture-Proof Packaging

As mentioned above, it is important to maintain the water content or AW. It is necessary to pack a cake that originally contains much water so as not to evaporate the water and a cake that contains less water so as not to absorb moisture. For this purpose, a packaging method of intercepting the air is needed.

Since a moisture-proofing effect can be obtained with it to a certain extent, seal-packaging, with plastic films such as polyethylene and polypropylene, films is used for a short-term distribution. When a higher standard of moisture-proof packaging is required, laminated materials such as other plastic films of higher moisture-proof effect are used. Table V shows the moisture permeability of various packaging materials.

As shown in Table V, suitable materials for highly moisture-proof laminated materials include polyvinylidene chloride (PVDC) film, PVDC-coated polypropylene film, and aluminum metallized film laminated packaging materials.

Packaging with desiccant is popular to provide moisture-proof packaging. In this packaging method, a desiccant, such as calcium oxide and silica gel, is packed with food with an air-permeable packaging material. This method is especially effective for dry cakes, since a desiccant maintains the water content of the product by absorbing and getting rid of the water that comes in through packaging materials.

B. Gas Barrier Packaging

Oxygen causes various kinds of quality deterioration of cakes, such as oxidation of oil and fat and deterioration of nutritious components, and growth of aerobes and harmful insects.

Chapter 24. Cakes and Snack Foods

Table V Degree of Humidity Permeability of Various Packaging Materials[a]

Packaging Materials	Water Vapor Transport Ratio (g/m² · 24h)
Oriented vinyl alcohol 15 μm	6–10
Ethylene–vinylalcohol copolymer 15 μm	80–100
Polyvinylidene chloride 30 μm	1–2
PVDC-coated oriented polypropylene 23 μm	3–5
Polyester 12 μm	25–40
Moisture-proof cellophane number 300	30–50
Oriented polypropylene 20 μm	7–10
Nylon (polyamide) 15 μm	150–200
Cast polypropylene 30 μm	8–10
Low-density polyethylene 40 μm	18–25
Polyester 12 μm/polyethylene 40 μm	10
Moisture-proof cellophane number 300/polyethylene 40 μm	10
Oriented polypropylene 20 μm/cast polypropylene 30 u	6
Oriented vinyl alcohol 15 μm/polyethylene 40 μm	3
Aluminum metallized polyester 12 μm/polyethylene 60 μm	Under 0.5
Polyester 12 μm/aluminum foil 12 μm/polyethylene 60 μm	Under 0.5

[a] From Yoshii et al. (1976).

Consequently, for cakes that contain higher levels of oil and fat, gas barrier packaging as well as moistureproof packaging is needed. Gas barrier packaging is needed for the cakes with AW of over 0.65 as a method of preventing the growth of aerobes, mainly composed of mold. There are several means in gas barrier packaging, including gas substitution packaging, which will be explained in the next chapter.

1. Seal-pack with high gas barrier packaging materials. Table VI shows the gas barriers of various, films and Table VII high gas barrier multilayer packaging materials.
2. Get rid of oxygen within products by means of vacuum packaging or gas substitution packaging, using high gas barrier packaging materials.
3. Seal-pack with oxygen absorbent, which absorbs oxygen, and gas substitute agent, which replaces oxygen with carbon dioxide, using gas barrier packaging materials.

Method 1 is a popular method of intercepting the water and the oxygen by using the high moistureproof gas barrier packaging materials. Methods 2 and 3 are also popular these days. Packaging with oxygen absorbent and gas substitution agent are examples of the effective gas barrier packaging that has been rapidly developed since the first half of 1977. Packaging methods 2 and 3 will be detailed in the following sections.

Table VI Gas Barrier of Various Films

Film	Gas Permeability ($cm^3/m^2 \cdot 24\ h \cdot atm$)[a]			Water Vapor Transport Ratio[a] ($g/m^2 \cdot 24h$), 40°C, 90% RH
	Carbon Dioxide	Nitrogen Gas	Oxygen Gas	
Low-density polyethylene	42,500	2,800	7,900[c]	24–48
High-density polyethylene	9,100	660	2,900[c]	22
Cast polypropylene (CPP)	12,600	760	3,800[c]	22–34
Oriented polypropylene (OPP)	8,500	315	2,500[c]	3–5
PVDC-coated oriented polypropylene[b]	8–80	8–30	16[c]	5
Plain cellophane	6–90	8–25	3–80[c]	720
Humidity-proof cellophane	—	—	40[d]	8–16
Polyester	240–400	11–16	95–130[c]	20–24
Cast nylon	160–190	14	40[c]	
Oriented nylon	—	—	30[d]	90
PVDC-coated oriented nylon[b]	—	—	10[d]	4–6
Polyvinyl chloride	320–790	30–80	80–320[c]	5–6
Polyvinylidene chloride, polyvinyl chloride, copolymer	60–700	2–23	13–110[c]	3–6
Polystyrene	14,000	880	5,500[c]	110–160
Polycarbonate	17,000	790	4,700[c]	170
Ethylene–vinyl alcohol copolymer	—	—	2[d]	30
Oriented vinylon	—	—	3[d]	4
Polyvinylidene chloride	—	—	10[d]	2
Polyacrylonitrile	—	8	3[d]	20

[a] The degree of gas permeability and water vapor transport ratio are shown by converting to a thickness of 25 μm.
[b] Value of the PVDC-coated film differs depending upon the coating agent, and its kind and quantity.
[c] At 25°C, 50% RH, ASTM D 1434-66.
[d] At 27°C, 65% RH, same-pressure oxygen electrode method.
[e] Value shown is that of OPP/polyvinylidene/CPP.

Table VII Examples of High Gas Barrier Multilayer Packaging Materials[a]

Film	Standard Thickness (μm)	Oxygen Permeability ($cm^3/m^2 \cdot 24\ h \cdot atm$, 20°C)	Water Vapor Transport Ratio ($g/m^2 \cdot 24\ h$)
PVDC-coated cellophane 25 μm/polyethylene 40 μm	65	1–13	9–11
PVDC-coated oriented polypropylene 23 μm/polyethylene 40 μm	63	10–20	3–5
PVDC-coated oriented nylon 18 μm/polyethylene 40 μm	58	4–8	6–8
PVDC-coated polyester 15 μm/polyethylene 40 μm	55	4–10	4–6
Moisture-proof cellophane/aluminum foil 7 μm/polyethylene 40 μm	67–70	0–[b]	0–10
Polyester/aluminum foil 7 μm/polyethylene 40 μm	59–63	0–[b]	0–10
Oriented polypropylene/ethylene–vinyl alcohol copolymer/polyethylene 60 μm	95–100	0.4–0.8	3–4
Oriented vinylon/polyethylene 80 μm	95–100	0.1–0.2	2.5–3.5

[a] From Suda (1980).
[b] Differs depending upon pinholes.

V. VACUUM PACKAGING AND GAS SUBSTITUTION PACKAGING

Vacuum packaging and gas substitution packaging of cakes are intended to prevent oxidation of oil and fat, control insects, etc.

A. Antioxidation Effect

Microbes can seldom grow in cakes with an AW of less than 0.6, but chemical changes such as oxidation or discoloration of oil and fat are serious problems. The speed of oxidation is affected by two factors. One is derived from the food itself, such as type of oil and fat, AW, pH, existence of oxygen, quantity of metallic iron, etc. The other is derived from the preservation and distribution, such as oxygen content in the packages, moisture, radiation, etc. Table VIII shows factors that affect the preservation of food containing oil and fat.

Table VIII Factors Affecting the Preservative Nature of Foods with Oil and Fat[a]

Used oil and fat
  Kind, fatty Acid composition
  Brand, extraction treatment
  Preservation condition
  Antioxidizing agent
  Metal
  Deterioration
Base
  Ingredient
  Properties (surface area)
  Food additives
  pH
Frying condition
  Heating condition (temperature, hours)
  Oil supply condition
  Oil absorption quantity
Aftercare
  Cooling method
  Existence of a coating
    Addition to frying oil
    Atomizing
    Dipping
    Treated wrapping paper
    Indirect addition
  Addition of antioxidizing agent
  Food additives (purity of salt)
Preservation condition
  Temperature
  Wavelength and strength of light radiation
  Open air: partial pressure of oxygen, air flow
  Packaging material: kind (light/radiation permeability, air permeability); thickness, combination
  Various bacteria: dependence on oxygen

[a] From Suda (1980).

In general, it is important to keep the oxygen content in the packages less than 1–2%. It is possible to keep it at 1–2% by means of vacuum packaging and nitrogen gas substitution packaging, using high gas barrier packaging materials as mentioned in Section IV, or packaging with oxygen absorbent and packaging with gas substitution agent, which will be explained in Section VI. Figure 1 shows the change of per oxide value (POV) of the oil and fat of fried dough cakes by means of nitrogen gas substitution. The graph proves that transparent barrier packaging materials like PVDC-coated cellophane number 300/polyethylene 40μm are effective enough to prevent oxidation if the oxygen content in the packages is low.

B. Restraining the Growth of Microbes

In the case of the cakes with AW of over 0.6, vacuum packaging and gas substitution packaging are intended the prevention of oxidation and dis-

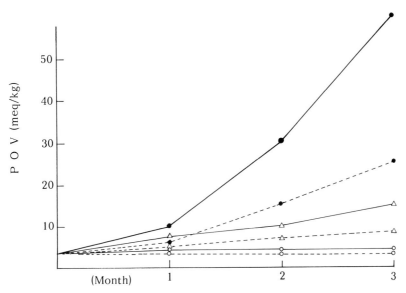

Figure 1. Change of POV of the oil and fat of fried dough cakes. Solid line: 40°C, 90% preserved in dark place. Dotted line: preserved at room temperature in light places. ● Non-gas-controlled packaging; ○ nitrogen substitution (remaining oxygen 10.1%); △ nitrogen substitution (remaining oxygen 3.4%). (From Suda, 1980.)

coloration, and to restrain the growth of microbes, especially mold and yeast.

Since it is possible to make the oxygen content less than 0.1% only by means of nitrogen substitution packaging, depending upon the capability of a packaging machine and characteristics of cakes, there are many cases in which the growth of mold, including *Aspergillus* and *Penicillium*, cannot be restrained. Table IX shows examples of the gas substitution method. Figure 2 shows the effects of the oxygen content against the growth of mold. As seen in Table IX, the rate of substitution is high in the chamber method. On the other hand, it is rather low in the nozzle and flash methods; therefore, it is necessary to pay close attention depening upon the purpose of using the method.

Carbon dioxide is effective to restrain the growth of mold only with gas substitution packaging. It is generally said that a carbon dioxide content of more than 30% has a restraining effect on the growth of mold. No mold will be generated with carbon dioxide at more than 90%. It is therefore utilized for cakes like Castella, whose taste is hardly affected by the carbon dioxide.

Due to the synergistic effect of low oxygen content and bacteriostatic effect of carbon dioxide, a restraining effect on the aerobic bacteria can be expected; however, there is no restraint on some kinds of yeast and aerobic bacteria. For cakes that need these prevented, such effective measures are needed as decreasing the primary viable count by means of heat sterilization, clean-room production, low-temperature distribution, etc.

Table IX Examples of Gas Substitution Method

Substitution Method	Type	Capacity	Remaining Oxygen	Method
Nozzle type	Single-headed	10–15 bags/min	2–5%	Intermittent
Chamber type	Single-headed	4–10 cycles/min	0.1–0.5%	Intermittent
	Many-headed	20–40 bags/min	0.1–0.5%	Intermittent
	Rotary	30–40 bags/min	0.1–0.5%	Consecutive
Tunnel type	Horizontal pillow	50–60 bags/min	2–5%	Consecutive
(flash type)	Vertical pillow	40–50 bags/min		

[a] From Suzuki (1981).

Chapter 24. Cakes and Snack Foods

C. Other Effects

Nitrogen gas substitution packaging and carbon dioxide gas substitution packaging are extremely effective to prevent and restrain discoloration due to oxygen, destruction of nutritious components, growth of harmful insects, etc. Most harmful insects are reported to die within hours in the low oxygen content of less than 1%. The carbon dioxide also has a large effect.

VI. PACKAGING WITH OXYGEN ABSORBENT AND PACKAGING WITH GAS SUBSTITUTION AGENT

Packaging with oxygen absorbent and packaging with gas substitution agent have been rapidly developed lately as the methods to adjust gas

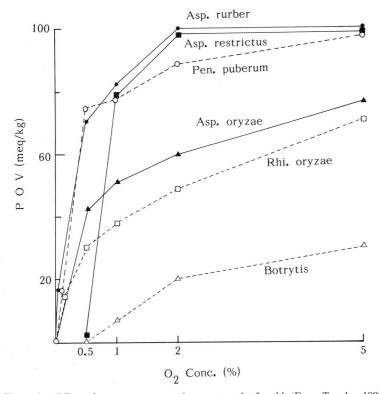

Figure 2. Effect of oxygen concentration on growth of mold. (From Tanaka, 1982.)

composition without using vacuum packaging machines and gas substitution packaging machines. An oxygen absorbent is a chemical (mainly composed of iron powder), which absorbs and removes oxygen, packed individually in air-permeable materials. A gas substitution agent is a substance (mainly composed of L-ascorbic acid), which can absorb oxygen and substitute it with carbon dioxide, also packed individually in air-permeable materials. Toppan Freshness Keeping Agent Type C Series, manufactured by Toppan Printing Co., Ltd., is a good example.

Table X shows these characteristics comparing with the gas substitution packaging. The common features of both are that they do not need large equipment and that they can maintain extremely low oxygen content. The difference is that the oxygen absorbent only absorbs oxygen, but gas substitution agent absorbs oxygen approximately 20% out of the air and at the same time generates almost the same quantity of carbon dioxide. The gas substitution agent does not change the externals of products and restrains the growth of microbes by carbon dioxide.

An explanation was given in Chapter 16 on gas substitution packaging regarding why oxygen absorbent and gas substitution agent are used in order to maintain the quality of cakes. The effects are to prevent oxidation of oil and fat due to oxygen, restrain the growth of microbes mainly composed of mold, prevent discoloration, and prevent the generation of harmful insects, and an advantage is that large equipment is not required (Suda, 1980; Watanabe, 1980; Igawa, 1978; Yoshii et al., 1976).

Comparisons of packaging methods using gas substitution packaging, packaging with oxygen absorbent, and packaging with gas substitution agent are given next.

A. Oxidation-Preventive Packaging of Peanut Butter

Figure 3 and Table XI and XII show examples of the oxidation-preventive packaging of peanut butter. In case of peanut butter packaging, it is important to prevent oxidation of oil and fat, and preserve the taste and flavor. Packaging with gas substitution agent (freshness keeping agent Type C) has brought excellent results both in preventing the increase of per oxide value (POV) and in flavor tests.

Packaging methods include (1) nitrogen gas substitution packaging, with remaining oxygen content 0.5%; (2) packaging with gas substitution agent, such as Toppan Freshness Keeping Agent Type C-150; or (3) packaging without using gas. Packaging materials are PVDC-coated polypropylene 20 μm/cast polypropylene 60 μm. The condition for preservation is 25°C and 60% relative humidity (RH).

Table X Comparison among Gas Substitution Packaging, Packaging with Gas Substitution Agent, and Packaging with Oxygen Absorbent

Item	Gas Substitution Packaging	Packaging with Gas Substitution	Packaging with Oxygen Absorbent
Theory	Substitute air in the container with inert gas	Absorb and remove oxygen in the container to substitute it with carbon dioxide	Same as gas substitution
Remaining oxygen	0.1–6% Depending upon productivity, substitution method, and contents	0% Is attainable depending upon proper usage	Same as gas substitution
Hourly change of oxygen content	Gradually increase depending upon oxygen-interceptive nature of packaging material	0% Is attainable for long hours if there is extra capability to absorb oxygen	Same as gas substitution
Oxidation-preventive effect	Effect is higher with lower remaining oxygen content	Extremely effective	Same as gas substitution
Restraint effect of microbes	Static bacteria effect can be expected by using carbon dioxide	Large effect including bacteriostatic effect of carbon dioxide	Effective against aerobic molds
External change of packaging material	Little external change with nitrogen gas; adjust external change by mixed gas of oxygen and nitrogen.	Little external changes	Approx. 20% reduction in capacity; adjust it by the size of packaging material
Applicable gas	N_2, CO_2, Ar, etc.	Substitute to $N_2/CO_2 = 80/20$	$N_2 = 100\%$ Avoid mixed use of CO_2 gas
Productivity	Generally fair with flash method and suitable for mass production	Suitable for small lot, also for mass production by using automatic machine	Same as gas substitution
Cost	Cost of equipment is required, but gas fee per bag is low	Cost per bag is high, but cost of equipment is very little required	Same as gas substitution
Packaging material	Use high gas-barrier material	With high gas-barrier material, freshness keeping agent of proper absorbing capability can be used	Same as gas substitution
Packaging shape	Various shapes can be chosen depending upon the packaging material	Various shapes can be chosen, but inside partition may be used depending upon the contents	Same as gas substitution
Use of metal detector	Possible	Possible	Impossible

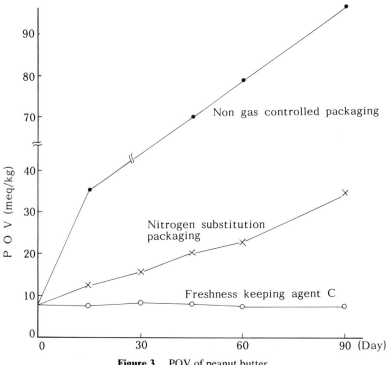

Figure 3. POV of peanut butter.

B. Packaging Preventing Bacteria and Molds for Moist Cakes

Tables XIII, XIV, and XV give examples of packaging moist cakes with AW of 0.92, including bean jam. Because of the effects of low oxygen content of less than 0.1% and carbon dioxide, packaging with gas substitution agent restrains the growth of molds and aerobic bacteria, and has made the shelf life longer.

Table XI   Changes in Water Content (%)

Packaging Method	Start	30 Days	60 Days	90 Days
Freshness keeping agent C−150	2.60	2.81	2.80	2.82
Nitrogen gas substitution	2.60	2.56	2.60	2.63
Non-gas-controlled packaging	2.60	2.61	2.61	2.73

Table XII Taste and Flavor Test[a]

Packaging Method	30 Days	60 Days	90 Days
Freshness keeping agent C − 150	81	87	90
Nitrogen gas substitution	76	70	68
Non-gas-controlled packaging	48	38	31
Product immediately after produced	95	105	108

[a] Criteria: Peanut butters that have been preserved under different conditions are ranked by 30 judges as first (4 points), second (3 points), third (2 points), and fourth (1 point).

Packaging methods include (1) packaging with gas substitution agent, such as Toppan Freshness Keeping Agent Type CW-500; (2) packaging with oxygen absorbent, such as produced by Company A; or (3) packaging without using gas. Packaging materials (400 ml volume tray) may be a tray and lid composition of polystyrene/polyethylene/polyvinyl chloride. The condition for preservation is 25°C and 60% RH.

C. Mold-Preventive Packaging for Castella

Table XVI shows an example of packaging a Castella with oxygen absorbent and nitrogen gas substitution packaging. The packaging with oxygen absorbent, which can maintain a lower oxygen content, showed excellent results in preventing molds and preserving good taste and flavor.

Packaging methods include (1) packaging with oxygen absorbent, such as produced by Company A; (2) nitrogen gas substitution packaging with a remaining oxygen content of 0.5%; or (3) packaging without using gas. Packaging materials may be PVDC-coated polypropylene film 20μm and polyethylene film 60 μm. The condition for preservation is 25°C and 60% RH.

Table XII Variation of Gas Composition

Test Item	Start	After 7 Days	After 14 Days	After 21 Days
Freshness keeping agent C	20.9/0.0	< 0.1/18.0	< 0.1/20.0	< 0.1/36.0
Oxygen absorbent	20.9/0.0	< 0.1/5.0	< 0.1/12.0	< 0.1–36.0
Non-gas-controlled packaging	20.9/0.0	11.8/29.0	< 0.1/62.0	—

[a] Gas composition: left value, oxygen concentration; right values carbon dioxide concentration (%).

Table XIV Bacteria Inspection

Test Item	Start		After 7 Days		After 14 Days		After 21 Days	
	Viable Count	Coliform Bacteria	Viable Count	Coliform Bacteria	Viable Count	Coliform Bacteria	Viable Count	Coliform Bacteria
Freshness keeping agent	1.7×10^3/g	(—)	5.3×10^4/g	(—)	1.7×10^5/g	(—)	3.7×10^5/g	(—)
Oxygen absorbent			6.1×10^4/g	(—)	8.0×10^5/g	(—)	1.0×10^6/g	(—)
Non-gas-controlled packaging			1.2×10^4/g	(—)	5.0×10^6/g	(—)	—	(—)

Chapter 24. Cakes and Snack Foods

Table XV   Variation of Outside Shape

Test Item	Start	After 7 Days	After 14 Days	After 21 Days
Freshness keeping agent	Fair	Fair	Fair	Container swelled due to gas generation
Oxygen absorbent	Fair	Container contracted due to loss of oxygen	Container contracted due to loss of oxygen	Container swelled due to gas generation
Non-gas-controlled packaging	Fair	Container swelled due to gas generation	Bag damaged due to gas generation Molds generated all over	

VII. PACKAGING WITH ALCOHOL-GENERATING AGENT

Gas substitution packaging, packaging with oxygen absorbent, and packaging with gas substitution agent prevent and restrain quality deterioration of food, including cakes, by means of reducing oxygen. On the other hand, packaging with an alcohol generating agent is a packaging method that diffuses ethyl alcohol vapor and demonstrates its effect of restraining the growth of microbes and its effect of sterilization, which ethyl alcohol originally has. This method has attracted public attention because a small quantity of ethyl alcohol is effective to prevent the growth of mold of food with low AW. Among alcohol-generating agents, there is a single alcohol, for which one individually packs an absorbent object, like silicon dioxide,

Table XVI   Preservative Effect of Castella Cakes[a]

Packaging Method	After 7 Days		After 14 Days		After 21 Days		After 40 Days	
Oxygen absorbent	(−)	○	(−)	○	(−)	○	(−)	○
Nitrogen gas substitution	(−)	△	(±)	△	(±)	△	(+)	X
Non-gas-controlled packaging	(+)	X	(⁺₊)	X	(⁺₊)	X		

[a] Key: (−) no molds; (±) molds generated partially; (+) molds generated; (⁺₊) many molds generated; ○ good taste and flavor; △ a little bad taste and flavor; X not edible.

which absorbs alcohol, in alcohol-permeable packaging materials, and there is an agent that simultaneously generates a gas other than alcohol or absorbs oxygen. Neither type of agent can be applied to food with high AW. The applicable AW range is 0.65–0.75. This packaging method has an advantage: it does not need high gas barrier packaging materials, which are used for gas substitution packaging. On the other hand, there is a big problem in that alcohol affects the food's taste and flavor. It is, however, expected that a new alcohol generating agent will emerge, and a method will be developed to combine the new agent with a new packaging method.

XIII. FUTURE PROSPECTS

From the standpoint of restraining the growth of microbes, the quality preservation packaging, which controls the gas atmosphere in the package, is not considered to be sufficient for the cakes of AW over 0.9. On the other hand, the cake industry's trend is to challenge this, by packaging cakes with higher AW. It is therefore necessary not only to develop packaging techniques for higher AW but also to realize the reduction of the primary viable count by means of introducing a new sterilization method, which has been already introduced in the industries of sliced ham and cut moist rice cakes.

REFERENCES

Igawa, F. (1978). *New Food Industry* **19,** 5.
Ishitani, T. (1978). *Quality Preservation of Processed foodstuffs*. Research Institute of Food Quality Reservation Technology.
Suda, H. (1980). "*Food Packaging and Materials,*" pp. 132–140.
Suzuki, S. (1981). *New Food Industry* **23:**12, 49–54.
Tanaka, A. (1982). *Japan Food Science* **11,** 34–45.
Watanabe, N. (1980). "*Science of Cakes,*" pp. 179–200. Dobun-shoin, Tokyo.
Yoshii, H., et al. (1976). "Food Microbiology," pp. 1–532. Gihodo, Tokyo.

PART VII

Physical Distribution and Food Packaging

CHAPTER 25

Physical Distribution of Packaged Foods

Yoshio Hasegawa
Japan Packaging Research Institute
Chiba, Japan

I. PACKAGING FOR PRESERVATION OF FOODS IN THE DISTRIBUTION PROCESS

Foods are packaged for preservation of their qualities. To get it done properly, first of all, we must investigate completely problems that packaged foods may come across in the distribution process. Possible problems are listed in Tables I–III.

As there are many hazards in the distribution process, it is necessary to protect foods from them. To protect foods from these hazards there are many means, and packaging is one of the most important methods. In Table I–III measures to take for protection from deterioration of foods by packaging are indicated for each problem.

When we look into protecting foods from deterioration by packaging, we must divide packaging into two fields: distribution packaging and consumer packaging. Distribution packaging is packaging for the distribution process, necessary for transportation and storage of foods. On the other hand, consumer packaging is packaging to pass into consumers' hands. In other words, the way goods are displayed on shelves at retailers is consumer packaging. A batch or group of goods in consumer packagings is packaged in one large distribution packaging. Then in the distribution process, the single distribution packaging is handled as a unit, but at the retailer, consumer packagings are taken out from distribution packaging and displayed on shelves. So at the retailer, one consumer packaging is handled as a unit. There are three ways to preserve the quality of foods: consumer packaging to carry out the principal role, distribution packaging to carry out the principal role, and both to carry out the same role.

Food Packaging Copyright © 1990 by Academic Press, Inc. All rights of reproduction in any form reserved.

Table I Physical Hazards in the Distribution Process

Hazards	Process of Physical Distribution	Examples	Measure to Take
Impact Vertical impact: dropping	Handling	1. When loading or unloading or transferring of packages is done by human power there is a chance that people might throw them occasionally down to the floor or onto other packages (identical or quite different). Its extent of effect varies depending on what the floor is made of and on the height from which it was dropped. If packages are carried on shoulders, a drop of about 120 cm is expected; however, it will be less than 60 cm when using both hands in carrying. Usually people do this if the distance of their trip is very short. Therefore we must take it into consideration that	Cushioning Don't drop from shoulder

there will be a chance of packages hitting the floor after slipping from a height of somewhere less than 40 cm (on average it will be 30 cm), when people do palletizing by hand.
2. Drop from vehicles. Uppermost packages stacked on the floor of truck face a drop to the ground from a height of more than 3 m. (Note: maximum allowable height of vehicle on the road is 3.8 mm in Japan.)
3. Faulty operation of cranes and fork lift trucks. If operator brings down its hook or forks too suddenly from raised position, then packages on them come to the ground with a thud, or part of them may jump out from the rest. One of ropes to the sling may break, taking too much

Careful loading
Careful handling

(continued)

Table I *Continued*

Hazards	Process of Physical Distribution	Examples	Measure to Take
		strain. If this happens the fall is from heigher points than any other cases described so far. But this last case belongs to a category we call accidents.	
Vertical impact by dropped packages	Handling	1. Packaged foods are dropped on other packages at loading. At consolidation, a corner of a wooden box may be dropped on corrugated fiber board box. 2. By mistake of crane handling, other freight is dropped on packaged foods. There are problems at loading on ships.	Careful handling Use of unit load system, for example, box pallet
Vertical motion of loading floor	Transportation	1. When a truck runs down rough roads without tightening freights, impact between freights and floor happens by jumping up of freights.	Cushioning Careful driving of truck

Horizontal impact between body of wagon and load	Transportation	When collecting and delivery of freights, as in trucks without tightening freights, vertical impact may happen. 2. Landing of airplane. 1. Impact by coupling of freight cars. 2. Impact between end structure of freight cars and loads when train stops suddenly.	Cushioning Cushioning for horizontal impact Careful loading
Horizontal impact between loads	Transportation	Impact between loads as mentioned above.	Similar to above
Horizontal impact of swinging crane	Handling	Freights strike on the hatch by swinging of crane handling. These happen by unskilled crane operation or under strong winds.	Careful handling
Horizontal impact by slide	Handling	Freights strike on other freights at end of slide	Careful handling
Horizontal strike	Handling	Freights are struck on sides by other freights.	Careful handling
Compound impact: rolling	Handling	1. Drums and barrels are rolling. 2. Moving heavy freight by rocking on their corners.	Appropriate strength of container Avoiding human power handling

(continued)

Table I *Continued*

Hazards	Process of Physical Distribution	Examples	Measure to Take
Compound impact: falling	Handling	Impact by falling down of unstable freights.	Stabilizing by packaging
Vibration			
Vibration in transportation	Transportation	1. When paper sacks rub against each other or with body of wagon, holes are made. This trouble may happen also in cans.	Careful loading
Freight car			
Truck			
Ship			
Airplane		2. Upper tier of packages of stack fall down by effects of vibration. Character of vibration varies by freight car, truck, ship, and airplane.	Careful loading
Compression	Storage	1. Total weight of upper packages stress the package in the bottom.	Use of container with appropriate strength
	Transportation		
	Stacking		

Fixing	2. In transportation, as there are effects of vertical vibration by freight car motion, more than above-mentioned weight stress on the package at the bottom.	
Transportation	1. Roping: loading freights are stressed by tightening with rope. 2. Binding up palletized load.	Use of corner pad
Handling	1. At crane handling by net, freights are stressed.	Use of board in net
Handling	2. Workers step on packages for handling.	Use of board on packages.

Table II   Climatic Hazards in the Distribution Process

Hazards	Process of Physical Distribution	Examples	Measure to Take
Temperature			
Natural high temperature	Storage Handling Transportation	1. Effects of direct sunshine when waiting at wharf and so on before loading. 2. Effects of high temperature in freight cars or holds.	Don't expose to the sunshine The freights that are affected by high temperature must be kept in insulated containers.
Natural low temperature	Storage Handling Transportation	1. Effects by low temperature in winter. 2. Transportation by airplane without air conditioning.	The freights that are affected by low temperature may be kept warm.
Natural change of temperature	Transportation	When freight cars or freight containers are exposed to the sun in the daytime, inside temperatures become higher than outside temperature, and in the evening, become lower than outside temperature.	Moisture-proof and waterproof packages.
Artificial high temperature	Storage, Transportation	When freights are placed near the boiler and so on.	Care of placement

Artificial low temperature	Storage Transportation	When freights are stored in the cold storage or transported in the refrigerated car, low temperature damage develops.	Care of temperature.
Artificial change of temperature	Storage Transportation	In storage and transportation, when the room temperature change up and down.	Care of temperature
Moisture High relative humidity	All physical distribution	Natural and artificial high relative humidity atmosphere.	Moisture protection of package
Low relative humidity	All physical distribution	Natural and artificial low relative humidity atmosphere.	Moisture protection of package (prevention from drying)
Liquid Fresh water: rain	Transportation Handling	1. Transportation by an open wagon. 2. Loading and unloading. 3. The roof leaks. 4. Snow blows into.	Care of raining Waterproof package
Fresh water: condensed	Transportation	Condensed water from atmosphere from rapid change of room temperature.	Waterproof package
Seawater in air	Transportation Handling	1. Open stacking at port. 2. Open loading in barge 3. Loading on deck.	Loading in freight containers.
Sunshine Ultraviolet rays	Transportation Handling	By ultraviolet rays, qualities of foods deteriorate.	Don't expose to the sunshine Prevention from passing of ultraviolet rays by package

Table III Miscellaneous Hazards in the Distribution Process

Hazards	Process of Physical Distribution	Examples	Measure to Take
Living things			
Microorganisms	Storage	Increase of bacteria, mold and other microorganisms.	Moisture-proof package. Package uses free oxygen scavenger. Storage in cold room.
Insects	Storage	Invasion by insects.	Seal up tightly
Rats	Storage, Transportation	Rats eat away foods in storage or in holds.	Improvement of environments.
Contamination			
Touching with other freights	Transportation	Contamination by touching with other freights.	Care of consolidation.
Contamination by leak from other freight	Transportation	Contamination from other freight loaded together or from trace of freight loaded there in the past.	Care of consolidation, clean up the floor and walls in the car.
Absorption of odor	Transportation Storage	Absorption of odor from other freight loaded together or from trace of freight loaded there in the past.	Care of consolidation. Clean up the floor and walls in the car.

Chapter 25. Physical Distribution of Packaged Foods

For example, when the quality of goods deteriorates by absorption of moisture, it is necessary to protect them from absorption of moisture in the distribution process and also at retail shops. At retail shops goods are taken out from distribution packaging and put on the shelves in only consumer packaging, and then consumer packaging must have the property of moisture-proofing. In these cases, distribution packaging may not be moisture-proof. For example when wheat flour is packed in large bags and distributed, the bags must be moisture-proof.

Protection against outside forces in the distribution process is a role of distribution packaging. After goods reach the retail shop and are displayed on shelves, they do not meet strong outside forces, and so consumer packaging does not require properties of protection against such force. Of course, there are cases when even consumer packaging must maintain protection against outside forces. For example, consumer packaging (pouches and bottles etc.) in distribution packaging must not burst when dropped in handling.

II. DISTRIBUTION OF PACKAGED FOODS AND TEMPERATURE ENVIRONMENT

To preserve the quality of foods that deteriorate at high temperature, keeping them in low temperature is necessary in addition to packagings. Suitable temperature differs by kind of foods, but there are roughly three groups to classify by suitable temperature range. The first group is frozen foods, and their suitable temperature is below $-18°C$. The second group is foods for which the suitable temperature is nearly $0°C$, for example fresh meats and fishes. The temperature for the third group is nearly $5°C$, for example, fruits and vegetables. But suitable temperatures of fruits and vegetables differ slightly by item. These temperatures must be maintained from the start to the end of the distribution. To keep suitable temperature, circumstances of loading and unloading must be suitably equipped for those perishables, and handling must be done quickly. To do quick handling, utilizing of unit load systems by mechanical handling is effective. The shortening of loading and unloading times can prevent a rise of goods temperature.

Utilizing of refrigerated vehicles that can control the temperature in the van is necessary. A refrigerated vehicle has fixed insulation on the walls to prevent heat transfer from outside, so the inner volume of the van is small. In particular, the inner width of the van become narrower than for a dry van, so when we want to utilize a unit load system, unit load dimensions differ from dimensions in a dry van, and special pallet sizes, for example, must be selected. Package sizes suitable to these unit load sizes differ from unit load sizes of dry cargo.

In the opposite direction, there are problems of protecting foods from cold injury in the distribution process. When many hours of transportation of foods in areas where the winter temperature falls lower than 0°C are necessary, foods must be protected from cold injury. In these cases, insulated vehicles are used and also a method that keeps warmth inside the van. The method of insulating each package must be avoided, and using insulated vehicles is desirable. This method is cheaper in total costs than using insulated packaging.

III. PHYSICAL DISTRIBUTION AND UNIT LOAD SYSTEM

A. Development of the Unit Load System

Besides the preservative aspects of the distribution packaging, distribution packaging must have properties that facilitate handling of goods. We must study these properties together with future direction for organizing physical distribution.

Cutting down human effort is also a direction for determining physical distribution, and as a method for cutting down of human effort, the unit load system must be utilized. This is the method that handles using machines and tools for the unit load of collected goods. Packaged foods are goods that can be suitable for this method.

The usual method that collects distribution packages into a unit is the use of a pallet. This method loads on a pallet distribution packages, arranged to prevent packages slipping out, and ships this unit load. This unit load is handled by the material handling machine, for example, forklift trucks, and so this method saves more human effort than methods that handle one-by-one distribution of packages by human efforts, and shorten handling times considerably.

When this method is used, goods suffer less shock than by human handling, so another advantage of this method is avoiding accidents in which goods are damage and injured. When goods are dropped from the shoulder by human handling, for example, fruits packed inside may get damaged, and decay starts. If we adopt the unit load system, these problems no longer exist.

Saving of human effort may not be necessary in areas where human labor is abundant, but one of the advantages of unit load system, that of preventing the damages and injuries that are inevitable when manpower does the job, must be recognized.

To introduce this system, there must be a large distribution quantity of goods. If there are masses of one kind of goods, this system can be introduced easily. When there are only small quantities of one kind of goods, it is necessary that many items of goods be collected and made into unit loads.

In many cases, we can transport, from factories to warehouses or distribution centers in consumption areas, one kind of goods making up unit loads. It is also acceptable to have unit loads of different items be loaded in one vehicle. When fruits and vegetables are transported, the same applies. From the packing plant in producing districts to markets in consumption areas, transportation by unit loads can be utilized. In these cases, if two items do not affect each other, unit loads can be made up with goods of different items combined.

However, distribution from warehouses or distribution centers to retailers etc. handles small quantities of each item, so smaller units are demanded. Besides, in retailers there are no forklift trucks available in many cases, so the roll box that can be handled by human power is utilized. When roll boxes are utilized, trucks for delivery must be equipped with a tailgate lifter.

B. Modular Coordination for a Physical Distribution System

When we promote the unit load system mentioned in the previous section, first unit load dimensions may be standardized, and then all facilities, machinery, and so on that relate to the distribution process may be coordinated by a modular system based on the unit load dimension.

In this case, the first problem is settling of the module, namely, its standard dimension. The numerical value of the module is difficult to change at present and also in the near future. The regulated largest outer width dimension of truck is the same in almost every country in the world, and this dimension is difficult to change. This dimension is 2500 mm, and many countries in the world use this dimension as regulations; and widening this outer width dimension is difficult in relation to facilities such as roads. Therefore we can consider that this dimension is more or less fixed in numerical value from the present to the near future. Then we may consider that we select this numerical value as the starting point. When we start from 2500 mm, which is the regulated largest outer width dimension of truck, inner dimension of truck will become about 2350 mm as the largest numerical value.

If we propose to place two unit loads in a truck of which the inner width dimension is 2350 mm, the total width of the two unit loads becomes a numerical value 2350 mm from which is subtracted the dimension of the gap necessary for easy loading and unloading. The dimension of the gap for easy loading and unloading is 50 mm in ordinary handling methods. Then, since $2350 - 50 = 2300$ mm, the dimension of one unit load is as follows:

$$2300 \text{ millimeter}/2 = 1150 \text{ millimeter}$$

This numerical value is a largest side dimension of unit load. When we consider that a square unit load is advantageous, the largest dimensions of the unit load become 1150×1150 mm. If this unit load is constituted on a pallet base, the dimensions of the pallet become 1100×1100 mm and the largest dimension of a unit load constituted on this pallet is necessarily smaller than 1150×1150 mm.

However, among the many kinds of foods, there are goods that demand low-temperature distribution. It is necessary that these goods be transported in a refrigerated truck. Insulation on the walls of the refrigerated truck is used to prevent heat penetration from the outside, so the inner width dimension of the refrigerated truck is narrower than that of the dry van. Therefore, above-mentioned dimension can not be applied here. So it is necessary that we consider other dimensions.

If we consider using refrigerater trucks that have good performance for transportation of frozen foods, the inner width dimension becomes 2180 mm. As the necessary dimension of the gap for loading and unloading in the refrigerator truck is larger than in the dry van, we settle its dimension at 80 mm:

$$2180 - 80 = 2100 \text{ millimeter}$$

Consequently the dimension of one unit load is $2100/2 = 1050$ mm. Then, if we consider loading in the dry van by combining width and length of pallet, the dimension of another side is as follows:

$$2300 - 1050 = 1250 \text{ millimeter}$$

So the largest side dimensions of the unit load for loading in the refrigerator truck become 1250×1050 mm. If this unit load is constituted on a pallet base, the dimensions of the pallet become 1200×1000 mm, and largest dimension of unit load constituted on this pallet is necessary smaller than 1250×1050 mm. This unit load dimension is adapted for low-temperature transportation and also for normal-temperature transportation. We illustrate these above mentioned details like this:

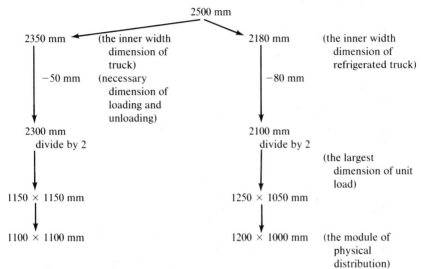

If we consider that 1100 × 1100 mm and 1200 × 1000 mm are the respective modules of physical distribution systems, we consider the modular coordination based on this module as follows.

IV. DISTRIBUTION PACKAGE DIMENSIONS BY MODULAR COORDINATION

A. Calculation Method for Distribution Package Dimensions

A calculation method for submultiples is shown in Fig. 1, giving distribution package dimensions based on the physical distribution module. In other words, the problem is how we determine distribution package dimensions that adapt to the 1100 × 1100 mm pallet or the 1200 × 1000 mm pallet. There are base dimensions (length × width) of parallelepipedic distribution packages that we may consider in this case.

Generally a container is used for putting goods in. So sizes of containers must be large enough to fit goods that are put in there. Then dimensions of containers may be settled by dimensions and the numbers of goods that are put in. So the dimensions of container are dependent on goods, and we can

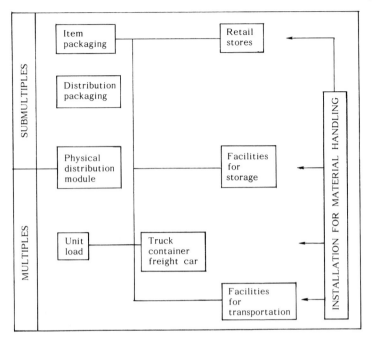

Figure 1. Modular coordination for physical distribution system.

not settle dimensions of containers without considering dimensions of goods.

However, as there are also goods that do not depend upon physical distribution for their shape and size, the idea of using containers that adapt to good efficiency of physical distribution may be considered. That is, dimensions of containers may first be settled and goods that adapt to these dimensions of containers in their manufacturing may be considered. The method in which the physical distribution module is first determined and then dimensions of containers are calculated, based on this module, is similar to the above-mentioned idea. When dimensions of goods or sizes of packages cannot be changed due to the character of goods, this method may be difficult to use. But fortunately foods are comparatively apt to change dimensions of goods or sizes of packages, so this method may be used easily. This will be testified to by examples mentioned subsequently.

1. Dimensions (Length × Width) Adapted to 1100 × 1100 mm

a. Combination of Numerical Values Dividing 1100 mm by Integral Numbers Numerical values dividing 1100 mm by integral numbers are shown in Table IV.

Chapter 25. Physical Distribution of Packaged Foods

Table IV Numerical Values of 1100 mm Divided by Integral Numbers

Submultiple	Dimensions of One Side (mm)
1	1100
2	550
3	366
4	275
5	220

The combinations by these numerical values make up 15 pairs:

1100×1100
$1100 \times 550 \quad 550 \times 550$
$1100 \times 366 \quad 550 \times 366 \quad 366 \times 366$
$1100 \times 275 \quad 550 \times 275 \quad 366 \times 275 \quad 275 \times 275$
$1100 \times 220 \quad 550 \times 220 \quad 366 \times 220 \quad 275 \times 220 \quad 220 \times 220$

b. Selection of Numerical Values Adapted to 1100 × 1100 mm by Combination As 1100 × 1100 mm is a square, the loading in Fig. 2 shows numerous dimensions that become length + width = 1100 mm. Selection is based on the standard of Table V. Combinations of length and width based on Table V are $a \times b$ (10), $a \times b/2$ (8), $b \times a/2$ (10), $a/2 \times b/2$ (8), and $b \times a/3$ (9), so there is a total of 45 combinations.

c. Other Combinations Other combinations that adapt to 1100 × 1100 mm are showed in Table VI. There are 69 combinations adapted to 1100 × 1100 mm.

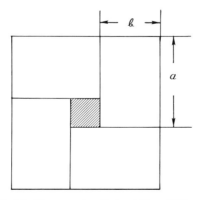

Figure 2. Packing method that becomes length + width = 1100 mm ($a + b = 1100$ mm).

Table V Selection of Numerical Values That Become Length + Width = 1100 mm

Aspect Ratio a/b	Dimensions of One Side (mm)				
	Length (a)	Width (b)	a/2	b/2	a/3
2/1 = 2.000	733	366	366	—	244
11/6 = 1.833	711	388	355	—	237
5/3 = 1.666	687	412	343	206	229
3/2 = 1.500	660	440	330	220	220
13/9 = 1.444	650	450	325	225	216
7/5 = 1.400	641	458	320	229	213
4/3 = 1.333	628	471	314	235	209
5/4 = 1.250	611	488	305	244	203
6/5 = 1.200	600	500	300	250	200
11/10 = 1.100	576	523	288	261	—

2. Dimensions (Length × Width) Adapted to 1200 × 1000 mm

a. Combinations of 1200 and 1000 mm Divided by Integral Numbers Numerical values for 1200 and 1000 mm divided by integral numbers are showed in Table VII. Combinations of these numerical values make up next 30 pairs:

1200 × 1000	1000 × 600	600 × 500
1200 × 500	1000 × 400	600 × 333
1200 × 333	1000 × 300	600 × 250
1200 × 250	1000 × 240	600 × 200
1200 × 200	1000 × 200	
500 × 400	400 × 333	333 × 300
500 × 300	400 × 250	333 × 240
500 × 240	400 × 200	333 × 200
500 × 200		
300 × 250	250 × 240	240 × 200
300 × 200	250 × 200	200 × 200

Table VI Other Combinations That Adapt to 1100 × 1100 mm

	l (mm)	w (mm)	l/w
1	412	275	1.498
2	293	220	1.333
3	275	206	1.333

Table VII Numerical Values Dividing 1200 and 1000 mm by Integral Numbers

Submultiple	Dimensions of One Side (mm)	
1	1200	1000
2	600	500
3	400	333
4	300	250
5	240	200
6	200	—

b. Combinations Adapted to 1200 × 1000 mm by Combined Loading By the loading shown in Fig. 3, the numerical values showed in Table VIII are obtained. If we omit combinations that were already included in (a) from combinations by c and d in Table VIII, new combinations are 600 × 400 mm and 400 × 300 mm.

c. Other Combinations of Length and Width Other combinations that adapt to 1200 × 1000 mm are shown in Table IX.

There are 40 combinations adapted to 1200 × 1000 mm. If they are shown both according to their dimension of length, they become Table X and Table XI.

If we select one of combinations in Table VIII, it becomes adapted to the 1100 × 1100 mm pallet, and if we select one of combinations in Table IX, it becomes adapt to the 1200 × 1000 mm pallet. As these combinations are shown as maximum dimensions, the practical dimensions of a container

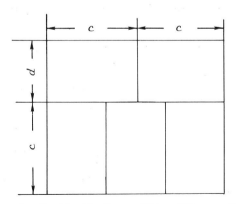

Figure 3. Combined loading on 1200 × 1000 mm pallet.

Table VIII Numerical Values Adapted to 1200 × 1000 mm by Combined Loading

Submultiple	Dimensions of One Side (mm)	
1	600	400
2	300	200
3	200	—

may be settled smaller than these numerical values in some cases. Compression bulge from stacking in plastic containers occurs rarely, and stacking irregularity does not arise as plastic container are fixed together top to bottom, so by using these dimensions themselves, unit load dimensions are settled at 1150 × 1150 mm. However, corrugated fiberboard boxes exhibit compression bulge and stacking irregularity, so smaller dimensions than used in these tables may be used for settling unit load dimensions within of 1150 × 1150 mm.

2. Comparison of This Method with ISO, ANSI, and SIS

ISO (International Organization for Standardization) deals with standardizations of packagings in TC 122, and the problem of how unit load dimensions should be decided has been discussed in the last 10 or more years. The reason that this problem could not be settled for many years is the insistence that one standard in the world should be settled in spite of different circumstances in different areas. If many countries persist in separate views, it may be difficult to establish international standards. Recently the technical committee of ISO succeeded in making a compromise, and it was settled that the following three types were adopted as

Table IX Other Combinations That Adapt to 1200 × 1000 mm

	l (mm)	w (mm)	l/w
1	500	233	2.145
2	475	250	1.900
3	433	333	1.300
4	380	240	1.583
5	333	216	1.541
6	316	250	1.264
7	300	233	1.287
8	266	200	1.333

Chapter 25. Physical Distribution of Packaged Foods

Table X Distribution Package Dimensions (1100 × 1100 mm)

Nominal Number	Length Width (mm)	Aspect Ratio (Length/Width)	Quantity of One Layer
11-1	1100 × 1100	1.000	1
11-2	1100 × 550	2.000	2
11-3	1100 × 366	3.000	3
11-4	1100 × 275	4.000	4
11-5	1100 × 220	5.000	5
11-6	733 × 366	2.000	4
11-7	711 × 388	1.833	4
11-8	687 × 412	1.666	4
11-9	687 × 206	3.333	8
11-10	660 × 440	1.500	4
11-11	660 × 220	3.000	8
11-12	650 × 450	1.444	4
11-13	650 × 225	2.888	8
11-14	641 × 458	1.400	4
11-15	641 × 229	2.800	8
11-16	628 × 471	1.333	4
11-17	628 × 235	2.666	8
11-18	611 × 488	1.250	4
11-19	611 × 244	2.500	8
11-20	600 × 500	1.200	4
11-21	600 × 250	2.400	8
11-22	576 × 523	1.100	4
11-23	576 × 261	2.200	8
11-24	550 × 550	1.000	4
11-25	550 × 366	1.500	6
11-26	550 × 275	2.000	8
11-27	550 × 220	2.500	10
11-28	523 × 288	1.818	8
11-29	500 × 300	1.666	8
11-30	500 × 200	2.500	12
11-31	488 × 305	1.600	8
11-32	488 × 203	2.403	12
11-33	471 × 314	1.500	8
11-34	471 × 209	2.253	12
11-35	458 × 320	1.431	8
11-36	458 × 213	2.150	12
11-37	450 × 325	1.384	8
11-38	450 × 216	2.083	12
11-39	440 × 330	1.333	8
11-40	440 × 220	2.000	12
11-41	412 × 343	1.200	8
11-42	412 × 275	1.498	10
11-43	412 × 229	1.799	12
11-44	388 × 355	1.092	8

(continued)

Table X *Continued*

Nominal Number	Length Width (mm)	Aspect Ratio (Length/Width)	Quantity of One Layer
11-45	388 × 237	1.637	12
11-46	366 × 366	1.000	9
11-47	366 × 275	1.333	12
11-48	366 × 244	1.500	13
11-49	366 × 220	1.666	15
11-50	343 × 206	1.666	16
11-51	330 × 220	1.500	16
11-52	325 × 225	1.444	16
11-53	320 × 229	1.400	16
11-54	314 × 235	1.333	16
11-55	305 × 244	1.250	16
11-56	300 × 250	1.200	16
11-57	300 × 200	1.500	20
11-58	293 × 220	1.331	18
11-59	288 × 261	1.103	16
11-60	275 × 275	1.000	16
11-61	275 × 220	1.250	20
11-62	275 × 206	1.334	21
11-63	250 × 200	1.250	24
11-64	244 × 203	1.201	24
11-65	235 × 209	1.124	24
11-66	229 × 213	1.075	24
11-67	229 × 206	1.111	25
11-68	225 × 216	1.041	24
11-69	220 × 220	1.000	25

unit load base dimensions, so an finally international standard may be settled before long:

1200 × 1000 mm
1200 × 800 mm
1140 × 1140 mm

These dimensions are maximum dimensions, and the permitted maximum deviation from them is −40 mm.

The 1200 × 1000 mm type has been used European countries for a long time. This unit load is derived from the 600 × 400 mm module used in several European countries.

The 1200 × 800 mm type is equal to the European pool pallet dimension, so this dimension can be used easily in European countries; this is also derived from the 600 × 400 mm module.

The 1140 × 1140 mm type is used for square unit loads. This dimension

Chapter 25. Physical Distribution of Packaged Foods

Table XI Distribution Package Dimensions (1200 × 1000 mm)

Nominal Number	Length Width (mm)	Aspect Ratio (Length/Width)	Quantity of One Layer
12-1	1200 × 1000	1.200	1
12-2	1200 × 500	2.400	2
12-3	1200 × 333	3.600	3
12-4	1200 × 250	4.800	4
12-5	1200 × 200	6.000	5
12-6	1000 × 600	1.666	2
12-7	1000 × 400	2.500	3
12-8	1000 × 300	3.333	4
12-9	1000 × 240	4.166	5
12-10	1000 × 200	5.000	6
12-11	600 × 500	1.200	4
12-12	600 × 400	1.500	5
12-13	600 × 333	1.801	6
12-14	600 × 250	2.400	8
12-15	600 × 200	3.000	10
12-16	500 × 400	1.250	6
12-17	500 × 300	1.666	8
12-18	500 × 240	2.083	10
12-19	500 × 233	2.145	10
12-20	500 × 200	2.500	12
12-21	475 × 250	1.900	10
12-22	433 × 333	1.300	8
12-23	400 × 333	1.201	9
12-24	400 × 300	1.333	10
12-25	400 × 250	1.600	12
12-26	400 × 200	2.000	15
12-27	380 × 240	1.583	13
12-28	333 × 300	1.110	12
12-29	333 × 240	1.387	15
12-30	333 × 216	1.541	16
12-31	333 × 200	1.665	18
12-32	316 × 250	1.264	15
12-33	300 × 250	1.200	16
12-34	300 × 233	1.287	17
12-35	300 × 200	1.500	20
12-36	266 × 200	1.333	22
12-37	250 × 240	1.041	20
12-38	250 × 200	1.250	24
12-39	240 × 200	1.200	25
12-40	200 × 200	1.000	30

is derived from the minimum internal width of the ISO series I for general-purpose freight container. This dimension is adopted by the convention of many countries that use ISO freight containers except countries on the Continent. As this dimension fits the ISO freight containers, it is very important for worldwide trade.

The dimensions mentioned in B-2 are 1150 × 1150 mm, but this dimension is 1140 × 1140 mm. This difference is due to the minimum width of ISO containers, which is 2330 mm. If we start from 2330 mm, the following figures are derived:

$$2330 - 50 = 2280 \text{ millimeter} \quad 2280/2 = 1140 \text{ millimeter}$$

As the permitted maximum deviation is −40 mm, we can stack goods on 1100 × 1100 mm pallets and make unit loads. The maximum dimension of these unit loads is permitted to be 1140 × 1140 mm. So the same packaging dimensions above mentioned can be adopted.

Packaging dimensions adapted to 1200 × 1000 mm are already established by ISO 3394 (Dimensions of rigid rectangular packages—Transport packages). In this international standard the dimensions given in Table XII are established. These are based on 600 × 400 mm and its multiples and submultiples.

These dimensions also fit the 1200 × 800 mm pallet. As these dimensions are based on 600 × 400 mm, it is natural that these dimensions fit 1200 × 800 mm. These dimensions differ somewhat from dimensions in Table IX.

There are methods like the one just discussed, based on 600 × 400 mm for a modular coordination system (DIN for Federal Republic of Germany, NF for France, and ISO), and also other methods based on dimensions of

Table XII   Dimensions of Distribution Packages in ISO 3394

Multiples	Submultiples	
1200 × 1000	600 × 400	600 × 133
1200 × 800	300 × 400	300 × 133
1200 × 600	200 × 400	200 × 133
1200 × 400	150 × 400	150 × 133
800 × 600	120 × 400	120 × 133
	600 × 200	600 × 100
	300 × 200	300 × 100
	200 × 200	200 × 100
	150 × 200	150 × 100
	120 × 200	120 × 100

the pallet (ANSI for USA, SIS for Sweden, JIS for Japan, etc.). The former ideas started from the packaging and the latter ideas started from the physical distribution. As there are no goods that do not depend upon physical distributions, starting from the physical distribution is proper.

V. EXAMPLES OF DISTRIBUTION PACKAGE DIMENSIONS SELECTED BY MODULAR COORDINATION

A. Sake in Japan

Sake is a unique Japanese alcoholic drink, and the amount of consumption is second in volume to beer in Japan. But with recent increases in the consumption of whisky and wines, the pattern of consumption has diversified and the consumption of sake has been reduced. In order to stimulate consumption of sake, its containers are beginning to play an important role and are getting various shapes.

Hitherto only glass bottles were used as containers of sake for maintaining the quality of sake. But other packaging materials were improved and began to be used. One of the diversified containers of sake is the paperboard container. The paperboard container began to be used for milk sold in supermarkets, and now the most milk is sold in paperboard containers, but containers of sake did not change to paperboard containers at the same time. Similarly, containers for whiskies and wines were glass bottles, and other packaging materials were not used. But as the amount of sake sold in supermarkets increased, the switch to paperboard containers began. The reason for this was the consumer requirement for light packages, as consumers themselves must carry back goods bought at supermarkets. When consumers buy sake at the retailers, the shops deliver to houses, so consumers are unconcerned with the weight of packaging.

At the same time that the conversion to paperboard packages begin, the rationalization of physical distribution began to be advanced, the introduction of unit load system was planned, and the modular coordination of pallet dimension and packaging dimensions was brought into these plans. A sample plan is as follows. The volume of container for sake is mainly 1.8 l. This is the Japanese traditional unit for liquid. The adoption of a paperboard container of the same volume was planned, and the shape of container was decided as in Fig. 4.

The total weight of sake in this paperboard container is shown in Table XIII, and this is 33% less than in the traditional glass bottle. A width of 80 mm and height of 289 mm were dimensions that adapt to the door pocket

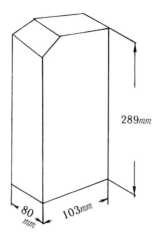

Figure 4. Paperboard container of sake.

of Japanese popular houshold refrigerators. The body diameter of the larger bottle for beer used in the Japanese home is 76 mm and the height is about 290 mm. The door pocket of the houshold refrigerator is designed to hold this bottle well, so the above mentioned dimensions were adopted. By deciding a width of 80 mm and a height of 289 mm, the length of 103 mm was calculated inevitably.

Six pieces paperboard containers were put in the plastic distribution container (Fig. 5).

$$103 \times 2 = 206 \text{ millimeter} \qquad 80 \times 3 = 240 \text{ millimeter}$$

The dimension of plastic container was thus 270 × 216 mm. These plastic containers are loaded on 1100 × 1100 mm pallets as shown in Fig. 6.

$$270 \times 4 = 1080 \text{ millimeter} \qquad 216 \times 5 = 1080 \text{ millimeter}$$

These plastic containers can be loaded very efficiently on a 1100 × 1100 mm pallet.

The total weight of filled plastic container is made up of

Filled paperboard containers $\qquad$ 1.9 kg × 6 = 11.4 kg

Table XIII Weight of Sake in Containers

Container	Total Weight	Percent of Glass Bottled Weight
Paperboard container	1900 g	67
Glass bottle	2850 g	100

Chapter 25. Physical Distribution of Packaged Foods

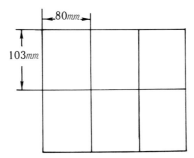

Figure 5. Paperboard containers for sake in the distribution container.

Plastic container 0.6 kg
Total weight 12.0 kg

The total weight of one unit load is 12 kg × 20 × 4 = 960 kg.
The dimensions of this plastic container correspond to no. 11-45

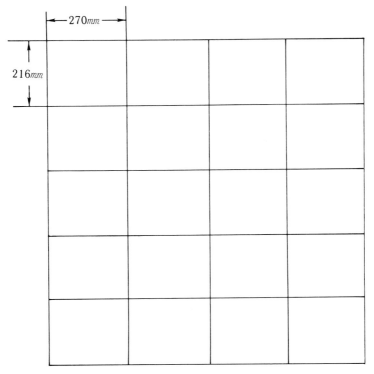

Figure 6. Plastic containers for sake loaded on 1100 × 1100 mm.

(275 × 220 mm) in "Distribution package dimensions by modular coordination." This example illustrates how the distribution package dimension and the item package dimension are decided by the idea of modular coordination.

B. Foodstuff Wholesalers Group in Sweden

The pallet size 800 × 1200 mm has now been approved as the only size accepted for European exchange in the pallet pool formed in 1961 by the International Railway Union. Most countries in Western and Northern Europe, among them Sweden, have acquired membership of this pool.

Since the railways do not want to exchange more than one single size of pallets and this size has been standardized to 800 × 1200 mm, the trade is bound to this size. This must consequently also form the basis of packaging dimensions within the food trade.

The base dimensions of transport packages that give compact pallet loads, correct as to size, on a pallet 800 × 1200 mm are standardized in Sweden (Table XIV). Most of the containers are also suitable for use with the 600 × 800 mm pallets. The outside dimensions for shipping containers shall be considered only as representative values. Inside dimensions, on the other hand, shall be considered as requirements, since they determine the outside dimensions of the consumer packages. Only outside dimensions are given for consumer packages, and these are to be considered as nominal or theoretical dimensions. The actual outside dimensions should, as a rule, be smaller than the dimensions given.

A pallet can either be loaded with shipping containers all of which have the same dimensions (uniform pallet load) or—under the conditions set forth in this section—with shipping containers of different dimensions (mixed pallet load). Uniform pallet loads are encountered most frequently in the transport and storage of goods on the manufacturer's premises and in trnsportation from manufacturer to wholesaler. Mixed pallet loads are used for the transport of small quantities of different types of goods from manufacturer to wholesaler or retailer, and especially for transport from wholesaler to retailer. Mixed pallet loads can be accomodated on a 800 × 1200 mm or a and 600 × 800 mm pallet.

The outside dimensions of consumer packages designed as rectangular parallelepipeds are given in the Table XV. All outside dimensions are theoretical, as they represent the total amount of space required by the package. In practice, allowance must be made for any space-consuming compartments, separators or dividers in the shipping container, bulges caused by the contents of the consumer package, etc. Allowance must also be made for situations where consumer packages must be easily accessible for insertion into or removal from shipping containers.

Table XIV Standard Dimensions of Shipping Containers in Sweden

	Outside Dimensions (mm)	Inside Dimensions (mm)
1M	1200 × 800	1140 × 760
2L	1200 × 400	1140 × 380
2M	800 × 600	760 × 570
3L	1200 × 270	1140 × 253
3M	800 × 400	760 × 380
4L	1200 × 200	1140 × 190
4M	800 × 300	760 × 285
4S	600 × 400	570 × 380
5M	800 × 240	760 × 228
6L	800 × 200	760 × 190
6M	600 × 270	570 × 253
6S	400 × 400	380 × 380
8L	600 × 200	570 × 190
8M	400 × 300	380 × 285
9M	400 × 270	380 × 253
10M	400 × 240	380 × 228
12M	400 × 200	380 × 190
12S	300 × 270	285 × 253
14L	400 × 170	380 × 162
15L	400 × 160	380 × 152
16L	400 × 150	380 × 142
16M	300 × 200	285 × 190
18L	400 × 130	380 × 126
18M	270 × 200	253 × 190
24L	400 × 100	380 × 95
24S	200 × 200	190 × 190

Based on these standards, the "Recommendation for distribution of consumer goods" was standardized. This standard recommends a uniform modular system for distribution of consumer goods in Sweden. The purpose of this standard is to illustrate the possibilities of simplifying and facilitating material handling and transportation throughout the distribution chain from producer to consumer by means of dimensional coordination.

The system is based on 800 × 1200 mm and 600 × 800 mm pallets. Because of their size, 800 × 1200 mm pallets can be used throughout the entire distribution chain only in experimental cases. Figure 7 shows goods leaving the producer on 800 × 1200 mm pallets. If the particular type of goods in question is shipped frequently, the pallet load should be divided into units having a bottom area of 600 × 800 mm even at this stage to reduce subsequent rehandling along the distribution chain. At the whole-

Table XV Standard Outside Dimensions of Consumer Packages in Sweden

	Shipping Containers and Inside Dimensions					
	1M 1140 × 760 mm		2L 1140 × 380 mm		2M 760 × 570 mm	
1	1140	760	1140	380	760	570
2	570	380	570	190	380	285
3	380	253	380	126	253	190
4	285	190	285	95	190	142
5	228	152	228	76	152	114
6	190	126	190	63	126	95
7	162	108	162	54	108	81
8	142	95	142	47	95	71
9	126	84	126		84	63
10	114	76	114		76	57
11	103	69	103		69	51
12	95	63	95		63	47
13	87	58	87		58	
14	81	54	81		54	
15	76	50	76		50	
16[a]	71	47	71		47	

[a] Others are omitted.

saler stage these 600 × 800 mm units can easily be transferred to live pallets of the same size, so that they can be easily rolled into a place in the retailer's shop or in the institutional or commercial premises.

Shipping containers should be adapted to fit on the 600 × 800 mm pallet.

The Erfa group of foodstuff wholesalers has been set up with a view to making the best possible use of the combined experience of the major food distributors in Sweden for the execution and coordination of certain necessary rationalization measures. Since a continued coordination of dimensions is an essential prerequisite of the more rational handling that can limit the rise in costs, the group favors the implementation of the recommendations issued by the Swedish Standard Association (SIS) for the distribution of consumer goods.

C. Migros Cooperatives in Switzerland

Migros Cooperatives are the union of 12 livelihood cooperatives in Switzerland, and their total members exceed 1 million. Considering family members, these cover half of all Swiss people (about 6 million). Migros has stores in the whole country, and the number of stores is more than 400.

Chapter 25. Physical Distribution of Packaged Foods

<u>Recommendations for distribution of consumer goods</u>
(SWEDISH STANDARD SIS 84 70 02 E)

Introduction This standard is a recommendation for a uniform modular system for the Swedish consumer goods industry, based on current Swedish standards SIS 84 70 01, etc.

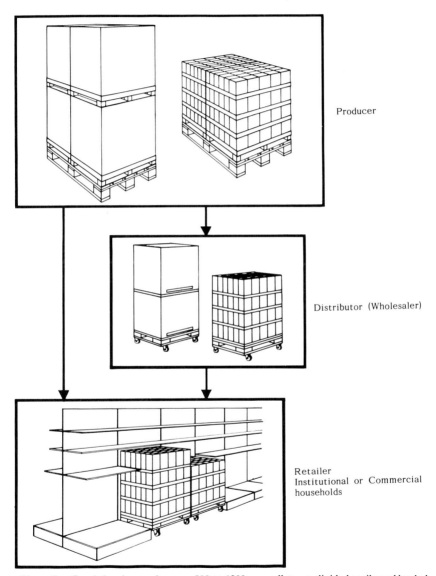

Producer

Distributor (Wholesaler)

Retailer
Institutional or Commercial households

Figure 7. Goods leaving producer on 800 × 1200 mm pallets are divided easily and loaded on 500 × 800 mm pallets at wholesaler and distributed to retailer.

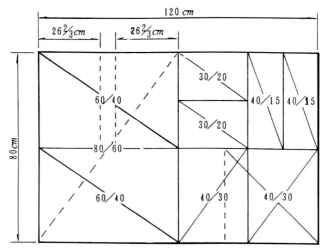

Figure 8. Standardized containers loaded on the 800 × 1200 mm pallet.

Table XVI   Base Dimensions of Transport Packages in SNV 25101

Outside Dimensions of Packages (mm)	Number of Packages in One Layer on Pallet
1200 × 1000	
1200 × 800	1
1200 × 600	
1200 × 400	2
800 × 600	2
600 × 400	4
600 × 200	8
600 × 133	
600 × 100	16
400 × 300	8
400 × 200	12
400 × 150	16
400 × 120	
300 × 200	16
300 × 133	
300 × 120	32
200 × 200	24
200 × 150	32
200 × 133	
200 × 120	40
200 × 100	48
150 × 133	
150 × 100	64
120 × 133	
120 × 100	80

Chapter 25. Physical Distribution of Packaged Foods

There are several big stores corresponding to hypermarkets, and sales of foods make up over 70% of total turnover.

To mechanize handling of a large quantity of goods, Migros planned coordinate dimensions of container by the module. First Migros decided the pallet size. Up to that time, Migros had used private size pallets, but on this occassion Migros adopted the Europallet (800 × 1200 mm, the pallet for the European pallet pool) that came in use generally at that time in Europe.

Next Migros decided dimensions of transport packagings adapted to this pallets (20 varieties, Fig. 8).

Switzerland has established the national standard SNV 25101 and has decided dimensions of rigid rectangular transport packages (Table XVI). This standard involves dimensions that adapt to both the 800 × 1200 mm pallet and the 1000 × 1200 mm pallet. The width and length dimensions that Migros adopted are chosen from this standard. The height dimensions was one that Migros itself decided, divided from 160 cm, which was the total height of pallet load.

In the 8 years after Migros decided this plan, 55% of total goods passing through 12 distribution centers came to follow this plan. Since then, standardized packages have been increasing, but all goods do not adapt to this method—that is to say, there are goods or packages that can not alter dimensions, so we cannot think that 100% of goods adapt to these methods.

INDEX

Accumulating and packaging system for multisized products, optimum, computer-controlled, 169–170
Acrylonitrile–butadiene–styrene (ABS) copolymer for containers, 125
Acrylonitrile-styrene copolymer for containers, 124–125
Activation energy of thermal death, 20
Air-containing packaging of meat by-products, 328–329, 330
Air interface area, lipid oxidation and, 38–39
Alcohol-generating agent, packaging cakes with, 377–378
Alcohol generation type of oxygen absorber, 243
Aluminum foil
　laminated to paper/paperboard, 57
　trays of, dual oven usage of, 264
Aluminum for containers, 90–93
Anodic undermining corrosion of cans, 103
Antioxidants, 40
Antioxidation effect of vacuum and gas substitution packaging for cakes, 367–369
Apple packaging for freshness preservation and storage, 298
Aseptic packaged foods, 213–227
　commercialized, 225–226
　complete, 221, 223–225
　definition of, 213
　future trends in, 227
　for institutional uses, 226–227
　manufacturing methods of, 221, 223–227
　packaging systems for, 217–221
　　food sterilizers in, 217–218
　　new aseptic fill/packaging machines in, 218–221, 222
　recent trends in
　　in Europe, 215–216
　　in Japan, 216–217
　　in the United States, 214–215
Aseptic packaging
　behavior of microorganisms and, 181–182
　of meat by-products, 329–332
Autooxidation of lipids, 25–30

Bacon, vacuum-packaged, 289, 291
Bacteria
　foods and, 14–16
　packaging preventing, for moist cakes, 374–375, 376–377
Bag-in-box, 72–73
Beef
　boxed, 313–317
　　advantages of, 313
　　hygienic care of, 316–317
　　packaging materials for, 313–314, 315
　　packaging process for, 316–317
　　packaging system for, 314–316
　corn, boiled, vacuum-packaged, 289, 291, 292
Boil-and-steam cooking of meat by-products, 326–327, 328
Bottles, glass
　color of, 108, 109
　forming, 105–106
　impact strength of, 107
　internal pressure strength of, 107
　lightweight, 109–113
　　chemical tempering of, 112
　　dual coating of, 109–110, 111
　　NNPB forming process for, 111–112
　　nonreturnable, strength of, 112–113
　prelabled, 113–115
　properties of, 107–108
　returnable, energy of, 158
　thermal shock resistance of, 107–108
　trends in, 108–115
Box, rigid, 63, 64
Butadiene–styrene copolymer for containers, 122–124
Butter, soft, aseptic packaged, 225–226

Index

Cakes
  fat content of, 360–361
  moist, packaging preventing bacteria and molds for, 374–375, 376–377
  packaging of, 357–378
    with alcohol-generating agent, 377–378
    classification of, 358
    future prospects for, 378
    gas barrier, 364–367
    gas substitution, 367–371
    with gas substitution agent, 371–377
    moisture-proof, 364, 365
    with oxygen absorbent, 371–377
    quality required for, 358, 359
    vacuum, 367–371
  quality deterioration of, 358–359, 362, 364
    from other factors, 362, 364
    related to water content, 358–359, 362
  water activity of, 362
  water content of, 360–361, 362
Can(s)
  cemented side-seam, 94–96
  composite, 66
  corrosion of, internal, 102–103
  draw and wall ironing (DWI), 97–98, 99
  drawn and redrawn (DRD), 99–101
  hygienic problems of, 156
  soldered side-seam, 93–94
  three-piece, 93–97
  two-piece, 97–100
  welded side-seam, 96–97
Carbon dioxide absorption type of oxygen absorber, 241–243
Carbon dioxide for gas-exchange packaging, 270
Carbon dioxide generating type of oxygen absorber, 240–241
Carrier cartons, 64–66
Carton making, 83
Carton(s)
  carrier, 64–66
  collapsible, 62
  flip-top, 63, 65
  folding, 59–62
  liquid, with spout, 68, 69
  lock-bottom, 60, 61
  manufacturing processes for, 74–83
    creasing in, 80–82
    die-cutting in, 80–82
    printing in, 74–79
    surface coating in, 79–80
  milk, 67
  multipack, 64–66
  oven-proof, 64
  seal-end, 60, 61
  with spout, 64
  straight, 60, 61
  tray, 61–62
  tucked-end, 60
Castella, mold-preventive packaging for, 375, 377
Cathodic detachment of organic coating of can, 103
Cemented side-seam can, 94–96
Chamber-type vacuum packaging machine, 280
Cheese
  fresh, packaging of, 351–353
  processed, packaging of, 353–355
Chemical tempering of lightweight glass bottles, 112
Citrus fruit packaging for freshness preservation and storage, 298–299
Cold sterilization, 19–20
Color of glass bottles, 108, 109
Composite cans, 66
Composite containers, 66–74
Computer-controlled optimum accumulating and packaging system for multisized products, 169–170
Computer-controlled packaging machines, 165–169
Conductivity, thermal, of foods, 11–13
Constant differential pressure method in retorting system, 199
Consumer packs, new aseptic fill/packaging machines for, 219–220
Corn beef, boiled, vacuum-packaged, 289, 291, 292
Corrosion of cans, internal, 102–103
Counterpressure method in retorting system, 199, 203
Creasing of cartons, 80–82
Cup
  hot-filling, 71, 72
  paper, 66
Cylindrical container, 71

Dairy products, packaging of, 349–355
  aseptic, 221, 223–224
Deep-draw type vacuum packaging

Index 417

machine for, 280–281
for meat, 319, 321
Deoxidizers, 249–251; *see also* Oxygen absorber
Deterioration, elimination of, oxygen absorber in, 234
Die-cutting of cartons, 80–82
Distribution of packaged foods, 381–413
  climatic hazards in, packaging for, 388–389
  food preservation in, packaging for, 381–391
  miscellaneous hazards in, packaging for, 390
  module dimensions for, 393–413; *see also* Distribution package dimensions
  physical hazards in, packaging for, 382–387
  temperature environment and, 391–392
  unit load system and, 392–395
Distribution package dimensions, 393–413
  by modular coordination, 395–413
    calculation method for, 395–405
      comparison of, with ISO, ANSI, and SIS, 400, 402, 404–405
      with dimensions adapted to 1200 × 1000 mm, 398–400, 401–402, 403
      with dimensions adapted to 1100 × 1100 mm, 396–398
    examples of, 405–413
      for foodstuff wholesalers group in Sweden, 408–410
      for Migros Cooperatives in Switzerland, 410–413
      for sake in Japan, 405–408
Double sterilization packaging of meat by-products, 325–326
Draw and wall ironing (DWI) can, 97–98, 99
Drawn and redrawn (DRD) can, 99–101
Drinks, fruit, aseptic packaged, 224, 225

Electrolytic chromium-coated steel (ECCS) for containers, 88, 90
Energy
  food packaging and, 149–161
  for manufacture
    of basic packaging materials, calculation of, 152–153
    of container, 154
  for transportation of reference containers, calculation of, 151–152, 153
Energy consumption of containers, total, 154–159
Environment, temperature, distribution of packaged foods and, 391–392
Environmental conditions, growth of microorganisms and, 17–19
Enzymes, lipid oxidation and, 38
Europe, aseptic packaged food trends in, 215–216
Extrusion coating of cartons, 80

Fatty acids, composition of, lipid oxidation and, 36
Filling and sealing equipment for retortable packaging, 192–198
  defective seals and, 194–196
  residual air and, 197, 198
  types of, 192–194, 195, 196, 197, 198
Film lamination of cartons, 80
Fish
  by-products of, packaging of, 341–348
  canned, packaging of, 339–340
  dried, packaging of, 338–339
  fresh, packaging of, 336–337
  frozen, packaging of, 337–338
  gas-exchange packaging for, 274–276
  salted, packaging of, 338–339
Fish ham, processing of, 344–345, 346
Fish paste products, packaging of, 341–344
Fish sausage, processing of, 346, 347
Flame sealer in carton making, 83
Flavor, oxygen absorber and, 234–235
Foamed polystyrene sheets, 121–122
Foil, aluminum
  laminated to paper/paperboard, 57
  trays of, dual oven usage of, 264
Folded forming paperboard trays, 258–259
Folding carton, 59–62
Food preservation technology, new trends in, 177–292; *see also* Preservation of food
Free-oxygen absorber, 41
Free-oxygen scavenging agent, enclosed, packaging with, behavior of microorganisms and, 180–181
Free oxygen scavenging packaging, 229–252; *see also* Oxygen absorber

Free oxygen scavenging packaging (*continued*)
  characteristics of, 234–235
  technology related to, advances in, 244–252
Frozen food, 253–267; *see also* Oven-proof trays
Fruit drinks, aseptic packaged, 224, 225
Fruits
  packaging of, 295–301
    for freshness preservation and storage, 295–299
    shock-absorbing, 299–301
  preservation of, controlled atmosphere method of, 295–296
Fungi
  foods and, 15, 16
  growth of, in anaerobic conditions, 248

Gamma rays, sterilization by, 19
Gas barrier, vacuum packaging materials as, 282–284
Gas barrier packaging for cakes, 364–367
Gas control agents for packaging of vegetables, 306
Gas-exchange packaging, 269–277
  aim of, 269–270
  deep-draw package for, 271
  for fish, 274–276
  foods packed by, 272–277
  form and system of, 270–271
  gases used in, properties of, 270
  for green tea, 276–277
  for ham, 272–273
  packaging materials for, 272
  pillow-type package for, 271
  pouch and bag for, 270
  for red meat, 273–274, 275
  for sausage, 272–273
  tray package for, 271
Gas pack, behavior of microorganisms and, 180
Gas packaging for meat, 319
Gas substitution packaging of cakes, 367–377
Gases for gas-exchange packaging, properties of, 270
Glass containers, 105–115; *see also* Bottles, glass
  history of, 48
Glass for food packaging, history of, 48

Gravure printing press for cartons, 76–79
  ink feeding structure for, 77, 78
  printing structure for, 77
  web-fed rotary press for, 77–79

Ham
  fish, processing of, 344–345, 346
  gas-exchange packaging for, 272—273
  roast and boneless, packaging of, 332–333
Heat
  specific, of foods, 8–11
  sterilization by, 20–21
    of microorganisms, 182
Heat pipe, heat-sealing device using, 170–173
Heat-sealing device using heat pipe, 170–173
Heating medium in retorting system, 201
Hematin compounds, lipid oxidation and, 38, 39
Hot-filling cup, 71, 72
Humidity-controlling materials for packaging of vegetables, 306
Hydroperoxides, formation mechanisms of, 27–28
Hysteresis in foods, 3–5

Impact strength of glass bottles, 107
In-mold bead forming for expandable polystyrene containers, 121
Industrial robots, 173–175
Institutional packs, new aseptic fill/packaging machines for, 221, 222
Internal pressure strength of glass bottles, 107
Irradiation, sterilization by, 19–20

Japan
  aseptic packaged food trends in, 216–217
  packaging machine development in, recent, 165–175
  sake in, distribution package dimensions for, 405–408
Japanese rice wine, aseptic packaged, 224–225

Kamaboko
  packaging of, 341–344

Index 419

processing of, 342-344
specially packaged, 348

Laminate materials, paper/paperboard in, 54, 57
Letterpress printing machines for cartons, 75
Light, lipid oxidation and, 37
Lightweight glass bottles, 109-113
    chemical tempering of, 112
    dual coating of, 109-110, 111
    NNPB forming process for, 111-112
    nonreturnable, strength of, 112-113
Lipid-oxidized products, reactions of, with other food components, 41-42
Lipids
    autooxidation of, 25-30
    oxidation of, 25-34; see also Oxidation of lipids
Lipoxygenase, 32, 34
Liquid carton with spout, 68, 69

Manufacture
    of basic packaging materials, energy for, calculation of, 152-153
    of container, energy for, 154
Materials, food packaging, 45-145
    basic manufacture of, energy for, calculation of, 152-153
    classification of, 50-51
    definitions of, 49
    functions of, 49-50
    for gas-exchange packaging, 272
    glass as, 105-115; see also Glass containers
    history of, 47-49
    introduction to, 47-51
    metal as, 85-103; see also Metal containers
    paper/paperboard as, 53-83; see also Paper/paperboard containers
    plastic as, 117-145; see also Plastic containers
    for retortable packaging, 190-191
Meat
    by-products of, packaging of, 323-333
        air-containing, 328-329, 330
        aseptic, 329-332
        boil-and-steam cooking, 326-327, 328

double sterilization, 325-326
oxygen-absorbing agent, 329, 331
retort sterilized, 327-328, 329
of roast and boneless ham, 360-361
simple wrapping in, 324
vacuum, 324-325
of wiener sausage, 333
fresh, packaging of, 309-321
    consumer, 311-312
    by gas packaging, 319, 320
    retail, 318-321
    traditional, 310-311
    by vacuum packaging in deep-draw or skin-pack form, 319, 321
    wholesale, 312-318
        of beef, 313-317
        of pork, 317
        of poultry, 318
red, gas-exchange packaging for, 273-274, 275
world production of, 309-310
Meat products, processed, aseptic packaged, 226
Metal containers, 85-103; see also Can(s)
    advantages of, 85-86
    of aluminum, 90-93
    easy-open ends and, 101-103
    of electrolytic chromium-coated steel, 88, 90
    features of, 85-86
    history of, 48
    three-piece can as, 93-97
    of tinplate, 86-88, 89
    two-piece can as, 97-100
Metals
    for food packaging, history of, 48
    lipid oxidation and, 37
Microbes, growth of
    minimum water activity for, 363
    restraining, vacuum and gas substitution packaging for cakes in, 369-370, 371
Microbiology
    of foods, 14-21
    new knowledge in, oxygen absorbers and, 252
Microorganisms
    behavior of
        aseptic packaging and, 181-182
        food packaging technology and, 179-182
        gas pack and, 180

Microorganisms (*continued*)
  packaging with enclosed free-oxygen scavenging agent and, 180–181
  vacuum packaging and, 179–180
 control of, in retort food, 204–208
 foods and, 14–16
 growth of, environmental conditions and, 17–19
 in packaged foods
  control of, 182–184
  sterilization of, 182–183
 sterilization and, 19–21
Microwaves, sterilization of microorganisms by, 182–183
Milk cartons, 67
Modular coordination for physical distribution system, 393–395
Moisture-proof packaging for cakes, 364, 365
Moisture sorption equilibrium of foods, 3–5
Molded formed paperboard trays, 287–288
Molds
 foods and, 16
 packaging preventing
  for Castella, 375, 377
  for moist cakes, 374–375
Monohydroperoxides of triglycerides and phospholipids, 34
Multifunction film for packaging vegetables, 306

Narrow neck press and blow (NNPB) forming process for lightweight glass bottles, 111–112
Nitrogen gas for gas-exchange packaging, 270
Nozzle-type vacuum packaging machine, 279, 280
Nylon for containers, 137–140

Offset printing press for cartons, 76, 77
Oven-proof trays, 64, 253–267
 development of, 253–254
 dual
  paperboard, 255–263
   characteristics of, 257
   coating for, 260–261
   composition of, 261
   development of, steps of, 262–263
   folded forming, 258–259
   molded formed, 259–260
   pressed formed, 257–258
   requirements for, 255–257
  requirements for, 254
 foil, dual oven usage of, 264
 lidding of, 264–265
 usage of, 265–267
 varieties of, 254–264
Oxidation
 of foods, 25–42
 of lipids, 25–34
  factors affecting, 35–40
  photosensitized, 30, 32, 33
  polymerized products from, 35, 38
  secondary products of, 34–35
  volatile compounds from, 35, 36, 37
Oxidation-preventive packaging of peanut butter, 372, 374–375
Oxygen
 growth of microorganisms and, 18
 in package, food preservation and, 285–288, 289
Oxygen absorbent, packaging cakes with, 371–377
Oxygen absorber
 airtight containers and, 244
 automatic single-pack throw-in machine and, 245–246
 characteristics of packing with, 234–235
 classification of
  by function, 240–243
  by material, 235
  by reaction speed, 237–238, 239
  by reaction style, 236
  by use, 238–239
 effects of, 230–233
 fixing and separating, 247, 252
 in food preservation, 229–252; *see also* Free oxygen scavenging packaging
 paper eye and, 245
 principles of, 230
 technology related to, advances in, 244–252
 types of, 235–243
  alcohol generation, 243
  carbon dioxide absorption, 241–243
  carbon dioxide generating, 240–241
 uses of, 235
 varieties of, 249–251

Index

Oxygen-absorbing agent packaging of meat by-products, 329, 331
Oxygen barrier, vacuum packaging materials as, 282–284

Packaging
  inner, 49
  item, 49
  materials for, 45–145; see also Materials, food packaging
  outer, 49
  primary, 49
Packaging machines
  computer-controlled, 165–169
  recent development of, in Japan, 165–175
Paper
  for food packaging, history of, 47–48
  grain of, 58, 59
Paper/paperboard containers, 53–83; see also Box; Carton(s); Cup
  brick-shaped, small, 68–71
  characteristics of, 53–54
  composite, 66–74
  cylindrical, 71
  history of, 47–48
  kinds of, 54, 55–56
  of laminate materials, 54, 57
  selection of, 57–58
  structure of, 54, 55–56
  with Ultra-Lock, 73–74
Paperboard
  for food packaging, history of, 47–48
  oven-proof trays of, 255–263; see also Oven-proof trays, dual, paperboard
Peanut butter, oxidation-preventive packaging of, 372, 373–374
Peanuts, vacuum-packaged, 288–289, 290
Pears, packaging of, for freshness preservation and storage, 296–297
Persimmons, packaging of, for freshness preservation and storage, 297–298
pH
  control of microorganisms by, 184
  growth of microorganisms and, 17–18
Phospholipids, monohydroperoxides of, 34
Photosensitized oxidation of lipids, 30, 32, 33
Physical properties of foods, 3–13
Pitting corrosion of cans, 102–103

Plasti-Shield label for glass bottles, 113–114, 115
Plastic containers, 117–145
  of acrylonitrile–butadiene–styrene copolymer, 125
  of acrylonitrile–styrene copolymer, 124–125
  advantages/disadvantages of, 117
  of butadiene–styrene copolymer, 122–124
  history of, 49
  of nylon, 137–140
  of polycarbonate, 141–142
  of polyethylene, 129–132
  of polyethylene terephthalate, 142–145
  of polypropylene, 133–137
  of polystyrene, 118–122
    expandable, 121–122
    general-purpose grade, 118–120
    high-impact grade, 120–121
  of polyvinyl chloride, 125–128
  of saponified ethylene–vinyl acetate copolymer, 140–141
Plastics
  for food packaging, history of, 49
  laminated to paperboard, 54, 57
Platen die-cutter for cartons, 80–81
Polycarbonate (PC) for containers, 141–142
Polyethylene film, high-density, for packaging of vegetables, 304–305
Polyethylene (PE)
  for containers, 129–132
    high-density, 132
    low-density, 130–132
  for packaging of fruit, 295–296
Polyethylene terephthalate (PET)
  for containers, 142–145
  oven-proof paperboard trays coated with, 254–258, 261–263
  oven-proof trays of, 263
Polymerized products from lipid oxidation, 35, 38
Polypropylene (PP) for containers, 133–137
Polystyrene, 118–122
  expandable, 121–122
  general-purpose grade, 118–120
  high-impact grade, 120–121
Polyvinyl chloride (PVC) for containers, 125–128

Pork, packaging of, 317
Poultry, packaging of, 318
Preservation of food
   aseptic packaging in, 213–227; *see also* Aseptic packaged foods
   free oxygen scavenging packaging in, 229–252; *see also* Free oxygen scavenging packaging; Oxygen absorber
   frozen foods in, 253–267
   gas-exchange packaging in, 269–277; *see also* Gas-exchange packaging
   oven-proof trays in, 253–267; *see also* Oven-proof trays
   oxygen content in package and, 285–288, 289
   packaging for, in distribution process, 381–391
   retortable packaging in, 185–210; *see also* Retortable packaging
   technology for
     behavior of microorganisms and, 179–182
     new trends in, 177–192
   vacuum packaging in, 279–292; *see also* Vacuum packaging
Preservatives, control of microorganisms by, 184
Press coating of cartons, 79
Pressed formed paperboard trays, 257–258
Printing of cartons, 74–79
   gravure printing press for, 76–79
   machines for, 75–79
     letterpress, 75
   offset printing press for, 76, 77
Production systems of retort food, 188–204; *see also* Retort food(s), production systems of
PVDC coating of cartons, 79

Radical theory of autooxidation, 26–27
Retort food(s)
   microorganisms control in, 204–208
     HTST sterilization method for, 205, 206
     manufacturing standard for, in Japan, 204–205
   production systems of, 188–204
     filling and sealing equipment in, 192–198

     packaging design in, 191–192
     packaging materials in, 190–191
     retorting system in, 199–204
   shelf life of, 209–210
   sterility of
     commercial confirmation of, 208
     factors affecting, 205–207
   types of, 186–188
Retort sterilized packaging of meat by-products, 327–328, 329
Retortable packaging, 185–210; *see also* Retort food(s); Retorting system
   history of, in Japan, 186
   standards for, 208–209
Retorting system, 199–204
   constant differential pressure method in, 199
   counterpressure method in, 199, 203
   equipment for, 199, 200
   heating medium in, 101
   sterilization racks in, 202, 204
   temperature recorder with F_o value measurement in, 204
Rice wine, Japanese, aseptic packaged, 224–225
Robots, industrial, 173–175
Rotary die-cutter for cartons, 82

Sack gluer in carton making, 83
Safety of oxygen absorber, 234
Saponified ethylene–vinyl acetate copolymer (EVOH) for containers, 140–141
Sausage
   fish, processing of, 346, 347
   gas-exchange packaging for, 272–273
   wiener, packaging of, 332–333
Scavenging agent, free-oxygen, enclosed, packaging with, behavior of microorganisms and, 180–181
Seafood products, packaging of, 335–340; *see also* Fish
Shelf life of retort foods, 209–210
Single oxygen quenchers, 40–41
Skin-type vacuum packaging
   machine for, 280
   for meat, 319, 321
Sleeve, 60, 61
Soldered side-seam can, 93–94
Special gluer in carton making, 83
Specific heat of foods, 8–11

Steam, superheated, sterilization by, 21
Steel, electrolytic chromium-coated, for containers, 88, 90
Stereoisomerization, 30, 31
Sterility of retort foods
 commercial, confirmation of, 208
 factors affecting, 205–207
Sterilization, 19–21
 of microorganisms, 182–183
 racks for, in retorting system, 202, 204
 of retort foods, high-temperature, short-time (HTST) method for, 205, 206
Sterilizers in aseptic food packaging systems, 217–218
Superheated steam, sterilization by, 21
Surface coating of cartons, 79–80
Sweden, foodstuff wholesalers group in, distribution package dimensions for, 408–410
Switzerland, Migros Cooperatives in, distribution package dimensions for, 410–413

Taste, oxygen absorber and, 234
Tea, green, gas-exchange packaging for, 276–277
Temperature
 growth of microorganisms and, 17
 lipid oxidation and, 36
 low, control of microorganisms by, 183
Temperature environment, distribution of packaged foods and, 391–392
Temperature recorder with F_o value measurement in retorting system, 204
Thermal conductivity of foods, 11–13
Thermal death, activation energy of, 20
Thermal diffusivity of foods, 13
Thermal properties of foods, 8–13
Thermal shock resistance of glass bottles, 107–108
Thermoprocessing for food preservation, 213; *see also* Retort food(s); Retortable packaging
Tinplate for containers, 86–88, 89
Transportation of reference containers, energy for, calculation of, 151–152, 153
Tray carton, 61–62
Trays, oven-proof, 64, 253–267; *see also* Oven-proof trays

Triglycerides, monohydroperoxides of, 34

Ultra-Lock, container with, 73–74
Ultraviolet rays, sterilization by, 19
 of microorganisms, 182–183
Unit load system
 development of, 392–393
 distribution of packaged foods and, 392–395
 modular coordination for, 393–395
United States
 aseptic packaged food trends in, 214–215
Unmanned operation, robots for, 174–175

Vacuum packaging, 279–292
 application of, to foods, 288–292
 behavior of microorganisms and, 179–180
 for cakes, 367–371
 machinery for, 279–281
  chamber-type, 280
  deep-draw type, 280–281
  nozzle-type, 279, 280
  skin-type, 280
 materials for, 282–292
  as gas barrier, 282–284
  as vapor barrier, 284–285
 for meat, 319, 321
 for meat by-products, 324–325
 for vegetables, 304
Vapor barrier, vacuum packaging materials as, 284–285
Varnish over-printing on cartons, 79
Vegetables, packaging of, 303–307
 future trends in, 307
 gas control agents in, 306
 high-density polyethylene film in, 304–305
 humidity-controlling materials in, 306
 multifunction film in, 306
 new concepts in, 305–306
 partial-vacuum, 304
 vacuum, 304
Vinyl coating of cartons, 79
Volatile compounds from lipid oxidation, 35, 36, 37

Water activity
 of cakes/cookies, 362

Water activity (*continued*)
  of foods, 5–8
    growth of microorganisms and, 18–19
  for growing microbes, 363
  lipid oxidation and, 39–40
Wax coating of cartons, 79–80
Web-fed die-cutter for cartons, 81, 82
Welded side-seam can, 97–97

Wiener sausage, packaging of, 332–333
Wine, rice, Japanese, aseptic packaged, 224–225

Yeasts, foods and, 16
Yogurt, packaging of, 349–351